CHROMOSOMES TODAY

CHROMOSOMES TODAY

Volume 14

Edited by

MICHAEL SCHMID

Department of Human Genetics,
University of Würzburg, Germany

and

INDRAJIT NANDA

Department of Human Genetics,
University of Würzburg, Germany

SPRINGER-SCIENCE+BUSINESS MEDIA, B.V.

A C.I.P. Catalogue record for this book is available from the Library of Congress.

ISBN 978-90-481-5855-3 ISBN 978-94-017-1033-6 (eBook)
DOI 10.1007/978-94-017-1033-6

Printed on acid-free paper

Contents

Color Plates ix

Part 1. Sex Chromosomes

Chapter 1.1. Mammalian Sex Chromosome Evolution – The Rise and Fall of the Y Chromosome
Jennifer A. Marshall Graves and Paul D. Waters 3

Chapter 1.2. Avian Sex Chromosomes and Sex-Linked Genes
I. Nanda, Z. Shan, C. Steinlein, M. Schartl, T. Haaf and M. Schmid 15

Chapter 1.3. *DMRT* Genes and Sex Determination in Medaka
Ute Hornung, Indrajit Nanda, Mariko Kondo, Atsushi Shimizu, Shuichi Asakawa, Jean-Nicolas Volff, Christoph Winkler, Zhihong Shan, Thomas Haaf, Nobuyoshi Shimizu, Akihiro Shima, Michael Schmid and Manfred Schartl 27

Chapter 1.4. Evolution of the Y Chromosome in Primates
Werner Schempp 39

Chapter 1.5. The Male Hypermethylation (MHM) Region on the Chicken Z Chromosome: Female-Specific Transcription and its Biological Implication
Mika Teranishi and Shigeki Mizuno 45

Chapter 1.6. Neo-X and Neo-Y Chromosomes in *Drosophila miranda*
S. Steinemann and M. Steinemann 55

Chapter 1.7. The X Chromosome Plays a Special Role During Speciation
Hildegard Kehrer-Sawatzki and Horst Hameister 65

Chapter 1.8. Sex Chromosome Evolution: Evidence from Fish, Fly and Moth Species
Walther Traut 73

Part 2. Meiosis

Chapter 2.1. Expression and Chromatin Binding Specificity of Ki-67 in Male and Female Meiosis of the Mouse
Heinz Winking, Johannes Gerdes and Walter Traut 85

Chapter 2.2. Conservation and Variability of Meiotic Processes – Lessons from the Unconventional Meiosis of *C. elegans*
Josef Loidl, Pawel Pasierbek and Ann M. Rose 93

Chapter 2.3. Meiotic Chromosomes and Meiotic Mechanisms
Alberto J. Solari 103

Chapter 2.4. The Nuclear Envelope at the Attachment Sites of Mammalian Meiotic Telomeres
Ricardo Benavente, Manfred Alsheimer and Elisabeth von Glasenapp 119

Chapter 2.5. Interphase Cytogenetics in Understanding Chromosome and Telomere Dynamics During Prophase I: Implications for Meiotic Telomere Movements
Harry Scherthan 127

Part 3. Cancer Cytogenetics

Chapter 3.1. Molecular Cytogenetics in the Study of Cancer
Jorn Koch 151

Chapter 3.2. Chromosomal Translocations in Leukaemia: Emerging Networks
Stefan K. Bohlander 157

Part 4. Chromosome Structures

Chapter 4.1. The EVO-DEVO of Pericentromeric DNA in the Mouse
Silvia Garagna, Ernesto Capanna, Maurizio Zuccotti and Carlo Alberto Redi 171

Chapter 4.2. New Developments in Multicolour Fluorescence *in situ* Hybridization
Christine Fauth, Jürgen Kraus, Sabine Langer and Michael R. Speicher 187

Chapter 4.3. Artificial Chromosomes – Past, Present and Future
Howard J. Cooke 197

Chapter 4.4. Differential Demethylation of Paternal and Maternal Genomes in the Preimplantation Mouse Embryo: Implications for Mammalian Development
Thomas Haaf, Wei Shi, Reinald Fundele, Katharine L. Arney, M. Azim Surani and Sheila C. Barton 207

Chapter 4.5. Integrated Comparative Genome Maps and Their Implications for Karyotype Evolution of Carnivores
Fengtang Yang and Alexander S. Graphodatsky 215

Chapter 4.6. Complex Relationships Between DNA Methylation Status and Chromosome Compaction and Cohesion
J. Bernardino-Sgherri, D. Flagiello and B. Dutrillaux 225

Chapter 4.7. Constitutive Heterochromatin of *Microtus agrestis*: Molecular Organization and Genetic Activity in Mitotic and Meiotic Cells
Karl Sperling, Sabine Henschel, Ilka Schulze and Heidemarie Neitzel 235

Part 5. Plant Cytogenetics

Chapter 5.1. FISHing Repeated DNA Sequences in *Beta* Genomes
T. Schmidt, C. Desel, D. Dechyeva, B. Fleischer, F. Gindullis, A. Schmidt, J. S. Heslop-Harrison and R. L. Doudrick 249

Chapter 5.2. Interphase Cytogenetics in Plants
I. Schubert, Z. Jasencakova, A. Meister, P. Fransz and M. Lysak 267

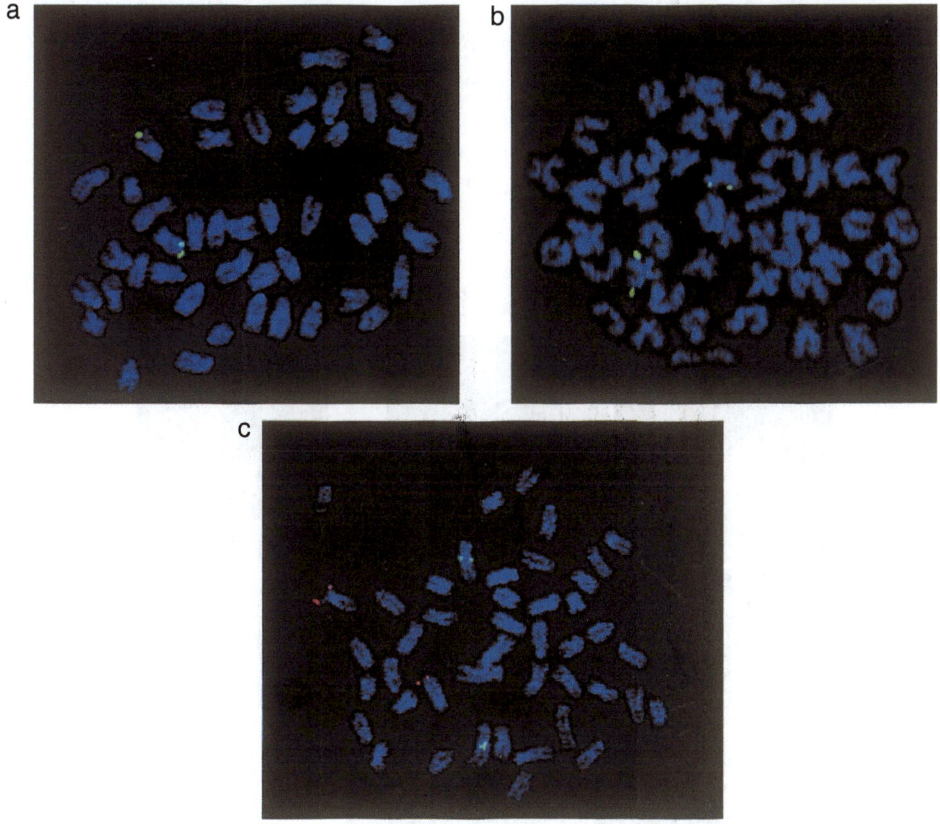

Figure 3. *From chapter 1.3, page 33.*

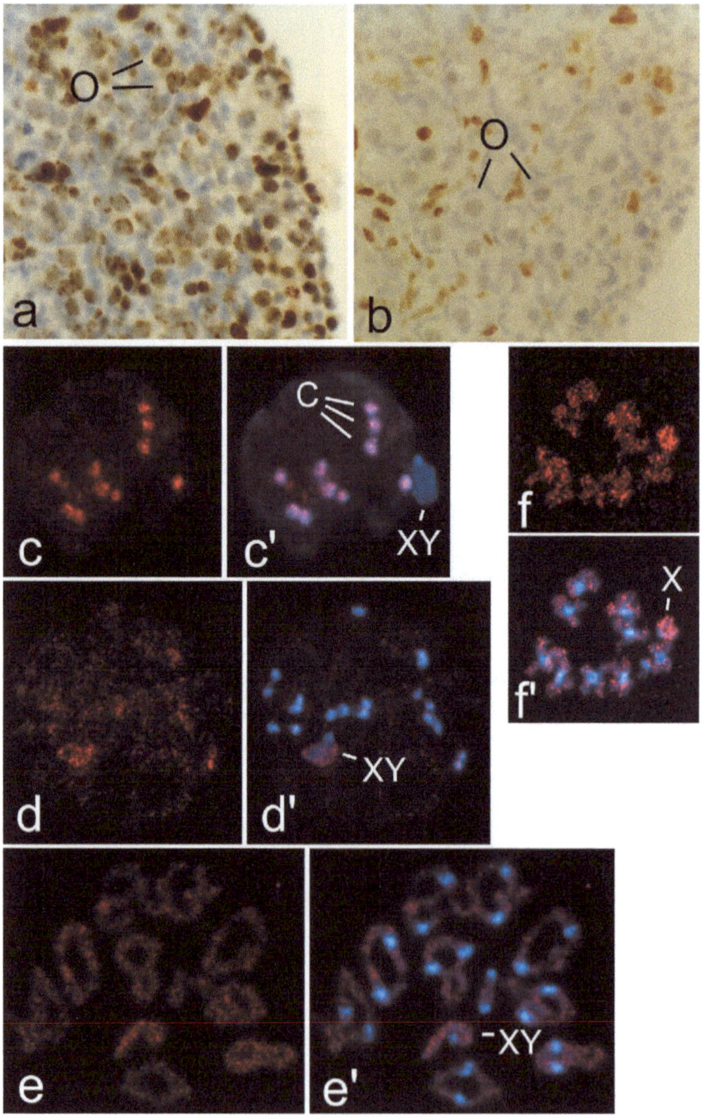

Figure 1. *From chapter 2.1, page 87.*

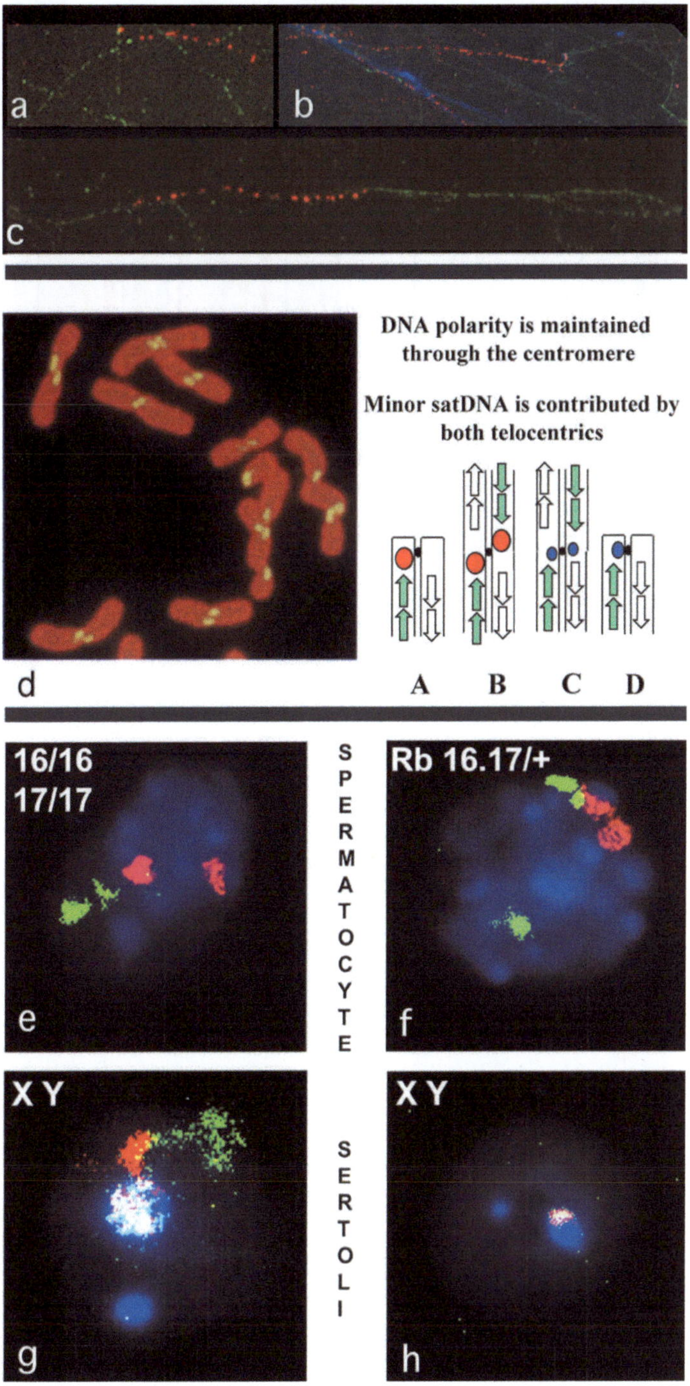

Figure 1. *From chapter 4.1, page 174.*

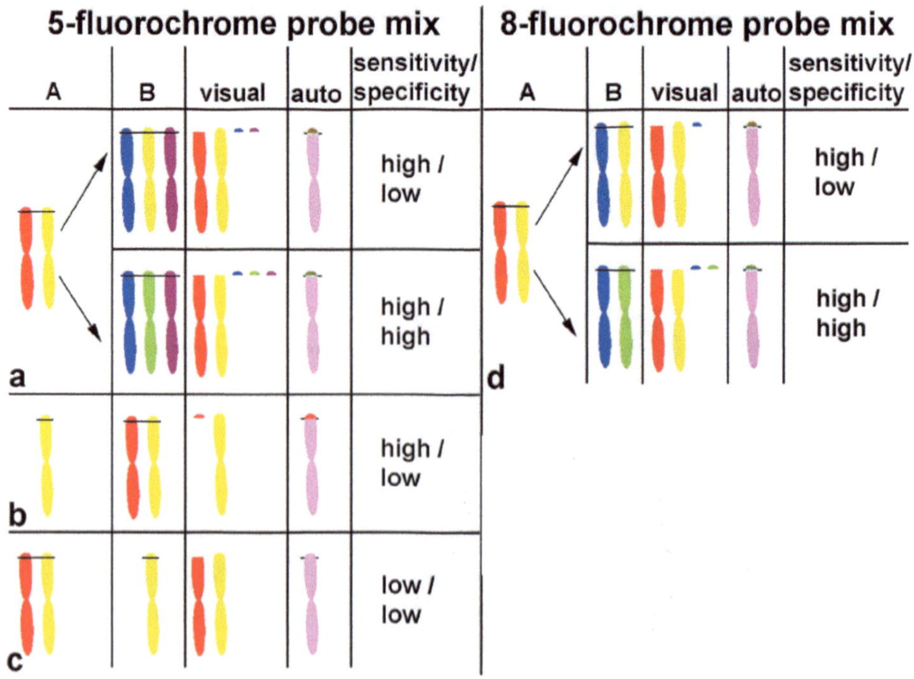

Figure 1. From chapter 4.2, page 190.

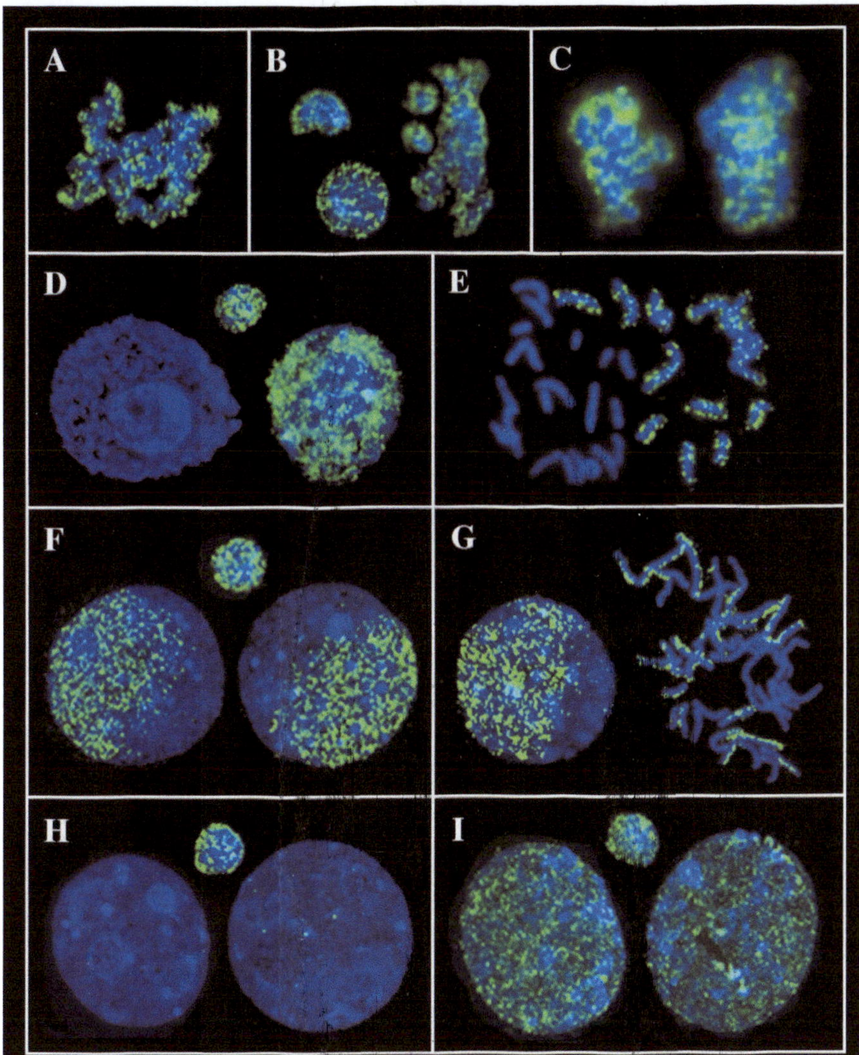

Figure 1. *From chapter 4.4, page 210.*

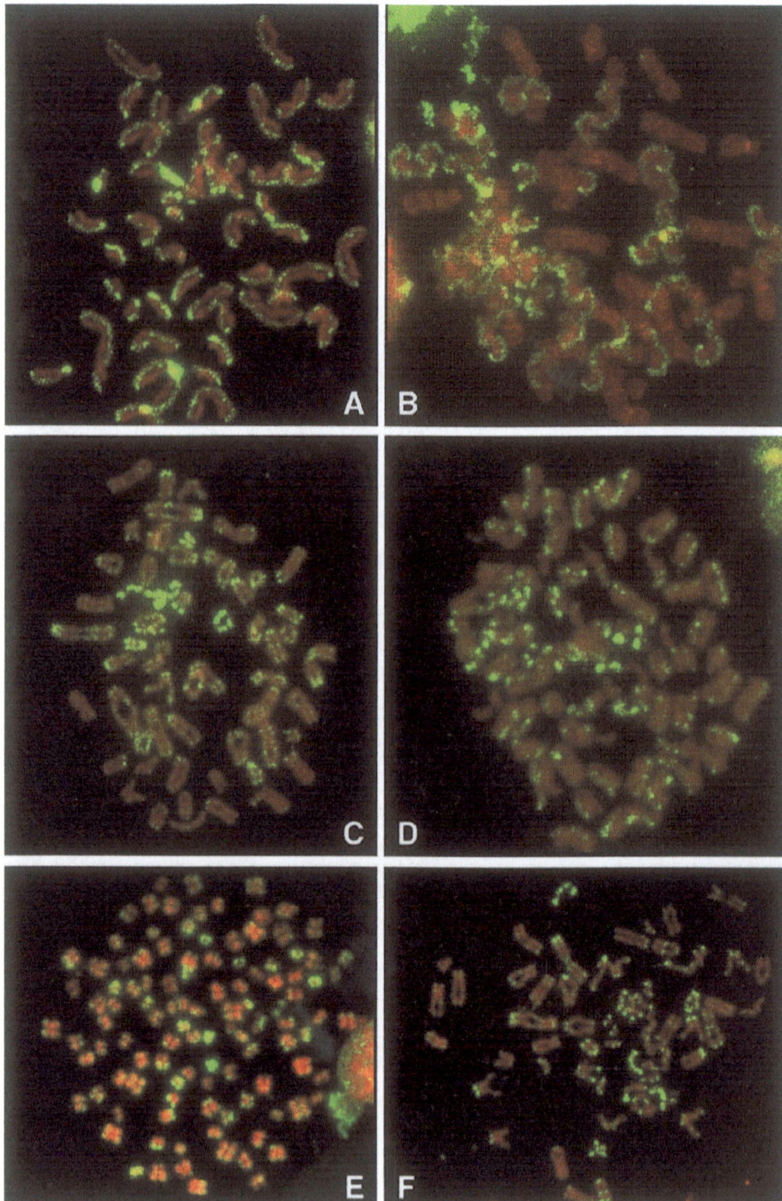

Figure 2. *From chapter 4.6, page 229.*

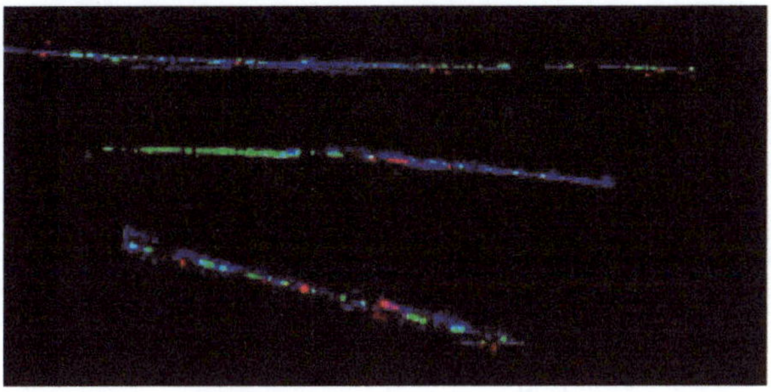

Figure 3. *From chapter 4.7, page 238.*

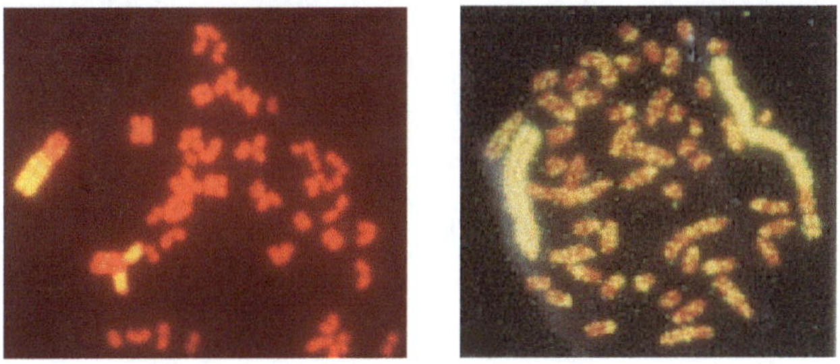

Figure 4. *From chapter 4.7, page 238.*

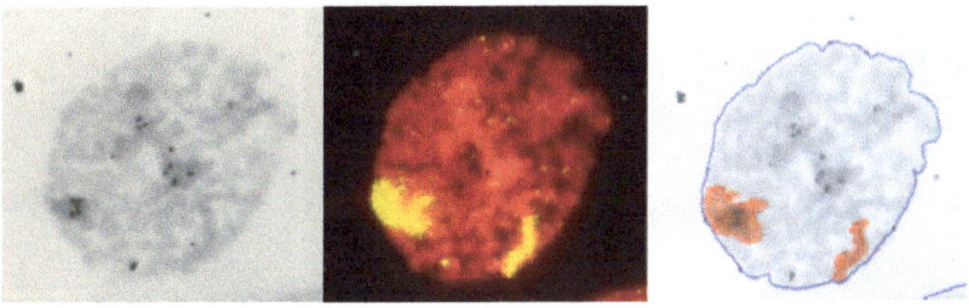

Figure 9. *From chapter 4.7, page 242.*

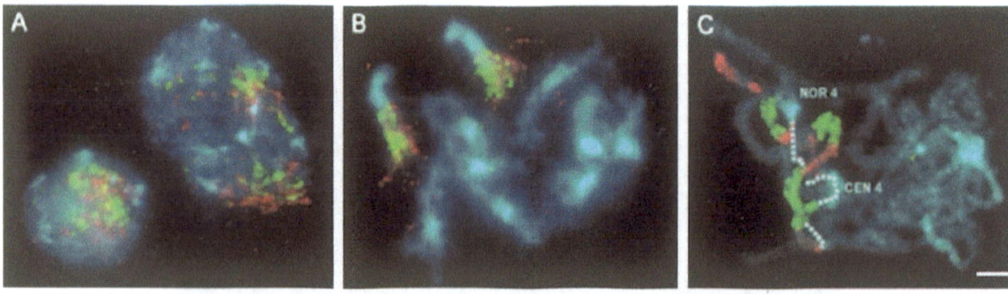

Figure 1. *From chapter 5.2, page 268.*

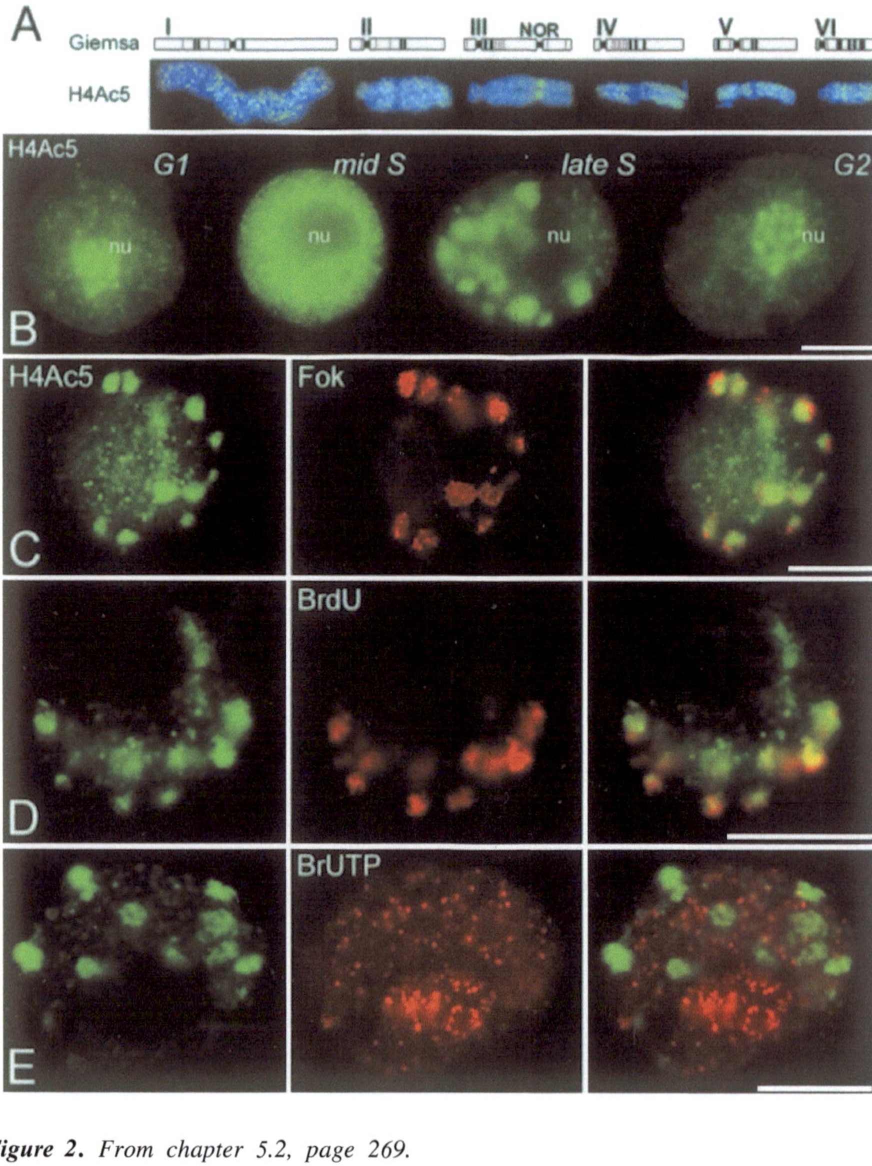

Figure 2. *From chapter 5.2, page 269.*

1. Sex Chromosomes

1.1. Mammalian Sex Chromosome Evolution – The Rise and Fall of the Y Chromosome

Jennifer A. Marshall Graves and Paul D. Waters
Research School of Biological Sciences, Australian National University, Canberra, ACT 2601, Australia

Sex determination and *SRY*

In mammals, females have two X chromosomes (XX) and males a single X and a Y (XY). The observations that XO individuals are female and XXY are male meant that the Y has a male dominant action. This was attributed to the action of a testis determining factor (TDF) on the Y, which instigates the pivotal genetic event in sex determination, the switching of the indifferent gonad (the genital ridge) to testis differentiation.

TDF was shown by deletion mapping to reside on the short arm of the human Y. *SRY* was positionally cloned from this region (Sinclair, 1990) and shown unequivocally to control sex determination by mutation analysis in man and transgenesis in mouse (reviewed by Koopman, 1995). Other species of eutherian mammals were found to have an equivalent *SRY* gene on the Y chromosome.

SRY is a small, single exon gene encoding an 80 amino acid DNA-binding motif shared by the high-mobility group (HMG) proteins that act as chromatin architectural factors. *SRY* defined the *SOX* gene family (for SRY-like HMG box containing) whose more than 20 members are highly conserved in all animal species and have many important functions in development. *SOX* genes include transcriptional activators and repressors (Uchikawa *et al.*, 1999), and seem to act in large complexes. *SRY* is a member of the intronless *SOX*B gene family.

SRY is thought to act by binding of the HMG box to a 6-bp target sequence in DNA, bending DNA through a specific angle (Harley *et al.*, 1993). Bending may promote association of regulatory elements bound to distant regions of DNA, forming a complex that controls the activity of other genes. It is not yet clear whether *SRY* control acts via activation or repression of other genes in the sex determining

M. Schmid and I. Nanda (eds.), Chromosomes Today, Volume 14, 3–14.

pathway, but a double repression has been suggested (McElreavey *et al.*, 1992; Graves 1998) by *SRY*.

SRY must act in a pathway with many other genes, several of which are known from studies of human sex reversal, and from mouse mutation and manipulation. One of these is a related *SOX* gene, *SOX9*, which is intimately involved in sex determination in all vertebrates (Foster *et al.*, 1996; Wagner *et al.*, 1996). There are likely to be many more – probably hundreds – as well as hundreds or thousands of genes required for germcell differentiation and male fertility. Only one needs to be on the Y, and for humans and mice at least, this trigger gene is *SRY*.

It might be imagined that a gene as important for reproduction as *SRY* would remain highly conserved through evolution. However, *SRY* is poorly conserved even within the HMG box. Outside the box, sequence cannot even be aligned between human and mouse. This lack of conservation is highly unusual for *SOX* genes, and suggests that, unlike other *SOX* genes, *SRY* action is all contained in the HMG box. This conclusion is reinforced by the finding that almost all of the known amino acid substitutions found in mutant SRY proteins from XY females lie within this region (Hawkins *et al.*, 1992). The mouse *Sry* gene seems to have acquired a 3′ region coding for a glycine-rich domain that is essential for its action (Bowles *et al.*, 1999).

The unusual structure, variation and evolution of the *SRY* genes turns out to be typical of genes on the Y chromosome. These strange properties of Y-borne genes are a result of the unique course of Y chromosome evolution.

Structure and gene content of the human Y chromosome

Mammalian X and Y chromosomes are morphologically very distinct and have quite different gene contents. The human X is large (about 160 Mb, amounting to 5% of the total haploid length). It bears about 1500 genes with a mixture of housekeeping and specialist functions. Most of these genes are not specifically concerned with sex, although there does seem to be a concentration of sex and reproduction-related genes on the X (reviewed by Graves and Delbridge, 2001). To ensure dosage equality between the sexes, only one X is genetically active in female cells. The set of genes on the X chromosome is almost completely conserved between different species of eutherian ("placental") mammal, probably because breaking it up would disrupt this chromosome-wide X inactivation mechanism (Ohno, 1967).

The human Y chromosome is much smaller (about 65 Mb). It is homologous to the X only over a short "pseudoautosomal region" (PAR), within which the sex chromosomes regularly pair and undergo crossing over. There is a 2.6 Mb PAR1 at the tips of the short arms and a second, shorter (0.5 Mb) PAR2 at the ends of their long arms. Y chromosomes in other eutherian species are quite variable in size, principally because of different amounts of heterochromatin.

Deletion analysis shows that the male-specific Y chromosome plays a pivotal role in sex determination and fertility, but has few other recognizable phenotypes. Deletions of *SRY* cause sex reversal, and deletions in several regions lead to azoospermia. An inventory of genes on the human and mouse Y started with the first candidate testis determining factor *ZFY* (Page *et al.*, 1987). Exhaustive screening of a human testis cDNA library with Y-specific YACs brought the number of recognizable gene sequences known on the human Y to a grand total of 37 (Lahn and Page, 1997). Of these, nine lie within the PAR1 and four in PAR2, leaving only 24 genes on the male-specific differential region of the Y. Many of these are inactive "pseudogenes" like *STS* (Yen *et al.*, 1988) which have been mutated or partially deleted so that they cannot be translated. Compared even to the X (which is moderately gene poor) the Y chromosome is therefore a genetic wasteland, containing only a few genes in a desert of repetitive sequence.

Uniquely, the Y chromosome gives the appearance of having a "coherent function" (Lahn and Page, 1997), specializing in male sex determination and reproduction. About half of the genes on the non-recombining, Y-specific (differentiated) region of the Y are expressed only in the testis and include several candidate spermatogenesis genes. Lahn and Page claimed that these male-specific genes formed a distinct "Class II" of genes that are testis-specific, multi-copy and have no homologue on the X. They were distinct from Class I genes, which were single copy, ubiquitously expressed and have homologues on the X.

Sex chromosome differentiation

How and when did the mammalian Y chromosome get its start? Why is it genetically depauperate, and why does it specialize in male-specific genes? Answers come from comparisons of sex chromosomes between different mammal and vertebrate groups.

Sex before the Y

The mammalian Y chromosome is unique to mammals. Our closest vertebrate relatives, birds and reptiles have a variety of sex-determining mechanisms, but none have an equivalent Y chromosome. Many reptiles such as alligators and turtles have no sex chromosomes at all, determining sex via environmental cues such as temperature. Birds and snakes have well-differentiated sex chromosomes, but the system is based on female heterogamety. In this ZZ male: ZW female chromosomal sex determination system the Z is large and gene-rich, and the W small and heterochromatic. The bird Z chromosome is highly conserved, as demonstrated by fluorescent *in situ* hybridization using a whole flow-sorted chicken Z chromosome as probe (chromosome painting). The chicken Z paints the Z chromosome of distantly related ratite birds, and even paints a similar sized autosome in turtles (Shetty *et al.*, 2000; Graves and Shetty, 2000). On the other hand, the W is very variable, being small and heterochromatic in carinate birds, but almost equivalent to the Z in ratites.

Although superficially similar to the large gene-rich X and small heterochromatic Y, the bird Z and W chromosomes are quite unrelated to the mammal sex chromosomes. Genes that map to the human X are found on chicken chromosomes 1 and 4, and genes on the chicken Z are mostly on human chromosome 9 (Nanda *et al.*, 1999). Most importantly, the sex determining genes that define the sex chromosomes are different in the two groups. Birds have no sex-specific *SRY* gene (Griffiths, 1991). Instead, a gene (*DMRT1*) on the Z chromosome appears to determine sex by its differential dosage of two copies in males and one in females (Raymond *et al.*, 1998;

Smith *et al.*, 1999), probably exaggerated by a controlling locus on the W that affects *DMRT1* methylation and activity in females. Human *DMRT1* lies on human chromosome 9 (Raymond *et al.*, 1999) and was known to be involved in sex determination because its deletion produced a sex reversed phenotype.

The conservation of the Z in birds and reptiles, and the conservation of *DMRT1* function across all vertebrates suggests that this gene has an ancient role in sex determination. Indeed, *DMRT1* has homologues involved in invertebrate sex determination (*dsx* in *Drosphila melanogaster* and *mab3* in *Caenorhabditis elegans*; hence its name, for *dsx-mab3* related transcription factor 1). This means that *SRY* has not always been the master switch that controls sex determination, but evolved in the lineage that led to eutherian mammals some time during the 300 million years since they diverged from reptiles.

Y chromosome differentiation

Ohno (1967) put forward a far-reaching hypothesis, initially to explain the peculiarities of snake sex chromosomes. He suggested that the variably sized W chromosome in different snake families represent stages in the gradual breakdown of the W, starting from a pair of autosomes. This hypothesis also explains the constant Z and the variable W chromosome in carinate and ratite birds. A process of W degradation has therefore taken place to different extents independently in different bird and snake lineages.

The same process is proposed to have occurred independently to differentiate the X and Y in mammals. Although the X and Y chromosomes are very different in size and gene content, homology is obvious in the pseudoautosomal regions as well as the "Class I" genes shared between the X and Y chromosomes. We conclude that the mammalian X and Y chromosomes also evolved from a pair of autosomes, albeit a different pair.

Why have the snake and bird W, and the mammalian Y chromosome, apparently degenerated? It is hard to imagine how gene deletion and loss of function could confer a selective advantage, and there has been vigorous debate about the forces which drive degradation of the Y (Charlesworth, 1991). It is thought that other genes with an advantage in males accumulate near the original sex determining locus. It then becomes an advantage to keep these genes together by suppressing recombination between the X and Y in this region. Genetic isolation of regions of the Y then leads to progressive degradation as the result of genetic drift ("Muller's ratchet"), or hitch-hiking of damaging mutations with a favourable mutation.

This theory views the snake and bird W as a degraded Z and the mammal Y as a degraded mammal X. Despite its male-dominant action, the Y is, after all, a wimp (Graves, 2000).

The mammalian proto-sex chromosomes

So what characteristics distinguished the original mammalian proto-X and Y chromosomes? Was this autosomal pair "chosen" for any particular reason? Was *SRY* the original sex determining locus, whose evolution initiated the differentiation of the mammal X and Y?

To deduce the gene content and function of the autosome that became the Y, we can define the regions of sex chromosomes that are shared between distantly related mammals. Comparisons of eutherians with marsupials (pouched mammals that diverged 130 million years ago) and monotremes (egg laying mammals that diverged 170 million years ago) have provided the variation to study the origin of mammalian sex chromosomes and sex determining genes.

The marsupial genome is much the same size as the human genome, but is divided up into a few very large autosomes. However, the X and Y are unusually small. The basic X constitutes only about 3% of the haploid genome, and the basic Y is only about 10 Mb. No synaptonemal complex and no chiasmata are observed during meiosis (Sharp, 1982) and no X–Y shared pseudoautosomal region is detected by painting the microdissected Y onto other chromosomes (Toder *et al.*, 2000).

Genes have been mapped to the marsupial X by somatic cell genetic analysis and fluorescent *in situ* hybridization (FISH) using cloned marsupial genes (reviewed Graves, 1995). Lining up the gene maps of the human and marsupial X showed that all the genes on the long arm of the human X and the pericentric region were also on the X in marsupials, implying that this region represented an ancient mammalian X that is at least 130 million years old. The same set of genes was also mapped to the X in monotremes, pushing back the age of this region to 170 million years. This region (called the X conserved region, XCR) is therefore equivalent to the ancient mammalian X. This has been beautifully confirmed using a probe made from a flow-sorted tammar X chromosome to hybridize *in situ* to human chromosomes. It painted only the XCR, the long arm of the X and the region around the centromere (Glas *et al.*, 2000).

The tiny marsupial Y chromosome also shares genes with the eutherian Y. Four genes from the human Y (*SRY, RPS4Y, SMCY, RBMY*) and one from the mouse Y (*Ube1y*) also map to the marsupial Y, so these must have been on the Y for at least 130 million years (Waters *et al.*, 2001). These genes define a Y conserved region (YCR) equivalent to XCR.

However, genes on the short arm of the human X all mapped to autosomes in marsupials and monotremes. Most cluster on the short arm of chromosome 5 in the tammar wallaby, and on chromosome 1 in monotremes. Since marsupials and monotremes diverged independently from eutherians, this must mean that an autosomal region was transferred to the X chromosome in the eutherian lineage. Thus the human X contains an X added region (XAR), which is still located on autosomes in non-eutherian mammals (Graves, 1995).

The human Y chromosome contains an equivalent added region. This is not surprising, since many genes on the human Y have partners within the XAR. The addition was directly demonstrated by cloning wallaby homologues of eight genes on the human Y with partners within the XAR (*DBY, USPY, ZFY, AMELY, STSP, ANT3* and *UTY*, and the pseudodautosomal *CSF2RA*). They all mapped to autosomes, most on the wallaby chromosome 5p, co-locating with the XAR genes (Waters *et al.*, 2001). These genes therefore define a region added to the Y (YAR) that shares homology with the X added region XAR and includes the pseudoautosomal region.

A piece of an autosome was therefore added to both the X and Y chromosomes early in the eutherian radiation. This probably occurred when the X and Y were only partly differentiated, when addition to an ancient PAR of one partner could result in its recombination onto the other partner (Graves, 1995).

Thus gene mapping reveals that the small marsupial X and Y chromosomes represent the ancient mammal X and Y chromosomes in a common ancestor of all mammals 170 million years ago. One or more autosomal additions occurred in the eutherian lineage. A different (as yet unknown) addition must also have occurred in the monotreme lineage to make up the large X and Y chromosome. In marsupial Family Macropodidae (the kangaroos and wallabies), too, ribosomal genes and associated heterochromatin was added to both the X and Y, then largely degenerated on the Y (Toder *et al.*, 1997). In other marsupials (e.g. the rare dasyurid, the dibbler, Young *et al.*, 1982), as in many eutherian species, large lumps of heterochromatin have been added to the Y, perhaps as a kind of ballast or an aid to pairing.

The tiny (10 Mb) marsupial Y, representing the last relic of the original Y, has its equivalent on the human Y, and we can now translate these data back to deduce the origin of the human Y chromosome. The four YCR genes shared with the human Y (*SRY, RPS4Y, SMCY* and *RBMY*) map to two small regions at either end of the human Y, defining a 10 Mb Y conserved region. This YCR was probably originally contiguous at the border of the large heterochomatic region but was separated by an inversion in the hominid lineage. Thus most of the extant human Y chromosome derives from a relatively recent addition to the sex chromosomes.

What was the sex determining gene that originally defined the Y? The *SRY* gene that now has the function of the sex-determining trigger, is proposed to have originated from its X homologue *SOX3*, which is expressed largely in the developing central nervous system. It has been suggested that *SRY* evolved from a prior system in which males were heterozygous for a null mutation of *SOX3*, and the dosage of active *SOX3* determined sex (Graves, 1998). The absence of *SRY* in monotremes, and the autosomal location of *SOX* genes suggests that there was an even more ancient sex determining gene.

One candidate for the sex determining gene that originally defined mammalian sex chromosomes is *ATRY* (Pask *et al.*, 2000. This gene has a widely expressed X-borne copy *ATRX* and a gonad-specific copy *ATRY* in marsupials, but in eutherians there is only an X-specific copy, the Y copy having been lost. The involvement of *ATRY* in testis determination is suggested by sex reversing mutations in human *ATRX*).

The decline and fall of the mammalian Y

What has the effect been of progressive degradation, and where will it all end? Again, comparative genetics has provided some answers.

Gene loss from the Y

Ohno's hypothesis predicts that loss of functional genes and insertion of non-coding repetitive DNA progressively degraded the proto-Y. The corollary of this hypothesis is that genes on the Y, including those essential for male-specific functions, are degraded versions of genes on the X. There is a great deal of evidence that the human Y represents just such a depauperate relic.

The high content of repetitive sequence on the Y chromosome is well documented. The 30 Mb on the distal region of the long arm is composed largely of two simple repeats. Even the euchromatic region on Yp and proximal Yq has very large repetitive regions. Some of these have a recent origin from illegitimate crossing over with the X (Affara and Ferguson-Smith, 1994), or amplification of degenerate gene sequences such as *TSPY*.

Since the original proto-Y is represented by the conserved region of the X, it originally contained about 1000 genes. Only 4 remain on the human Y (*SRY, RPS4Y, SMCY* and *RBMY*), and an overlapping four on the mouse Y (*Sry, Smcy, Rbmy* and *Ube1y*). The marsupial Y contains homologues of these five genes, plus *ATRY*, which has X- and Y-borne copies in marsupials, but not in human or mouse (Pask *et al.*, 2000). The other 990-odd genes have all been inactivated and deleted at least within the last 300 million years since mammals diverged from birds and reptiles. This gives a minimum loss rate of about 3.3 genes/Myr.

The autosomal region added to the Y 80–130 million years ago has also been largely stripped of its genes, although a greater proportion survive on the Y. The original autosomal region is represented intact as the XAR, which contains about 400-odd genes, about 12 of which have survived on the Y. The greater survival probably means only that the region has had less time to degrade (80–130 Myr). In fact the overall rate of loss of genes (~390 genes/130 Myr) is much the same as for te YCR (about 3/Myr). The continuing degradation of these genes is obvious by the recent inactivation of many sequence present on the Y only as pseudogenes (e.g. *STS, KAL, ARSD, ARSE, PRKY*).

Degradation of genes on the Y

Genes on the human Y – both the YCR and YAR – illustrate a spectrum of degradation and loss. Pseudoautosomal genes have homologues on the X and Y that pair and recombine at meiosis. Some genes (e.g. *SMCY*) still maintain active homologues on the Y, and their X homologues escape X inactivation in females. Other Y-borne genes (e.g. *RPS4Y* from the original Y and *AMELY* from the added region) are only partially active. Many genes, like *STS*, are represented only by pseudogenes on the Y that have been mutated and partially deleted so that they no longer code for a sensible protein. The overwhelming majority of genes have been completely deleted from the Y and their X homologues have become subject to X inactivation in females. Some genes are present on the Y in one species and absent on another. Different stages of degradation and loss are represented by the same gene in different species. For instance, *UBE1Y* is pseudoautosomal in monotremes, active but male-specific in mouse and marsupial, present only as pseudogene fragments in several primates, and absent from the human Y (Mitchell *et al.*, 1999).

The differences in extent of degradation is likely to be related to the time of isolation of the X and Y homologues from each other, which is a function both of the location on the original or the added segment, as well as events such as major inversions that isolated regions. These events have been timed by comparing the sequences of X and Y homologues. The finding that genes fall into five clusters have been suggested to indicate five "geological layers" that separated at different times (Lahn and Page, 1999a).

The genes on the Y that are of the most interest are those with a male-specific function. Such genes were classified as "Class II" genes by Lahn and Page (1997) on the basis of their testis-specific action, presence of multiple copies and the apparent absence of a homologue on the X. Are Y-borne genes with functions in sex determination or fertility a special class that did not arise from the X, but were recruited by a "selfish Y" because of their male-specific function? This is likely to be true of at least two genes recently transposed to the Y from autosomes (Lahn and Page, 1999b).

However, the division into Classes I and II genes breaks down when comparisons are made across species. For instance, *ZFY* is a single copy, ubiquitously expressed gene in humans, but its mouse homologue *Zfy* is testis-specific and there are multiple copies in several Old World mouse species (Nagamine, 1994). Even more damaging to the hypothesis is the observation that two genes with critical male-specific functions, originally classified as Class II by Lahn and Page (1997), do indeed have X homologues. *RBMY*, a multicopy candidate spermatogenesis gene in one of the AZF deletion regions, has been shown to have a ubiquitously expressed homologue *RBMX* on the X in human (Delbridge *et al.*, 1999) and mouse (Mazeyrat *et al.*, 1999), so undoubtedly arose from a widely expressed gene with functions in both sexes. Even *SRY*, the testis determining factor itself, has a homologue (*SOX3*) on the X, from which it evolved (Foster and Graves, 1994). It seems more likely that "Class I" and "Class II" genes simply represent opposite ends of a spectrum of Y gene degradation. At one extreme are genes that remain intact and at the other are genes that have been inactivated and lost.

How did widely expressed genes on the proto-Y acquire male-specific functions after differentiation of the Y? Each gene was moulded by mutations and become restricted in its transcription pattern. For instance, the candidate spermatogenesis gene *RBMY* has undergone mutation and exon amplification from the original *RBMX* gene, then been amplified into a gene family, only a few members of which are active. From a ubiquitous expression profile, its transcription has become restricted to germcells. The small *SRY* gene, which has all its biochemical activity within the DNA binding and bending HMG box, seems to have been derived by truncation of the *SOX3*. This left nothing except the HMG box, and could have changed its function from an activator to an inhibitor (Graves, 1998). *SRY* is expressed largely in the developing gonad in mice, but has some expression in the brain (Mayer *et al.*, 2001), whereas its ancestor *SOX3* is expressed largely in the central nervous system, with minor expression in the gonad (Collignon *et al.*, 1993). Its evolution is therefore typical of genes on the Y chromosome.

It seems clear, therefore, that the "functional coherence" of the Y resulted from ruthless extermination of any gene that did not contribute a strong selective advantage. Since genes on the Y cannot be required for viability (since half the population lacks a Y), their selective advantage must lie either in a strict adherence to dosage equality, or must be be gained by acquiring a critical role in reproduction.

Fall of the Y chromosome

The entire mammalian Y seems to be headed for extinction. There is practically nothing left of the original mammalian Y chromosome. Autosomal additions may have prologed the useful life of the eutherian Y for a little while longer, but even the gene rich added region is disappearing fast. At the rate of gene loss of more than 3/Myr that occurred in the human Y over the last 300 Myr, the last 30-odd genes would disappear in another 10 Myr.

The marsupial Y chromosome is even more precarious. In the absence of autosomal additions, it has been whittled down to a small non-recombining package containing *SRY* and a few spermatogenesis genes, but little else. Some species of marsupials dispense with the Y during somatic growth by eliminating it from the embryo (Watson *et al.*, 1999), implying it no longer bears any genes required for general functions. In macropodid marsupials, addition of the nucleolus organizer region to the macropodid X and Y increased the size of the sex chromosomes, but in most species only nucleolar heterochromatin remains on the Y.

The end of the line for the Y chromosome seems, indeed, to have come for two rodent species. Two mole vole species of the genus *Ellobius* have completely lost the Y chromosome. Animals of both sexes have an XX or XO sex chromosome constitution. A third species *E. fuscocapillus* has a normal Y and a normal *SRY* gene. However, no *SRY* gene homologue can be detected in either species lacking a Y (Just *et al.*, 1996), even using a probe amplified from the closely related *E. fuscocapillus*. Evidently some new gene has usurped *SRY* function in triggering the male developmental pathway in these species.

Mole voles are evidently leading the way to a new round of sex chromosome evolution, initiated by a new sex determining gene. There are no outward signs of heteromorphism, but somewhere in the *Ellobius* genome, a new sex chromosome system is just beginning the cycle of degradation, initiated by the evolution of a new sex determining factor. This is the same process as presumably occurred in the early evolution of mammals from reptiles, but over the last 300 million years the trail has gone rather cold. It will therefore be important to understand the *de novo* sex determining system of mole voles, in order to understand the process by which new mammal sex chromosomes may take over from old.

References

Affara NA, Ferguson-Smith MA (1994). DNA sequence homology between the human sex chromosomes. In: Wachtel SS, ed. *Molecular Genetics of Sex Determination.* San Diego, CA: Academic Press, pp. 255–266.

Bowles J, Cooper L, Berkman J, Koopman P (1999). *SRY* requires a CAG repeat domain for male sex determination in *Mus musculus. Nat Genet* 22: 405–408.

Charlesworth B (1999). The evolution of sex chromosomes. *Science* 251: 1,030–1,033.

Collignon J, Sockanathan S, Hacker A, Cohen-Tannoudji M, Norris D, Rastan S, *et al.* (1996). A comparison of the properties of *SOX*3 with *SRY* and two related genes, *SOX*1 and *SOX*2. *Development* 122: 509–520.
Delbridge ML, Disteche CM, Graves JAM (1999). The candidate spermatogenesis gene *RBMY* has a homologue on the human X chromosome. *Nat Genet* 22: 223–224.
Foster JW and Graves JAM (1994). An *SRY*-related sequence on the marsupial X chromosome: implications for the evolution of the mammalian testis-determining gene. *Proc Natl Acad Sci USA* 91: 1,927–1,931.
Foster JW, Dominguez-Steglich M, Guioli S, Kwok C, Weller PA, Stevanovic M, Weissenbach J, Mansour S, Young ID, Goodfellow PN, Brook JD, Schafer AJ (1996). Campomelic dysplasia and autosomal sex reversal caused by mutations in an SRY-related gene. *Nature* 372: 525–530.
Glas R, Graves JAM, Toder R, Ferguson-Smith MA, O'Brien PC (1999). Cross-species chromosome painting between human and marsupial directly demonstrates the ancient region of the mammalian X. *Chromosome Res* 10: 1,115–1,116.
Graves JAM (1995). The origin and function of the mamalian Y chromosome and Y-borne genes – an evolving understanding. *BioEssays* 17: 311–320.
Graves JAM (1998). Interactions between *SRY* and *SOX* genes in mammalian sex determination. *BioEssays* 20: 264–269.
Graves JAM (2000). The human Y chromosome, sex determination and spermatogenesis – a feminist view. *Biol Reprod* 63: 667–676.
Graves JAM, Delbridge ML (2001). The X – a smart and sexy chromosome. *BioEssays* (in press).
Graves JAM, Shetty S (2001). Sex from W to Z – evolution of vertebrate sex chromosomes and sex determining genes. *J Exp Zool* 281: 472–481.
Griffiths R (1991). The isolation of conserved DNA sequences related to the human sex-determining region Y gene from the lesser black-backed gull (*Larus fuscus*). *Proc Roy Soc Lond* 224: 123–128.
Gubbay J, Collignon J, Koopman P, Capel B, Economou A, *et al.* (1990). A gene mapping to the sex-determining region of the mouse Y chromosome is a member of a novel family of embryonically expressed genes. *Nature* 346: 245–250.
Harley VR, Jackson DI, Hextall PJ, *et al.* (1992). DNA binding activity of recombinant *SRY* from normal males and XY females. *Science* 255: 453–456.
Hawkins JR, Taylor A, Berta P, Levilliers J, van der Auwera B, Goodfellow PN (1992). Mutational analysis of *SRY*: nonsense and missnse mutations in XY sex reversal. *Hum Genet* 88: 471–474.
Just W, Rau W, Vogel W, Akhverdian M, Fredga K, Graves JAM, Lyapunova E (1995). Absence of *SRY* in species of the vole Ellobius. *Nat Genet* 11: 117–118.
Koopman P (1995). The molecular biology of *SRY* and its role in sex determination in mammals. *Reprod Fertil Dev* 7: 713–722.
Koopman P, Gubbay J, Vivian N, Goodfellow P, Lovell-Badge R (1991). Male development of chromosomal female mice transgenic for *SRY*. *Nature* 351: 117–121.
Lahn B, Page DC (1997). Functional coherence of the human Y chromosome. *Science* 278: 675–680.
Lahn BT, Page DC (1999a). Four evolutionary strata on the human X chromosome. *Science* 286: 964–967.
Lahn BT, Page DC (1999b). Retroposition of autosomal mRNA yielded testis-specific gene family on human Y chromosome. *Nat Genet* 21: 429–433.

Mayer A, Lahr G, Swaab DF, Pilgrim C, Reisert I (1998). The Y-chromosomal genes *SRY* and *ZFY* are transcribed in adult human brain. *Neurogenetics* 1: 281–288.

Mazeyrat S, Saut N, Mattei M-G, Mitchell MJ (2000). *RBMY* evolved on the Y chromosome from a ubiquitously transcribed X-Y identical gene. *Nat Genet* 22: 224–226.

McElreavey K, Rappaport R, Vilain E, Abbas N, Richaud F, Lortat-Jacob S, *et al.* (1992). A minority of 46,XX true hermaphrodites are positive for the Y-DNA sequence including *SRY*. *Hum Genet* 90: 121–125.

Mitchell MJ, Wilcox SA, Watson JM, Lerner JL, Woods DR, Scheffler J, Hearn JP, Bishop CE, Graves JAM (1998). The origin and loss of the ubiquitin activating enzyme gene on the mammalian Y chromosome. *Hum Mol Genet* 7: 429–434.

Nagamine CM, (1994). The testis-determining gene, *SRY*, exists in multiple copies in old world rodents. *Genet Res* 64: 151–159.

Nanda I, Shan Z, Schartl M, Burt DW, Koehlar M, *et al.* (1999). 300 million years of conserved synteny between chicken Z and human chromosome 9. *Nat Genet* 21: 258–259.

Ohno S. (1967). Sex chromosomes and sex linked genes. Berlin: Springer Verlag.

Page DC, Mosher R, Simpson EM, Fisher EMC, Mardon G, Pollack J, McGillivray B, de la Chapelle A (1987). The sex determining region of the human Y chromosome encodes a finger protein. *Cell* 51: 1,091–1,104.

Pask A, Renfree MB, Graves JAM (2000). The human sex-reversing gene *ATRX* has a homologue on the marsupial Y chromosome. *Proc Natl Acad Sci USA* 97: 13,198–13,202.

Raymond CS, Shamu CE, Shen MM, Seifert KJ, Hirsch B, Hodgkin J, Zarkower D (1998). Evidence for evolutionary conservation of sex-determining genes. *Nature* 391: 691–695.

Raymond CS, Parker ED, Kettlewell J, *et al.* (1999). A region of human chromosome 9p required for testis development contains two genes related to known sexual regulators. *Hum Mol Genet* 8: 989–996.

Sharp P (1982). Sex chromosome pairing during male meiosis in marsupials. *Chromosoma* 86: 27–47.

Shetty S, Griffin D, Graves JAM (1999). Comparative chromosome painting reveals strong chromosome homology over 80 million years of bird evolution. *Chromosome Res* 7: 289–295.

Sinclair AH, Berta P, Palmer MS, Hawkins JR, Griffiths BL, Smith MJ, Foster JW, Frischauf AM, Lovell-Badge R, Goodfellow, PN (1990). A gene from the human sex-determining region encodes a protein with homology to a conserved DNA-binding motif. *Nature* 346: 240–244.

Smith CA, McClive PJ, Western PS, Reed KJ, Sinclair AH (1999). Conservation of a sex determining gene. *Nature* 402: 601–602.

Toder R, Wienberg J, Voullaire L, O'Brien PCM, Maccarone P, Graves JAM (1997). Shared DNA sequences between the X and Y chromosomes in the tammar wallaby – evidence for independent additions to eutherian and marsupial sex chromosomes. *Chromosoma* 106: 94–98.

Toder R, Wakefield M, Graves JAM (2000). The minimal mammalian Y chromosome – the marsupial Y as a model system. *Cytogenet Cell Genet* 91: 285–292.

Uchikawa M, Kamachi Y, Kondoh H (1999). Two distinct subgroups of Group B *SOX* genes for transcriptional activators and repressors: their expression during embryonic organogenesis of the chicken. *Mech Dev* 84: 103–120.

Wagner T, Wirth J, Meyer J, Zabel B, Held M, Zimmer J, Pasantes J, Bricarelli FD, Keutel J, Hustert E, Wolf U, Tommerup N, Schempp W, Scherrer G (1994). Autosomal sex reversal and campomelic dysplasia are caused by mutations in and around the *SRY*-related gene *SOX9*. *Cell* 79: 1,111–1,120.

Waters P, Duffy B, Frost CJ, Delbridge ML, Graves JAM (2001). The human Y chromosome derives largely from a single autosomal region added 80–130 million years ago. *Cytogenet Cell Genet* 92: 74–79.

Watson CM, Margan SH, Johnston PG (1998). Sex-chromosome elimination in the bandicoot *Isoodon macrourus* using Y-linked markers. *Cytogenet Cell Genet* 81: 54–59.

Yen PH, Marsh B, Allen E, Tsai SP, Ellison J, Connolly L, Neiswanger K, Shapiro LJ (1988). The human X-linked steroid sulfatase gene and a Y-encoded pseudogene: evidence for an inversion of the Y chromosome during primate evolution. *Cell* 55: 1,123–1,135.

Young GJ, Graves JAM, Barbieri I, Woolley PA, Cooper DW, Westerman M (1982). The chromosomes of dasyurids (Marsupialia). In: Archer M, ed. *Carnivorous Marsupials*, vol. 2. NSW: Royal Zool. Soc, pp. 783–795.

1.2. Avian Sex Chromosomes and Sex-Linked Genes

I. Nanda[1], Z. Shan[3], C. Steinlein[1], M. Schartl[2], T. Haaf[3] and M. Schmid[1]

[1]Departments of Human Genetics, [2]Physiological Chemistry I, University of Wuerzburg, Wuerzburg, Germany; [3]Max-Planck-Institute of Molecular Genetics, Berlin, Germany

Avian sex determination operates through a ZZ/ZW sex chromosome system. In contrast to mammals, female birds are the heterogametic sex, carrying one copy each of the Z and W sex chromosomes, whereas the homogametic males have two identical Z chromosomes. Given the very distinct nature of mammalian and avian sex chromosomes, the sex determination pathway in birds is unlikely to resemble that in mammals, where the presence of the *SRY* gene product exclusively in males regulates expression of a cascade of downstream genes that triggers male sexual development. In addition, the failure to detect an avian *SRY* homolog(s) argues against its role in avian sex determination. However, a recent study suggests that at least some genes associated with gonadal development are conserved among mammals and birds (Smith *et al.*, 1999).

Compared with the mammalian X and Y sex chromosomes, the function(s) of the Z and W in avian sex determination is poorly understood (Ellegren, 2001; Graves and Shetty, 2001). It is not even clear whether avian sex is determined by a dominant factor on the W chromosome or by a dosage-sensitive factor on the Z. The identification of Z- and W-chromosomal genes which are involved in sexual differentiation can provide new insights into the evolution of independent sex determination mechanisms in vertebrates. Since the W chromosomes of modern birds are mainly composed of constitutive heterochromatin (Pigozzi, 1999), many studies on avian sex-determination mechanisms have focused on the Z. Construction of a comprehensive genetic map of the chicken (*Gallus gallus*), the most important avian animal model, has greatly benefited from the dramatic advances in human genome research, particularly in acquiring information regarding synteny of the chicken genome with cognate genes of mammalian species. In this chapter, we review recent progress in

M. Schmid and I. Nanda (eds.), Chromosomes Today, Volume 14, 15–26.

the research on avian sex chromosomes, in particularly on their gene content. Comparative gene mapping is one important tool for the identification of candidate sex-determining genes in birds. In addition, it should help to elucidate the origin of heteromorphic avian sex chromosomes and their evolutionary relationship to mammalian sex chromosomes.

Avian sex chromosomes and sex-determining mechanisms

In modern birds, especially in the subclass Carinatae, the Z and W sex chromosomes are highly differentiated and can be easily identified by standard cytogenetic procedures. The Z is usually a medium-sized macrochromosome, representing 7–10% of the entire genome. Considering the variable morphology of avian Z chromosomes, they must have undergone considerable structural changes during evolution. As a result, gene order may considerably differ between the Z chromosomes of different species. The average W chromosome of modern birds is much smaller than the Z and largely heterochromatic (Figure 1). In contrast, in so-called "primitive" birds the W resembles the Z in size and morphology and most likely also in gene content (Fridolfsson *et al.*, 1998; Ogawa *et al.*, 1998). The W of galliform birds is characterized by the presence of two large repeat DNA families (Saitoh and Mizuno, 1992) and contains only few structural genes. During meiosis the highly differentiated Z and W are paired to some extent and at least one recombination nodule is formed in the synapsed region (Pigozzi, 1999). In somatic tissues the W appears to be late-replicating and in a more condensed heterochromatic state, termed W chromatin (Suka *et al.*, 1993). Thus, in many cytological aspects the avian W appears to resemble the mammalian Y.

The primary role(s) of Z and W for the developmental decision to form testes or ovaries remains unknown. Because the W is only found in the heterogametic females, it has been speculated that avian sex is controlled by a dominant ovary-determining gene on the W. In this context, it is important to emphasize that ovariectomy of the left

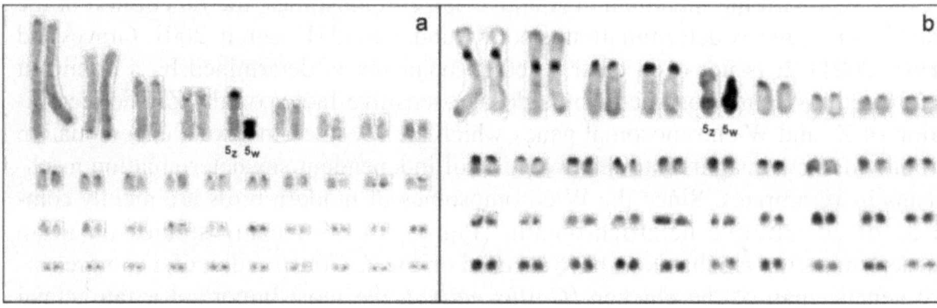

Figure 1. *Karyotypes of chicken (a) and quail (b) displaying C-band-positive heterochromatin in the sex chromosomes. Note the large amounts of constitutive heterochromatin in the W.*

gonad in young chickens induces differentiation of the right gonad into a testis. Similarly, treatment with aromatase inhibitors during early development causes a male phenotype in genetically female (ZW) birds (Elbrecht and Smith, 1992; Clinton and Haines, 2001). Triploid ZZW chickens first appear to be female and then progressively change to a male phenotype (Thorne and Sheldon, 1993). None of these findings is consistent with a dominant ovary-determining factor on the W.

Alternatively, avian sex may be determined by the dosage of a Z-linked gene(s). In this scenario, male birds with two Z chromosomes develop testes, whereas one Z, either alone or along with a W chromosome, leads to female sexual differentiation. In most dosage-based sex determination systems, i.e., in *Drosophila* and *Caenorhabditis elegans*, the ratio between the number of (X) sex chromosomes and autosomes acts as the primary genetic switch. Unfortunately, since ZO aneuploids have not been described in birds, it is difficult to prove the importance of Z chromosome dosage for avian sex determination. It is possible that the expression of Z-orthologous genes on the W is essential for avian normal embryo development and, therefore, ZO embryos do not develop to term.

Absence of dosage compensation

In birds the homogametic (ZZ) male sex possesses a double dose of Z-linked genes compared with females. Highly diverged animals with differentiated sex chromosome systems including mammals, *Drosophila* and nematodes have evolved independent mechanisms to ensure comparable dosages of sex-linked gene products between males and females. However, several lines of evidence suggest the absence of dosage compensation in birds. First, the additional Z chromosome in normal males and aneuploid/triploid birds does not cause formation of a Barr body-like condensed structure during interphase and does not result in heteropycnotic behaviour during meiosis. Both Z chromosomes appear to adopt an active chromatin structure throughout the cell cycle. Secondly, bromodeoxyuridine-labeling experiments did not detect replication asynchrony, which is a cytological hallmark of X inactivation, between the two Z chromosomes in male birds (Schmid *et al.*, 1989). Thirdly, enzymatic activity of the Z-linked cytosolic aconitase was consistently higher in males than in females in a number of bird species, whereas activity of an autosomal gene did not differ between sexes (Baverstock *et al.*, 1982). Unfortunately, these classical experiments do not provide unequivocal evidence for the absence of Z dosage compensation. In fact, two recent reports, one quantifying mRNA levels in developing chick embryos with real-time PCR and the other one using FISH with antisense riboprobes to detect nascent transcripts, presented contradictory results. The PCR approach showed equivalent expression of at least six of nine tested Z-linked genes in both male and female embryos (McQueen *et al.*, 2001), whereas RNA FISH always revealed two hybridization signals (transcription sites) of Z-linked genes in males (Kuroda *et al.*, 2001). Although the *in situ* experiment excludes transcriptional silencing of one Z-linked allele in males, it cannot rule out the formal possibility that transcription of both Z alleles is downregulated in males. However, despite apparently conflicting results at the single gene level, it may be safe to conclude that in contrast to mammals, chromosome-wide inactivation (involving replication asynchrony and facultative heterochromatinization) does not exist in birds.

Conserved synteny between avian sex chromosomes and human chromosomes

Because of the economical importance of poultry breeding and the growing scientific interest in comparative genomics, international collaborative efforts have been made to construct a genetic map of the chicken genome, which up to now includes more than 2,000 loci (Schmid *et al.*, 2000). The Z chromosome spans almost 210 cM (Levin *et al.*, 1993). Several loci, mainly morphological traits had already been assigned to the chicken Z by classical linkage analysis (Bitgood, 1993). More recently 28 genes were physically and/or genetically mapped to the Z, four of which also map to the W (Figure 2). In addition to type I loci, several anonymous DNA sequences including repetitive DNAs and pseudogenes were identified on both Z and W (Ellegren, 2000).

By comparing the location of these Z-linked genes with the human genome, 18 Z-linked genes were found to have orthologs on human chromosome 9pter-q32, demonstrating that the chicken Z is largely homologous to human 9 (Figure 2). In addition, human chromosomes 5, 8 and 18 show smaller segments of conserved synteny with the chicken Z. Evidently, translocations and/or insertions that happened during evolution have moved genetic material from the last common ancestor of human 9 and chicken Z to other chromosomes. So far no human orthologs of Z-linked *ZOV3*, a member of an immunoglobulin gene superfamily, and *AVIDIN* have been identified. Although the overall homology between chicken Z and human 9 is extensive, the gene order has not been highly conserved (Nanda *et al.*, 1999). Thus, intrachromosomal rearrangements must have occurred much more frequently during vertebrate evolution than interchromosomal rearrangements. The results of comparative gene mapping indicate that an inversion has either moved *B4GALT1-ACO1* in the avian lineage to the chicken Zq or in the mammalian lineage to human 9p (Nanda *et al.*, 2000). Although the human orthologs of Z-linked *ATP5A1* and *PKCI (HINT)* are not located in the syntenic regions of the human genome, it is interesting to note that human 9 carries an *ATP5A1* pseudogene and a *HINT* paralogue. The presence of orthologous genes on interspecific homologous chromosome segments reflects conservation of synteny from the last common ancestor of mammals and birds, which diverged approximately 300 million years ago (Kumar and Hedges, 1998). It is evident that most of the avian Z and human 9 evolved from the same ancestral autosome. In contrast, comparative gene mapping data did not detect orthologous loci on the avian Z and mammalian X chromosomes. This makes it very unlikely that avian and mammalian sex chromosome systems have a common ancestral origin.

Identification of a Z-linked male-regulatory locus

Information on conserved chromosomal synteny in the chicken, a relatively poorly mapped species compared with humans, can greatly facilitate the assignment of additional chicken genes to a region of interest (Burt *et al.* 1999). The delineated

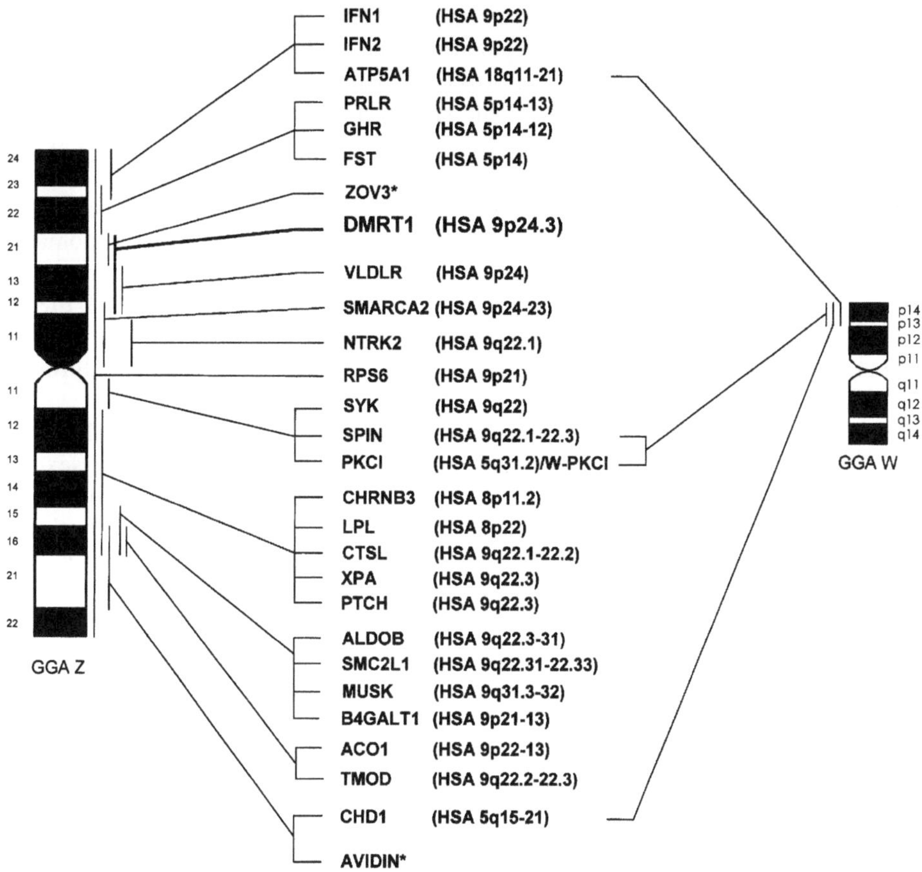

Figure 2. G-banded ideograms of the chicken Z and W chromosomes showing the cytogenetic locations of 28 Z- and four W-linked genes. The map position of individual genes may slightly differ when cytogenetic and genetic information are integrated. The chromosomal positions of CHD1 *and* ATP5A1 *in the W chromosome have not yet been resolved. Considering the heterochromatic nature of large parts of the W including the entire long arm, both genes are positioned tentatively on the short arm. The locations of human orthologs are shown in parentheses, illustrating conserved synteny between chicken Z and human 9. No human orthologs have been identified for two Z-linked genes (*). The presence of Z-linked homologs of the four W-linked genes demonstrates the homology between Z and W.*

homology between chicken Z and human 9 gained considerable attention, because several reports have associated human 9p deletions with gonadal dysgenesis and XY sex reversal. The critical region has been narrowed down to a relatively small interval on human 9p24. It contains two candidate genes, *DMRT1* (Doublesex and Mab-3 related transcription factor 1) and *DMRT2*, which are expressed in the adult testis.

Both *DMRT1* and *DMRT2* have the DNA-binding DM domain and share significant structural homology with male sex-regulatory genes from *Caenorhabditis elegans* (mab-3) and *Drosophila melanogaster* (dsx) (Raymond *et al.*, 1998, 1999a). The most likely explanation for male to female sex reversal in patients with monosomy 9p is haploinsufficiency of a dosage-sensitve gene in the critical region. Based on comparative chromosome maps, it was plausible to assume that a *DMRT* ortholog(s) on the Z may be implicated in avian sex determination. Screening of a chicken cDNA library with a human EST clone containing the highly conserved DM domain identified an avian *DMRT1* ortholog. The chicken DMRT1 has 95% similarity (at the amino acid level) with the DM domain of human DMRT1. As expected, FISH mapping of a chicken *DMRT1*-containing cosmid revealed its location on the short arm of the chicken Z (Figure 3) and on the Z chromosomes of other galliform birds (Figure 3).

Since evolutionary conservation is usually considered a good indicator of biological significance, these findings argue in favor of the notion that avian sex is determined by a gene-dosage effect: two *DMRT1* doses may be required for testis formation, whereas a single gene dose leads to female sexual differentiation. Additional evidence comes from studies on *DMRT1* expression in male and female chickens. Whole mount *in situ* hybridization revealed higher *DMRT1* mRNA levels in the genital ridges (undifferentiated gonads) of male embryos than in those of female embryos. After the onset of sexual differentiation *DMRT1* expression became testis-specific (Figure 4). There was no detectable expression in the developing ovary. Testis-specific expression in adult birds was confirmed by Northern blot and RT–PCR analysis (Figure 5). Most importantly, *DMRT1* expression was

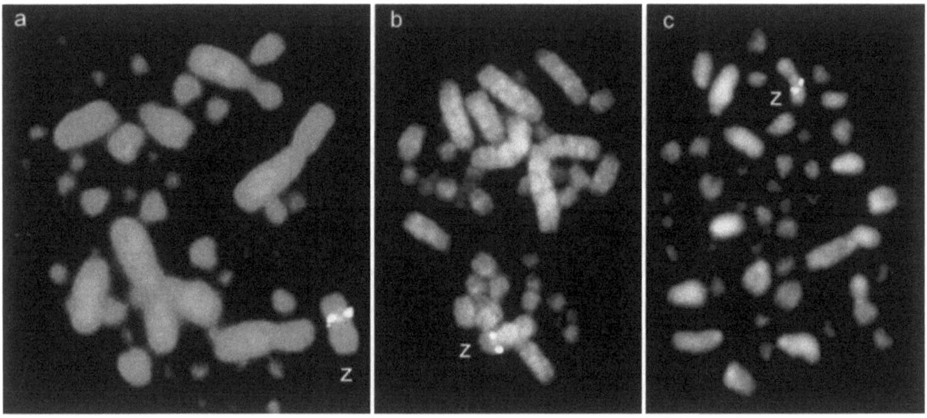

Figure 3. *Conserved location of* DMRT1 *on avian Z chromosomes. Hybridization of a chicken* DMRT1*-containing cosmid to the Z chromosomes of chicken (a), golden pheasant (b) and pheasant (c). The biotinylated probe DNA was detected by FITC-avidin. Chromosomes are counterstained with DAPI.*

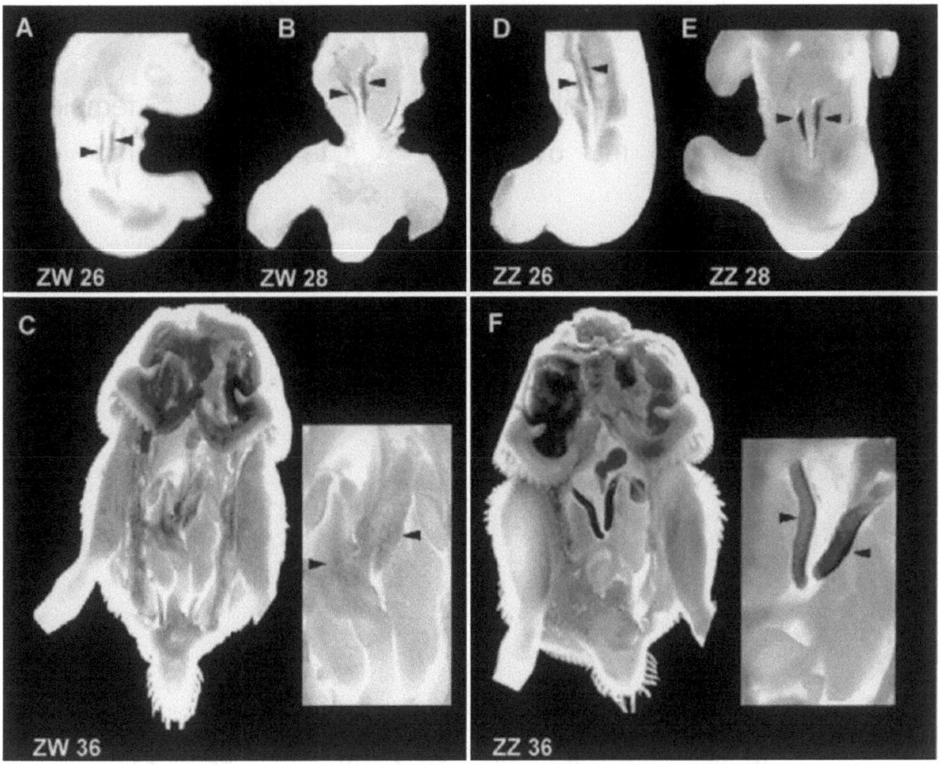

Figure 4. DMRT1 *expression during chicken embryogenesis. Chick embryos of both sexes were subjected to non-radioactive whole mount* in situ *hybridization with anti-sense RNA. (A) female embryo prior to sexual differentiation at stage HH26. (B) female embryo during sexual differentiation at stage HH28. Note the specific expression of* DMRT1 *in the genital ridges (arrows) of the developing urogenital system. (C) Female embryo after sexual differentiation at stage HH 36. The insert shows a magnification of the urogenital system. Note the absence of detectable* DMRT1 *expression in the ovaries (arrows). (D,E) Male embryos at stage HH 26 (D) and HH 28 (E).* DMRT1 *expression in the developing male gonads is stronger than in corresponding female embryos. (F) Male embryo with differentiated testis at stage HH 36. Insert shows strong* DMRT1 *expression in testis.*

detectable prior to *AMH*, up to now the earliest marker for testis differentiation in birds (Raymond *et al.*, 1999b; Shan *et al.*, 2000). Although several genes with female specific functions, i.e., *ZOV3, VLDLDR and AVIDIN*, are known to reside on the Z, *DMRT1* is the only Z-linked gene which plays a role in male sexual differention and, thus, is an excellent candidate for the primary testis-determining gene in birds.

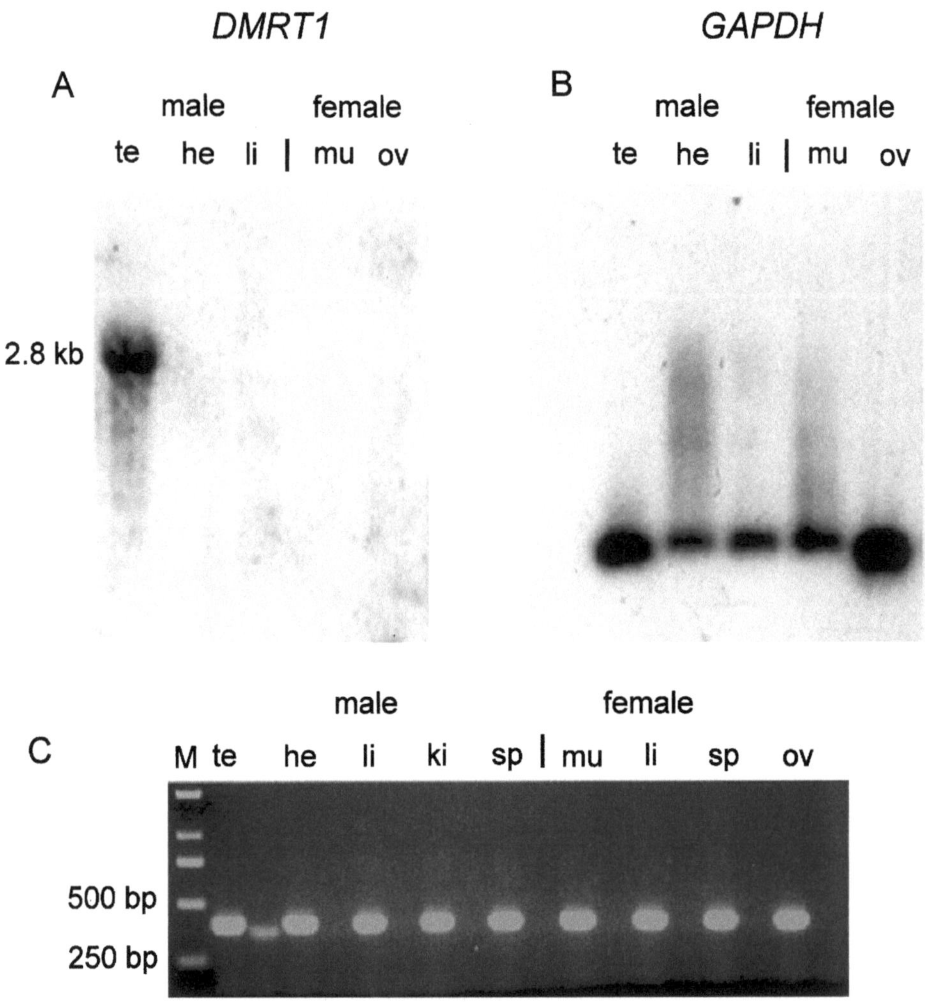

Figure 5. *Testis-specific expression of* DMRT1 *in adult chickens. (A) Northern blot with RNAs from three male and two female tissues hybridized with chicken* DMRT1 *cDNA. (B) The same blot rehybridized with chicken* GAPDH *cDNA. (C) RT–PCR reaction performed with RNA preparations from testis (te), heart (he), liver (li), kidney (ki), spleen (sp), muscle (mu), and ovary (ov). The first lane of each tissue represents the control amplification of a 387-bp* GAPDH *cDNA fragment), the second lane amplification with a primer pair specific for* DMRT1. *A 367-bp* DMRT1 *cDNA fragment is amplified exclusively from testicular RNA.*

W-linked genes

Since the W chromosome is much smaller than the Z and largely heterochromatic, it is not surprising that the hunt for W-linked genes has not been very rewarding. So far four genes that were assigned to the chicken W are known to have homologs on the Z.

The genomic structure of *ATP5A1, CHD1* and *SPIN* appears to be identical on Z and W, whereas Z- and W-linked *PKCI* display a different structural organization. The amino acid sequences of W-PKCI and PKCI Z are 65% identical. *PKCI Z* maps near the centromere of the Z long arm. *W-PKCI*, which corresponds to the independently isolated W-linked *ASW* locus (O'Neill *et al.*, 2000), lacks the characteristic HIT motif but contains a conserved alpha-helix and unique Leu- and Arg-rich regions. It maps to the non-heterochromatic region of the W and is highly conserved on the W chromosomes of Carinatae birds. *PKCI* encodes a functional protein that may play a role in avian sex determination, particularly in the female developmental pathway. Whole mount *in situ* hybridization gave a strong W-PKCI signal in the undiffferentiated female gonads but not in male embryos. It has been proposed that W-PKCI and PKCI Z form heterodimers and homodimers which can trigger female versus male sexual development (Hori *et al.*, 2000).

Conservation and evolution of avian sex chromosomes

Based on similar Z chromosome sizes and conserved Z linkage of common genetic traits such as albinism in different bird species, Ohno (1967) proposed that similar to the mammalian X, the Z sex chromosome has been conserved during avian evolution. Early biochemical studies on the Z-linked aconitase gene product (Baverstock *et al.*, 1982) in different bird orders confirmed Ohno's hypothesis of the conservation of Z-linked genes. Later, classical G-banding (Christidis, 1990) and DNA replication banding (Schmid *et al.*, 1989) demonstrated extensive structural and functional homology between different avian Z chromosomes. The partial meiotic pairing between Z and W in Carinatae birds (Pigozzi, 1999) suggests homology of an evolutionarily conserved segment between Z and W.

At least three genes, *IREBP, ZOV3* and *CHD1*, and one anonymous DNA marker, *EEO.6*, have been mapped to the Z chromosomes in widely different bird species, including the primitive Ratitae (Saitoh *et al.*, 1993; Ogawa *et al.*, 1998; Ellegren, 2000). Comparative hybridization of four Z-linked chicken probes directly visualized the orthologous loci on the quail Z (Suzuki *et al.*, 1999). Similarly, a chicken cosmid was used to localize the candidate testis-determining factor, *DMRT1*, to the Z of different galliform birds (Nanda *et al.*, 2000). Since these syntenic loci are distributed along the entire length of the chicken Z, it is unlikely that only particular Z segments have been conserved during avian evolution. In addition, a chicken Z chromosome-specific DNA library painted both the Z and W of the primitive emu (Shetty *et al.*, 1999), indicating that Z and W differentiated from an ancestral homologous chromosome pair. Despite its highly variable morphology, the Z has been conserved in its entirety during avian evolution. It is generally assumed that conservation of the mammalian X as a chromosomal unit is a consequence of dosage compensation by X inactivation. However, to the extent of our present knowledge, dosage compensation of sex-linked genes through chromosome-wide inactivation does not exist in birds. Thus, the mechanism responsible for synteny conservation of the avian Z remains to be elucidated. The observed extraordinary chromosomal stability may also apply to other macrochromosomes of the bird karyotype.

The few known W-linked genes have also been conserved during evolution. Most importantly, *W-PKCI* (*ASW*), up to now the strongest candidate for an ovary-determining factor, was found on the W chromosome of different modern birds (O'Neill *et al.*, 2000; Hori *et al.*, 2000). In addition, *CHD1* and *EEO.6*, which have homologs on the Z, were mapped to the W chromosome of many different birds. Their W-linked copies provide useful molecular markers for determining the sex of carinate birds by PCR or restriction analysis. The four genes and the other DNA sequences that have been mapped to both Z and W (Fridolfsson *et al.*, 1998; Ellegren, 2000) provide strong evidence for the common evolutionary history of Z and W. It is interesting to note that *CHD1, ATP5A1, SPIN*, and *W-PKCI* are W-linked in modern birds, however Southern blot hybridization to genomic DNA of primitive birds did not always produce sex-specific RFLPs. One possible explanation is the largely homomorphic nature of the sex chromosomes in primitive birds which may indicate a highly homologous DNA sequences. Interestingly, FISH analysis revealed Z-linked *ZOV3* alleles on both homologs of the fourth chromosome pair in female ostrich, whereas Z-linked *IREBP* was detected in only one homolog (Ogawa *et al.*, 1998). This reflects the early stage of sex chromosome differentiation between the ostrich Z and W. A small deletion including the *IREB* locus may have generated an active (Z) as well as a degenerate copy (W) of primitive avian sex chromosomes. Subsequent heterochromatinization and additional deletions of genetically inert material on the W, as well as extensive intrachromosomal rearrangements of the Z may have contributed to the evolution of highly differentiated sex chromosomes in modern birds.

Conclusion

Until recently very little was known about the mechanism(s) regulating the development of testis and ovary in birds. Comparative gene mapping and molecular analyses have identified two candidate genes for avian sex determination, *DMRT1* on the Z and *W-PKCI* on the W. This has left us with the baffling question, whether Z dosage, a dominant factor on the W or both act as a primary genetic switch(es). Considering the conserved role of *DMRT1* in male sexual differentiation among different phylogenetic groups, its Z-chromosomal location in birds and its expression pattern in early chick embryos and differentiated testes, *DMRT1* appears to be the best candidate for a primary sex determining gene in birds. Although mammalian *DMRT1* functions downstream in the hierarchy of sex-determining genes, testicular differentiation of both the avian and mammalian genital ridge involves a *DMRT1* dosage-sensitive mechanism.

The relatively low expression of *DMRT1* in the genital ridges of ZW embryos during early developmental stages and the subsequent complete cessation of expression in the ovary may be crucial for female sexual differentiation. The mechanism silencing the one *DMRT* gene in female embryos remains to be elucidated. It is possible that high *W-PKCI* expression from the W chromosome in the female developing gonad interferes with *DMRT1* expression, thus promoting development of an ovary. However, further experiments, i.e. the introduction of additional *DMRT1* copies in female embryos are necessary to assess the possibly interdependent roles of these two key sexual regulatory genes. The fact that the phenotype of ZZW triploids progressivley changes from female to male phenotype provides circumstantial evidence that Z chromosome dosage outweighs the influence of a dominant W-linked gene(s).

References

Baverstock PR, Adams M, Polkinghorne RW, Gelder M (1982). A sex-linked enzyme in birds – Z-chromosome conservation but no dosage compensation. *Nature* 296: 763–766.

Bitgood JJ (1993). The genetic map of the chicken and availability of genetically diverse stocks. In: Etches RJ, Gibbins AMV, eds. *Manipulation of the Avian Genome*. Boca Raton: CRC Press, pp. 61–79.

Burt DW, Bruley C, Dunn IC, Jones CT, Ramage A, Law AS, Morrice DR, Paton IR, Smith J, Windsor D, Sazanov A, Fries R, Waddington D (1999). The dynamics of chromosome evolution in birds and mammals. *Nature* 402: 411–413.

Christidis L (1990). Aves. In: John B, ed. *Animal cytogenetics, vol. 4: Chordata 3*, Berlin: Gebrüder Bortnträager.

Clinton M, Haines LC (2001). An overview of factors influencing sex determination and gonadal development in birds. In: Scheer G, Schmid M, eds. *Genes and Mechanisms in Vertebrate Sex Determination*. Basel: Birkhäuser Verlag, pp. 97–115.

Elbrecht A, Smith RG (1992). Aromatase enzyme activity and sex determination in chickens. *Science* 255: 467–469.

Ellegren H (2000). Evolution of the avian sex chromosomes and their role in sex determination. *Trends Ecol Evol* 15: 188–192.

Ellegren H (2001). Hens, cocks and avian sex determination; A quest for gene on Z or W. *EMBO Reports* 21: 192–196

Fridolfsson A-K, Cheng H, Copeland NG, Jenkins NA, Liu H-C, Raudsepp T, Woodage T, Chowdhary B, Halverson J, Ellegren H (1998). Evolution of the avian sex chromosomes from an ancestral pair of autosomes. *Proc Natl Acad Sci USA* 95: 8,147–8,152.

Graves JAM, Shetty S (2001). Sex from W to Z: evolution of vertebrate sex chromosomes and sex determining genes. *J Exp Zool* 290: 449–462.

Hori T, Asakawa S, Itoh Y, Shimizu N, Mizuno S (2000). Wpkci, encoding an altered form of PKCI, is conserved widely on the avian W chromosome and expressed in early female embryos: implication of its role in female sex determination. *Mol Biol Cell* 11: 3,645–3,660.

Kumar S, Hedges SB (1998). A molecular timescale for vertebrate evolution. *Nature* 392: 917–920.

Kuroda Y, Arai N, Arita M, Teranishi M, Hori T, Harata M, Mizuno S (2001). Absence of Z-chromosome inactivation for five genes in male chickens. *Chromosome Res* 9: 457–468.

Levin I, Crittenden LB, Dodgson JB (1993). Genetic map of the chicken Z chromosome using random amplified polymorphic DNA (RAPD) markers. *Genomics* 16: 224–230.

McQueen H, McBridge D, Milele G, Bird A, Clinton M (2001). Dosage compensation in birds. *Curr Biol* 11: 253–257.

Nanda I, Shan Z, Schartl M, Burt DW, Koehler M, Nothwang H, Grutzner F, *et al.* (1999). 300 million years of conserved synteny between chicken Z and human chromosome 9. *Nature Genet* 21: 258–259.

Nanda I, Zend-Ajusch E, Shan Z, Grutzner F, Schartl M, Burt DW, Koehler M, Fowler VM, Goodwin G, Schneider WJ, Mizuno S, Dechant G, Haaf T, Schmid M (2000). Conserved synteny between the chicken Z sex chromosome and human chromosome 9 includes the male regulatory gene DMRT1: a comparative (re)view on avian sex determination. *Cytogenet Cell Genet* 89: 67–78.

Ogawa A, Murata K, Mizuno S (1998). The location of Z- and W-linked marker genes and sequence on the homomorphic sex chromosomes of the ostrich and the emu. *Proc Natl Acad Sci USA* 95: 4,415–4,418.

Ohno S (1967). *Sex Chromosomes and Sex Linked Genes*. Berlin: Springer Verlag.

O'Neill M, Binder M, Smith C, Andrews J, Reed K, Smith M, Millar C, Lambert D, Sinclair A (2000). ASW: a gene with conserved avian W-linkage and female specific expression in chick embryonic gonad. *Dev Genes Evol* 10: 243–249.

Pigozzi MI (1999). Origin and evolution of the sex chromosomes in birds. *Biocell* 23: 79–95.

Raymond CS, Shamu CE, Shen MM, Seifert KJ, Hirsch B, Hodgkin J, Zarkower D (1998). Evidence for evolutionary conservation of sex-determining genes. *Nature* 391: 691–695.

Raymond CS, Parker ED, Kettlewell JR, Brown LG, Page DC, Kusz K, Jaruzelska J, Reinberg Y, Flejter WL, Bardwell VJ, Hirsch B, Zarkower D (1999a). A region of human chromosome 9p required for testis development contains two genes related to known sexual regulators. *Hum Mol Genet* 8: 989–996.

Raymond CS, Kettlewell JR, Hirsch B, Bardwell VJ, Zarkower D (1999b). Expression of *Dmrt1* in the genital ridge of mouse and chicken embryos suggests a role in vertebrate sexual development. *Dev Biol* 215: 208–220.

Saitoh Y, Mizuno S (1992). Distribution of *Xho*I and *Eco*RI family repetitive DNA sequences into separate domains in the chicken W chromosome. *Chromosoma* 101: 474–477.

Saitoh Y, Ogawa A, Hori T, Kunita R, Mizuno S (1993). Identification and localization of two genes on the chicken Z chromosome: implication of evolutionary conservation of the Z chromosome among avian species. *Chromosome Res* 1: 239–251.

Schmid M, Enderle E, Schindler D, Schempp W (1989). Chromosomal banding and DNA replication patterns in bird karyotypes. *Cytogenet Cell Genet* 52: 139–146.

Schmid M, Nanda I, Guttenbach M, Steinlein C, Hoehn M, Schartl M, Haaf T, Weigend S, Fries R, Buerstedde JM, Wimmers K, Burt DW, Smith J, A'Hara S, Law A, Griffin DK, Bumstead N, Kaufman J, Thomson PA, Burke T, Groenen MA, Crooijmans RP, Vignal A, Fillon V, Morisson M, Pitel F, Tixier-Boichard M, Ladjali-Mohammedi K, Hillel J, Maki-Tanila A, Cheng HH, Delany ME, Burnside J, Mizuno S (2000). First report on chicken genes and chromosomes 2000. *Cytogenet Cell Genet* 90: 169–218.

Shan Z, Nanda I, Wang Y, Schmid M, Vortkamp A, Haaf T (2000). Sex-specific expression of an evolutionarily conserved male regulatory gene, DMRT1, in birds. *Cytogenet Cell Genet* 89: 252–257.

Shetty S, Griffin DK, Graves JA (1999). Comparative painting reveals strong chromosome homology over 80 million years of bird evolution. *Chromosome Res* 7: 289–295.

Smith CA, Smith MJ, Sinclair AH (1999). Gene expression during gonadogenesis in the chicken embryo. *Gene* 234: 395–402.

Suka N, Shinohara Y, Saitoh Y, Saitoh H, Ohtomo K, Harata M, Shpigelman E, Mizuno S (1993). W-heterochromatin of chicken; its unusual DNA components, late replication, and chromatin structure. *Genetica* 88: 93–105.

Suzuki T, Kansaku N, Kurosaki T, Shimada K, Zadworny D, Koide M, Mano T, Namikawa T, Matsuda Y (1999). Comparative FISH mapping on Z chromosomes of chicken and Japanese quail. *Cytogenet Cell Genet* 87: 22–26.

Thorne MH, Sheldon BL (1993). Triploid intersex and chimeric chickens: useful models for studies of avian sex determination. In: Reed KC, Graves JAM, eds. *Sex Chromosomes and Sex-Determining Genes*. Chur: Harwood Academic Publishers, pp. 199–205.

1.3. *DMRT* Genes and Sex Determination in Medaka

Ute Hornung[1], Indrajit Nanda[2], Mariko Kondo[1], Atsushi Shimizu[3], Shuichi Asakawa[3], Jean-Nicolas Volff[1], Christoph Winkler[1], Zhihong Shan[4], Thomas Haaf[4], Nobuyoshi Shimizu[3], Akihiro Shima[5], Michael Schmid[2] and Manfred Schartl[1]

[1]Physiologische Chemie I, [2]Institute for Human Genetics, Biocenter, University of Würzburg, Germany; [3]Department of Molecular Biology, Keio University School of Medicine, Tokyo, Japan; [4]Max Planck Institute for Molecular Genetics, Berlin, Germany; [5]Department of Integrated Biosciences, Graduate School of Frontier Sciences, University of Tokyo, Bunkyo-ku, Tokyo 113–0033, Japan

Introduction

The vast majority of animal species are bisexual, and in many cases the decision, if an embryo develops to a male or female, is made by the genome. Sex determination genes initiate a series of developmental processes that establish the male or female phenotype. The genetic scenarios how sex can be determined are of extraordinary diversity and have provoked questions on the evolution and maintenance of bisexuality. One of the unsolved problem is, why such a simple and highly conserved biological phenomenon as the existence of males and females should be controlled by so different mechanisms like the ratio of X chromosomes to autosomes for instance in the fruit fly *Drosophila melanogaster* and the worm *Caenorhabditis elegans* or the presence of a Y chromosome in male mammals.

In most mammals, in several flies and in the worm *C. elegans* the genes that translate the chromosomal systems into molecular actions are known and reasonably well understood. However, the master regulator encoded on the Y chromosome of most mammalian species, *SRY*, is not functioning like that in some mammals (Just *et al.*, 1995). In non-mammalian species, which have also a XX/XY sex determination system *SRY* is not present at all. Neither the *sxl* gene of *Drosophila* nor *xol* of *C. elegans*, the genes, which are at the top of the sex determination cascade in these

M. Schmid and I. Nanda (eds.), Chromosomes Today, Volume 14, 27–37.

organisms, are not functioning in the same way in more distantly related species (for review see Marin and Baker, 1998). This indicates that on the molecular level another dimension of diversity is added to the complex situation of multiple genetic systems for sex determination. In between worms and flies on the one side and mammals on the other side, there is a large gap in our knowledge about sex determination genes. This makes it difficult to discuss the evolution of sex from a molecular biological perspective.

Fishes are an attractive group of organisms to study the evolution of sex determination because members of this class exemplify a broad range of various types of sexuality from hermaphrodism to gonochorism and from environmental to genetic sex determination (for review see Baroiller, 1999). This offers the possibility to compare the structure and expression of sex determination genes in such species and evaluate thus their relative importance in the respective organism, their position in the sex determination cascade and their conservation during evolution. Unfortunately, in both main fish models, the pufferfish (*Takifugu rubripes*) and the zebrafish (*Danio rerio*), no information exists on the mode of sex determination, the potential presence of sex chromosomes and the process of sex differentiation. Remarkably, none of the many markers of the zebrafish genomic map was found to be sex-linked (Amores *et al.*, 1998; Postlethwait *et al.*, 2000; Woods *et al.*, 2000).

The situation for a molecular analysis of sex determination is much more favourable in another recently emerging small aquarium fish model species, the medaka (*Oryzias latipes*) (for review see Wittbrodt *et al.*, 2002). Medaka has a XX/XY sex determination system like mammals. The Y chromosome carries by definition the male determining locus, which initiates the processes leading to male development of the undifferentiated primordial gonad. Male and female medaka are distinguished by a number of secondary sex characters (see Yamamoto, 1975) some of which – like the shape and size of the dorsal and anal fin – can be easily determined. In addition two loci for body coloration (*lf, R*) are closely linked to the sex determination locus. In the *drR* strain of medaka females are colourless white and males are orange, because of a dominant allele or *R* on the Y. In the *quart* strain the sex chromosomes express different alleles of the *lf* pigment marker (wild-type allele on the Y chromosome and a mutant allele on the X chromosome). The presence of leucophores in the male and their absence on the female allow to differentiate both sexes as early as at 2–3 days of embryonic development (Wada *et al.*, 1998).

Sex chromosomes and sex reversal in medaka

It is generally accepted that the gonosomes have evolved from a pair of autosomes and that the Y chromosome (or in the case of female heterogamety the W chromosome) undergoes a process of recombinational isolation (Charlesworth, 1991). This prevents crossing-overs in the sex determination region between both types of sex chromosomes and guarantees the persistence of the hemizygous chromosome. It is brought about for example by a degeneration process on the Y, which includes the loss of most functional genes and the accumulation of transposable elements and other repetitive DNA. In many species this process has progressed to a stage, where almost over their entire length the X and Y chromosomes do not pair in meiosis and consequently crossing over is absent. Only a hand-full of genes is encoded besides *SRY* in the non-recombining region of human Y. A number of them are multicopy and have no homologue on the X and some are spermatogenesis genes. This obviously represents the most advanced stage of Y chromosome evolution (see Lahn *et al.*, 2001).

The X and Y chromosomes of medaka are considered to be at a very early stage of evolution. They are morphologically indistinguishable (homomorphic) (Matsuda *et al.*, 1998) and crossing-overs are possible over the entire length. The linkage map of sex chromosomes (LG1, represented by the second largest pair of chromosomes) contains for most parts an even distribution of markers like the autosomes. Around the sex-determining locus, however, a pronounced clustering of markers indicates a region of reduced recombination (Kondo *et al.*, 2001; Matsuda *et al.*, 1999; Naruse *et al.*, 2000). Interestingly the evolution of sex chromosomes in medaka has not yet led to the inactivation of important genes. Unlike in higher vertebrates full sex reversals can be obtained for instance by treatment with sex steroid hormones during the larval period. By this and other means YY males, XY females, XX males and even YY females have been produced (see Yamamoto, 1975). Such experiments uncovered three important phenomena. First, there is a time window during development, the so-called "sensitive period" where sex can be artificially reverted. This is as long as the gonad is morphologically indifferent, which in genetically female fish is until hatching and in males even until two weeks after hatching. Second, the uncompromised viability of YY females points to the fact that no genes essential for the fish have been inactivated or eliminated during Y chromosome evolution. Third, the full fertility of XX males points to the fact that there are no genes that have a function in the development of the testes or for spermatogenesis are located within the Y-specific region.

Candidate sex determination genes

In the search for the primary sex determinator a number of genes, which are homologous to known sex determination genes from mammals, flies and worms were analysed. A homologue of *SRY* – as expected – was not found so far. Two genes, that are placed downstream in the male sex determination cascade of mammals, have been scrutinized in medaka, namely *Sox9* and *DMRT1*. Both genes belong to gene families with several, closely related members. In both cases the isolation of the medaka genes was attempted by "homology cloning" using heterologous probes from higher vertebrates for library screening under reduced stringency conditions and degenerate primers from highly conserved region for PCR cloning. It is almost unavoidable that in the first cloning steps not only the searched gene but also some other gene family members are isolated as well. The analysis is even more difficult due to the fact, that many gene families have more members in teleost fish than in mammals (Meyer and Schartl, 1999; Wittbrodt *et al.*, 1998). In addition a frequently observed phenomenon in this context is that even orthologous gene family members do not exert the identical function in fish and mammals and thus fail to fulfil the homology criterium.

Sox genes are a family of high-mobility group (HMG) domain containing DNA-binding proteins that act as transcription factors. One member is *Sox 9*, which has been associated with testis determination of mammals, birds and reptiles (Foster *et al.*, 1994; Kent *et al.*, 1996; Wagner *et al.*, 1994; Western *et al.*, 1999). Mutations in *Sox9* cause campomelic dysplasia, a bone morphogenetic disease that is associated with XY sex reversals (Foster *et al.*, 1994; Mansour *et al.*, 1995; Wagner *et al.*, 1994). It is generally accepted that *Sox9* is the crucial sex determining gene immediately downstream to *Sry* driving the undetermined genital ridge along the male pathway (Koopman, 2001). Several *Sox* genes have been cloned and characterized from medaka (Fukada *et al.*, 1995). It revealed that the *Sox* gene family has undergone an expansion in the teleost lineage. Two genes with clearly highest similarity to mammalian *Sox 1* and three genes that could represent either *Sox2, 3* or *14* were found. None of these sequences maps to the sex chromosomes (Tanaka, pers. comm.). The orthologue(s) of *Sox 9* has (have) not been reported to date.

DMRT1 is a candidate downstream sex determination gene in mammals and appears to be involved in a certain type of XY sex reversal in humans. Gene hierarchy studies in the worm and the fly revealed that the corresponding *DMRT1* homologues, *mab-3* and *DSX*, are placed at the bottom of the sex determination cascade. The position of *DMRT1* in the mammalian gene cascade is unknown at present, but it is supposed to be far downstream. It encodes a protein with a DNA-binding motif (Zhu *et al.*, 2000), called DM domain, and it is conserved in a wide range of animals with diverse sex-determining mechanisms, including *C. elegans, Drosophila*, fish, reptiles, birds and mammals (Guan *et al.*, 2000; Marchand *et al.*, 2000; Nanda *et al.*, 1999; Raymond *et al.*, 1998, 1999a; Smith *et al.*, 1999).

Interestingly, *DMRT1* is Z-linked and absent from W in chicken and it has been suggested that *DMRT1* in birds has been recruited as an upstream regulatory sex-determining factor (Nanda *et al.*, 2000). In chicken, *DMRT1* expression precedes expression of all other potential SD genes, is stronger in male than in female gonads and is evident before the sex differentiation of the gonad anlage starts (Raymond *et al.*, 1999b; Shan *et al.*, 2000; Smith *et al.*, 1999). In the turtle, *DMRT1* expression is higher in the gonads of embryos incubated at the male determining temperature prior to sexual differentiation (Kettlewell *et al.*, 2000). In the mouse gonad specific expression of *DMRT1* is detected at E9.5, while *Sry* expression begins around E10.5 (Raymond *et al.*, 1999b). Also in humans the simultaneous onset of *DMRT1* and *Sry* expression suggested a role of *DMRT1* in early events of sex determination (Moniot *et al.*, 2000).

DMRT genes form a family of genes, which share the highly conserved DM domain but have no similarity outside this motif. Four different *DMRT* genes have been cloned and characterized in medaka so far. A comparative phylogenetic analysis of all known DM domain genes revealed a very complex situation (Figure 1). The four medaka genes fall each in a cluster of orthologous sequences. *OlaDMRT-1, -2* and *-3* were named according to the priority nomenclature of the mammalian genes. The remaining one was designated *DMRT-4* to keep the systematic. It is placed in the phylogenetic tree together with the previously described *DMO* (DM domain gene expressed in ovary) gene of tilapia (Guan *et al.*, 2000). However, as the ovary specific expression of *DMO* appears to be a species-specific feature (Kondo *et al.*, in press), this name was abandoned.

Analysis of overlapping cosmids from medaka of Northern and Southern strains revealed that *OlaDMRT-1, -2* and *-3* are located in a gene cluster (Figure 2). This cluster organization is conserved in the pufferfish *T. rubripes* and humans. Moreover the whole genomic region around the *DMRT* cluster shows conserved synteny between these organisms. Genetic linkage analysis and fluorescent *in situ* hybridization (FISH) localization assigned the *DMRT-1, -2, -3* cluster to the subtelomeric region of a pair of autosomes (Figure 3) that represent linkage group 9 of medaka (Brunner *et al.*, 2001). As the sex chromosomes of medaka are linkage group 1, none of the three genes can be the primary sex-determining gene. The same holds true for *OlaDMRT-4*, which has

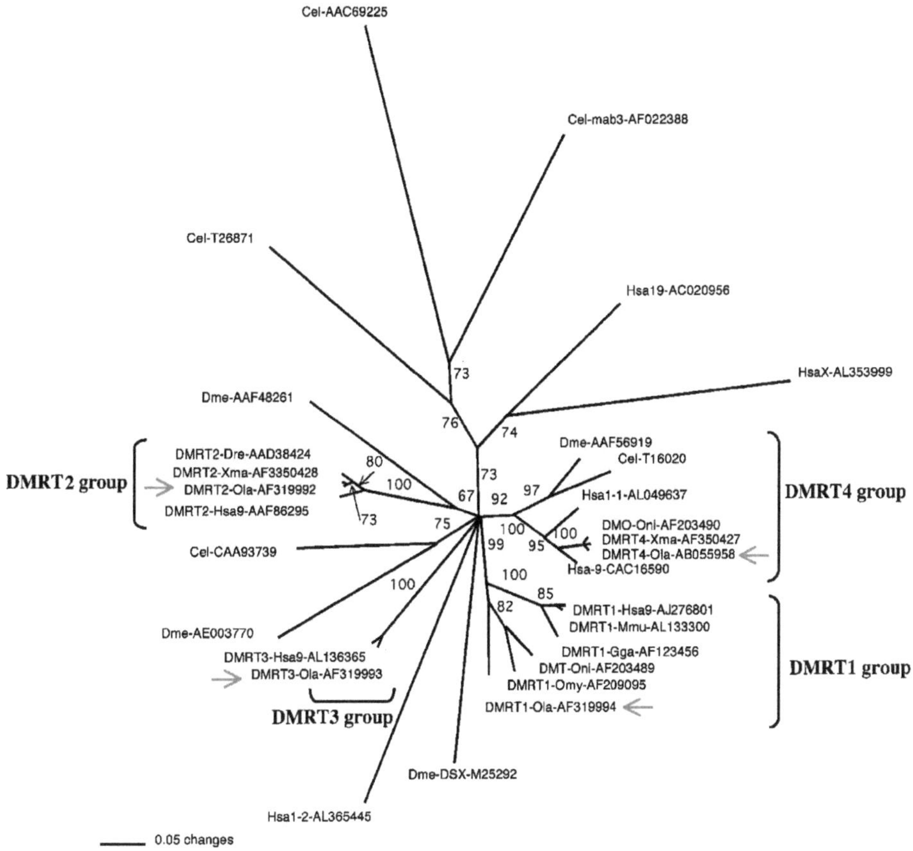

Figure 1. *Phylogenetic tree of invertebrate and vertebrate DM domain genes. Arrows highlight the medaka* DMRT *genes. Hsa, human, Gga, chicken, Mmu, mouse, Ola, medaka, Fru, fugu, Cel,* Caenorhabditis elegans, *Dme,* Drosophila, *Omy, rainbow trout, Oni, tilapia, Xma, platyfish; the accession number is given with each gene symbol. Numbers above the branches indicate bootstrap values (Neighbour Joining Analysis, 1,000 replicates).*

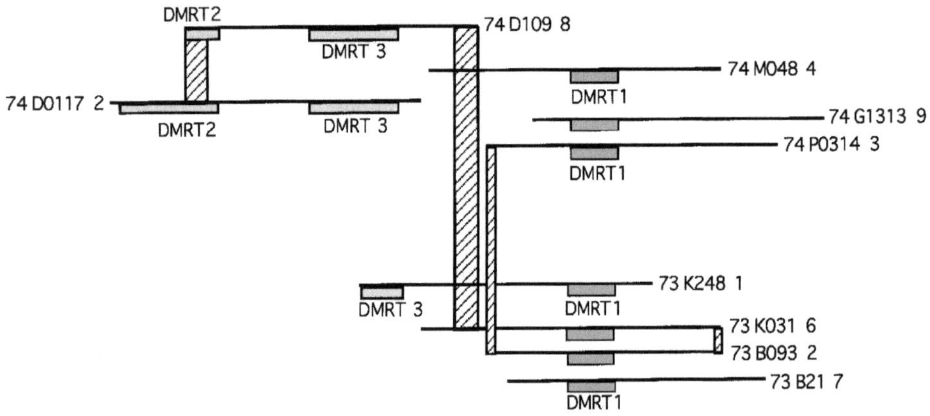

Figure 2. *Cosmid contigs from Northern (strain Kaga) and Southern (strain Carbio) medakas from the* DMRT-1, -2, -3 *cluster. The first two numbers of the cosmid ID indicate the library source: 73 = Carbio, 74 = Kaga. Shaded boxes indicate the location of the* DMRT *genes. Striped boxes indicate homologous regions determined by Southern analysis using cosmid end probes.*

an interstitial location (Figure 3) in the linkage group 18 equivalent chromosome pair (Kondo *et al.*, in press).

In adult medaka *DMRT-1* is only expressed in testis. No expression above background was seen in any other organ including ovary. *DMRT-2* is expressed in testis as well, but also at equal levels in ovary and gills of both sexes. *DMRT-3* is also expressed in testes. The abundance of the transcript, however, is lower than of the two other *DMRT* (Brunner *et al.*, 2001). *DMRT-4* shows the widest expression. It is most abundantly expressed in gills, and it is also found in testes, ovary, kidney, eye and brain (Kondo *et al.*, in press). Intriguingly all *DMRT* genes are expressed in testes, which may indicate a shared conserved or redundant function of all four transcription factors in this organ.

During embryogenesis expression of *DMRT-1* is barely detectable. Notably, there were no transcripts around the hatching stage which in medaka is the critical period of development of the primordial gonad and the time where the first differences between male and female development can be seen (Hamaguchi and Sakaizumi, 1992). For *DMRT-2*, maternal transcripts were detected. Zygotic *DMRT-2* expression commences with a strong expression at somitogenesis (stages 19+) and continues throughout embryonic development and early larval stages. *DMRT-3* is expressed during mid-embryogenesis starting at day 3 and decreasing thereafter until hatching (Brunner *et al.*, 2001). *DMRT-4* expression starts between midblastula and the primitive streak formation. After a strong initial expression it decreases until day 5 of development and thereafter resumes higher levels (Kondo *et al.*, in press). These data suggest a function for *DMRT-2, -3* and *-4* during early development.

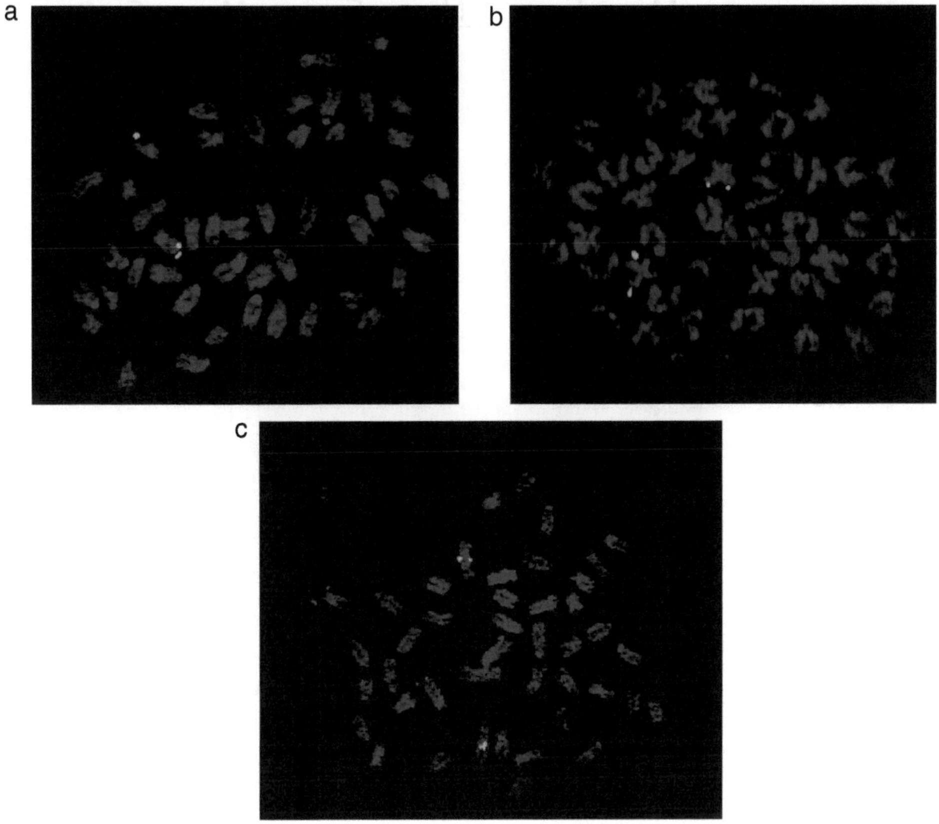

Figure 3. *Chromosomal localization of two different DMRT loci in medaka by FISH: (a) hybridization of cosmid 73K2481 from the* DMRT -1, -2, -3 *cluster (see Fig. 2) revealing its subtelomeric location on the chromosome that corresponds to linkage group 9. The biotinylated cosmid probe was detected with avidin-conjugated FITC (green). (b) Hybridization of a BAC clone (24C08) containing* DMRT-4 *to the long arm interstial region of a chromosome corresponding to linkage group 18. (c) Double colour FISH hybridization of the* DMRT-1, -2, -3 *cluster probe (red) and an anonymous marker locus (BAC 98C17) close to the centromere of the undifferentiated sex chromosomes, hybridized with a digoxygenin-labelled genomic BAC clone detected by anti-digoxygenin conjugated FITC (green). Note that the* DMRT-1, -2, -3 *cluster does not colocalize with the sex chromosome specific marker.*

A duplicated copy of *DMRT-1* at the sex determination locus on the Y chromosome

Southern blot analysis of male and female DNA revealed that males have an additional DMRT fragment that hybridized under conditions of high stringency to the *OlaDMRT-1* probe (Figure 4), but not to any of the other three genes. The corresponding genomic fragment was cloned and found to be a fully intact second copy of *DMRT-1*

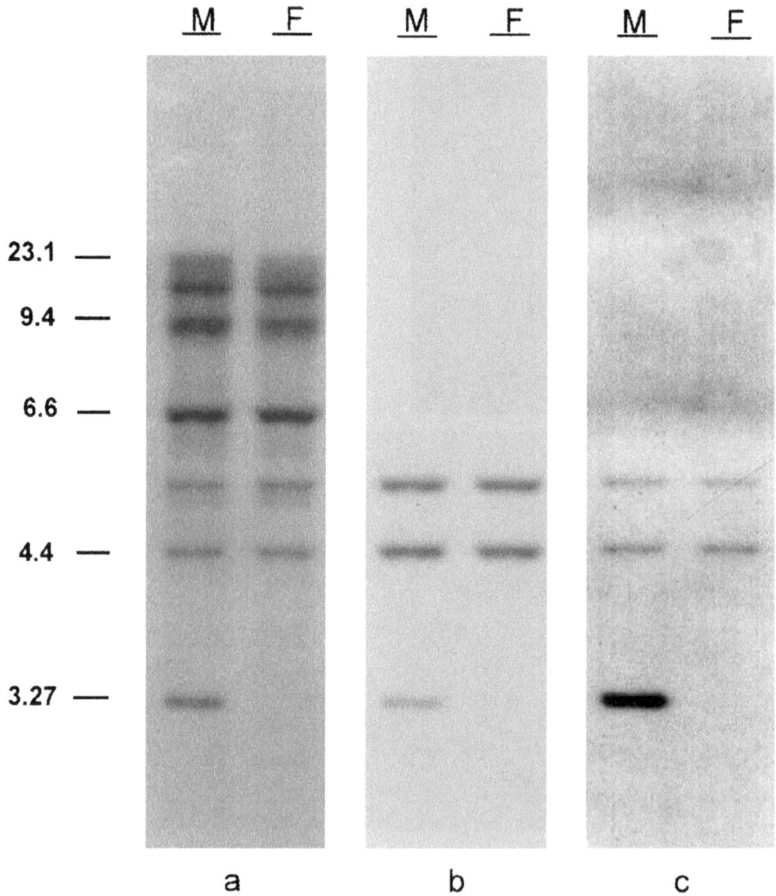

Figure 4. *Sex-specific hybridization pattern of a* DMRT *gene in medaka: Southern blot analysis of EcoRI digested male and female DNA first hybridized with the human* DMRT-1 *cDNA probe at low stringency (a) and after stripping rehybridized with cloned DM domains from* DMR-T1 *(b) and* DMRT-1Y *(c) at high stringency. Note the additional hybridization fragment only in the male.* DMRT-1 *and* DMRT-1Y *show cross hybridization to each other even under conditions of highest stringency due to the high sequence similarity.*

(Nanda *et al.*, submitted). This gene was mapped by linkage analysis employing a high-resolution sex-reversed mapping panel and allele-specific PCR that allowed to distinguish the two copies of *DMRT-1*. It mapped without recombination to the sex-determining locus on the Y chromosome. It was therefore designated *DMRT-1Y*. FISH with a *DMRT-1Y* containing BAC produced a single strong hybridization signal on the Y close to the centromere, but no signal on the X chromosome. A weak hybridization was detected with the subtelomeric region of a pair of autosomes, most probably due to cross hybridization with the *DMRT* cluster on linkage group 9.

A BAC contig was established that covers the *DMRT-1Y* containing region. Sequencing of three overlapping BACs revealed a Y-specific region of approximately

260 kb. Within this region *DMRT-1Y* is the single intact gene (Nanda *et al.*, submitted). Other genes identified in this region either by sequence similarity or by gene prediction programmes have large deletions or point mutations that compromise the respective gene's function. In addition a large number of transposable elements and repetitive sequences is present, reminiscent of the situation found in the platyfish, *Xiphophorus maculatus* (Volff and Schartl, in press).

DMRT-1Y is expressed from day 1 onwards exclusively in male embryos and hatchlings. In adults the transcript is present in the Sertoli cells of the testis (Nanda *et al.* submitted).

The location of *DMRT-1Y* as the single functional gene in the Y chromosome specific region and its expression pattern make this gene an excellent candidate for the primary sex determining gene of medaka. Experiments on the biochemical function of *DMRT-1Y* are needed to understand, how a gene that in other organisms acts as an effector downstream in the sex determination cascade could take up a position at the top of the male sex determination process and what role gene duplication of sex determination and differentiation genes may play in the evolution of sex determination systems. The fact that the Y-specific region is very small will allow to trace back the phylogenetic origin of all components of the sex-determining locus. The system is certainly at a very early stage of evolution and may help to understand how XY sex determination mechanisms have evolved and what happens in the initial period of Y chromosome evolution.

Acknowledgements

We thank G. Schneider, H. Schwind and P. Weber for breeding of the fish and G. Scherer, Freiburg, for the human *DMRT1* cDNA clone. We are especially indebted to the Sequencing Team of Keio University for BAC clone sequencing. This work was supported by grants supplied by the Commission of the European Community (FAIR CT 97-3796) and Fonds der Chemischen Industrie to M. Schartl, by a grant from the Deutsche Forschungsgemeinschaft (SCHM 484/18-1) to M. Schmid, grants from the Future Program (JSPS-RFTF96L00401) from the Japanese Society for the Promotion of Science to A. Shima, and a Grant-in-Aid for Scientific Research on Priority Areas (Area No. 813) from the Ministry of Education, Science, Sports and Culture to A. Shima and S. Asakawa, and a grant form the Biofuture Programme of the Bundesministerium für Bildung und Forschung to J.N. Volff.

References

Amores A, Force A, Yan YL, *et al.* (1998). Zebrafish hox clusters and vertebrate genome evolution. *Science* 282: 1,711–1,714.

Baroiller JF (1999). Endocrine and environmental aspects of sex differentiation in fish. *Cell Mol Life Sci* 55: 910–913.

Brunner B, Hornung U, Shan Z, *et al.* (2001). Genomic organization and expression of the double sex-related gene cluster in vertebrates and detection of putative regulatory regions for *DMRT1*. *Genomics* 77: 8–17.

Charlesworth B (1991). The evolution of sex chromosomes. *Science* 251: 1,030–1,033.

Foster JW, Dominguez-Steglich MA, Guioli S, *et al.* (1994). Campomelic dysplasia and autosomal sex reversal caused by mutations in an *SRY*-related gene. *Nature* 372: 525–530.

Fukada S, Tanaka M, Iwaya M, Nakajima M, Nagahama Y (1995). The SOX gene family and its expression during embryogenesis in the teleost fish, medaka (*Oryzias latipes*). *Dev Growth Differ* 37: 379–385.

Guan G, Kobayashi T, Nagahama Y (2000). Sexually dimorphic expression of two types of DM (Doublesex/Mab-3)-domain genes in a teleost fish, the Tilapia (*Oreochromis niloticus*). *Biochem Biophys Res Commun* 272: 662–666.

Hamaguchi S, Sakaizumi M (1992). Sexually differentiated mechanisms of sterility in interspecific hybrids between *Oryzias latipes* and *O. curvinotus. J Exp Zool* 263: 323–329.

Just W, Rau W, Vogel W, *et al.* (1995). Absence of Sry in species of the vole Ellobius. *Nat Genet* 11: 117–118.

Kent J, Wheatley SC, Andrews JE, Sinclair AH, Koopman P (1996). A male-specific role for SOX9 in vertebrate sex determination. *Development* 122: 2,813–2,822.

Kettlewell JR, Raymond CS, Zarkower D (2000). Temperature-dependent expression of turtle Dmrt1 prior to sexual differentiation. *Genesis* 26: 174–178.

Kondo M, Nagao E, Mitani H, Shima A (2001). Differences in recombination frequencies during female and male meioses of the sex chromosomes of the medaka, *Oryzias latipes. Genet Res* 78: 23–30.

Koopman P (2001). The genetics and biology of vertebrate sex determination. *Cell* 105: 843–847.

Lahn BT, Pearson NM, Jegalian K (2001). The human Y chromosome, in the light of evolution. *Nat Rev Genet* 2: 207–216.

Mansour S, Hall CM, Pembrey ME, Young ID (1995). A clinical and genetic study of campomelic dysplasia. *J Med Genet* 32: 415–420.

Marchand O, Govoroun M, D'Cotta H, *et al.* (2000). *DMRT1* expression during gonadal differentiation and spermatogenesis in the rainbow trout, *Oncorhynchus mykiss. Biochim Biophys Acta* 1493: 180–187.

Marin I, Baker BS (1998). The evolutionary dynamics of sex determination. *Science* 281: 1,990–1,994.

Matsuda M, Matsuda C, Hamaguchi S, Sakaizumi M (1998). Identification of the sex chromosomes of the medaka, *Oryzias latipes*, by fluorescence *in situ* hybridization. *Cytogenet Cell Genet* 82: 257–262.

Matsuda M, Sotoyama S, Hamaguchi S, Sakaizumi M (1999). Male-specific recombination frequency in the sex chromosomes of the medaka, *Oryzias latipes. Genet Res* 73: 225–231.

Meyer A, Schartl M (1999). Gene and genome duplications in vertebrates: the one-to-four (-to-eight in fish). rule and the evolution of novel gene functions. *Curr Opin Cell Biol* 11: 699–704.

Moniot B, Berta P, Scherer G, Sudbeck P, Poulat F (2000). Male specific expression suggests role of DMRT1 in human sex determination. *Mech Dev* 91: 323–325.

Nanda I, Shan Z, Schartl M, *et al.* (1999). 300 million years of conserved synteny between chicken Z and human chromosome 9. *Nat Genet* 21: 258–259.

Nanda I, Zend-Ajusch E, Shan Z, *et al.* (2000). Conserved synteny between the chicken Z sex chromosome and human chromosome 9 includes the male regulatory gene *DMRT1*: a comparative (re)view on avian sex determination. *Cytogenet Cell Genet* 89: 67–78.

Naruse K, Fukamachi S, Mitani H, *et al.* (2000). A detailed linkage map of medaka, *Oryzias latipes*: comparative genomics and genome evolution. *Genetics* 154: 1,773–1,784.

Postlethwait JH, Woods IG, Ngo-Hazelett P, *et al.* (2000). Zebrafish comparative genomics and the origins of vertebrate chromosomes. *Genome Res* 10: 1,890–1,902.

Raymond CS, Shamu CE, Shen MM, *et al.* (1998). Evidence for evolutionary conservation of sex-determining genes. *Nature* 391: 691–695.

Raymond CS, Parker ED, Kettlewell JR, *et al.* (1999a). A region of human chromosome 9p required for testis development contains two genes related to known sexual regulators. *Hum Mol Genet* 8: 989–996.

Raymond CS, Kettlewell JR, Hirsch B, Bardwell VJ, Zarkower D (1999b). Expression of *Dmrt1* in the genital ridge of mouse and chicken embryos suggests a role in vertebrate sexual development. *Dev Biol* 215: 208–220.

Shan Z, Nanda I, Wang Y, Schmid M, Vortkamp A, Haaf T (2000). Sex-specific expression of an evolutionarily conserved male regulatory gene, *DMRT1*, in birds. *Cytogenet Cell Genet* 89: 252–257.

Smith CA, McClive PJ, Western PS, Reed KJ, Sinclair AH (1999). Conservation of a sex-determining gene. *Nature* 402: 601–602.

Wada H, Shimada A, Fukamachi S, Naruse K, Shima A (1998). Sex-linked inheritance of the *lf* locus in the medaka fish (*Oryzias latipes*). *Zool Sci* 15: 123–126.

Wagner T, Wirth J, Meyer J, *et al.* (1994). Autosomal sex reversal and campomelic dysplasia arc caused by mutations in and around the *SRY*-related gene *SOX9*. *Cell* 79: 1,111–1,120.

Western PS, Harry JL, Graves JA, Sinclair AH (1999). Temperature-dependent sex determination: upregulation of *SOX9* expression after commitment to male development. *Dev Dyn* 214: 171–177.

Wittbrodt J, Meyer A, Schartl M (1998). More genes in fish? *Bioessays* 20: 511–515.

Wittbrodt J, Meyer A, Schartl M (2002). Medaka – a model organism from the far east. *Nat Rev Genet* 3: 53–64.

Woods IG, Kelly PD, Chu F, *et al.* (2000). A comparative map of the zebrafish genome. *Genome Res* 10: 1,903–1,914.

Yamamoto T (1975). *Medaka (Killfish): Biology and Strains.* Tokyo: Keigaku Publishing Co.

Zhu L, Wilken J, Phillips NB, *et al.* (2000). Sexual dimorphism in diverse metazoans is regulated by a novel class of intertwined zinc fingers. *Genes Dev* 14: 1,750–1,764.

1.4. Evolution of the Y Chromosome in Primates

Werner Schempp
Institute of Human Genetics and Anthropology, University of Freiburg, Breisacher Strasse 33, 79106 Freiburg, Germany

Introduction

The human Y chromosome is one of the smallest human chromosomes with an estimated average size of 50–60 Mb (Morton, 1991; Tilford *et al.*, 2001). Size variability is mostly caused by polymorphism in the length of the constitutive heterochromatin located in the long arm in Yq12. It is characterized by its bipartite structure and function. It is flanked by pseudoautosomal regions at the distal short arm (PAR1) and distal long arm (PAR2), the sequences of which are shared by, and undergo meiotic recombination with the X chromosome. Between these two PARs there is the non-recombining region of the human Y chromosome (NRY), which comprises about 95% of the chromosome. This NRY does not undergo sexual recombination and is thus male-specific.

Phylogenetically, the human Y chromosome is of recent origin having reached its present configuration during the divergent evolution of the branches leading to great apes and human. In a broader evolutionary context it is believed that the mammalian X and Y chromosomes have evolved from an autosomal pair (Ohno, 1967). The "proto" X retained and the "proto" Y chromosome gradually lost most ancestral genes. Only after the "proto" Y chromosome acquired a strict sex-determining function in the form of the *SRY* gene, a block of DNA surrounding the *SRY* gene may have stopped recombination. Thereafter, "evolutionary events" (e.g. large chromosomal inversions) are postulated that should suppress recombination between the "proto" sex chromosomes in a stepwise fashion (Lahn and Page, 1999). Each inversion is thought to have reduced the size of the PAR and enlarged the non-recombining portions of the X and Y chromosomes.

"Evolutionary breakpoints" on the primate Y chromosomes

Recently, we could show that the ancestral eutherian PAR might be represented by, for instance, that of the dog (carnivores), and the gene content of this ancestral

M. Schmid and I. Nanda (eds.), Chromosomes Today, Volume 14, 39–43.

eutherian PAR is still conserved in the prosimian lineage (Gläser *et al.*, 1999). Our molecular and fluorescence *in situ* hybridization (FISH) mapping data strongly indicate that only after branching of the prosimian lineages did the Y-chromosomal transposition of *SRY* into the ancestral eutherian PAR occur in a common ancestor of simian primates creating a new pseudoautosomal boundary on the Y chromosome (PABY). By means of that transposition event the close proximity of *SRY* to the PABY in simian primates resulted. By this event, several formerly pseudoautosomal genes including *STS* and *PRK* as well as the *KAL* and *ARSE* were excluded from the X–Y crossover process and thus became susceptible to rearrangements and/or deterioration on the Y chromosome in simian primates. Indeed, comparative mapping of human X–Y homologous genes and Y chromosome genes and gene families on chromosomes of simian primates by FISH has shown that Y chromosomal rearrangements are confined to the NRY portion outside the pseudoautosomal and sex-determining region in higher primates (Gläser *et al.*, 1997a, 1998a, 1999). Moreover, some of these pseudoautosomal rearrangements must have occurred species-specifically and might be helpful for a better definition of the phylogenetic branching in higher primates.

During our search for evolutionary breakpoints within a defined segment on the long arm of the Y chromosome, we performed FISH mapping of individual clones from a human Y-chromosomal PAC contig spanning 2.8 Mb in proximal Yq11.21 including the entire AZFa region, to chromosomes of humans, great apes, gibbon and several Old World monkeys (Wimmer *et al.*, in press). For a specific subset of clones proximal to AZFa it turned out that a transposition of an approximately 100 kb DNA fragment from chromosome 1 onto the Y chromosome must have occurred in a common ancestor of human, chimpanzee and bonobo.

Only the Y chromosomes of these three species contain the chromosome 1-derived fragment, it could not be detected on the Y chromosomes of gorillas and of other primates examined (Wimmer *et al.*, 2001).

Physical mapping of human NRY

Genes and gene families that play an important role in spermatogenesis are content of the large Y-specific region (NRY) where no recombination occurs. Thus, a genetic linkage map from this region cannot be constructed, and the localization of this genes and gene families within NRY depends on physical mapping.

Construction of DNA contigs as a mapping strategy

The construction of yeast artificial chromosomes (YACs) and cosmid contigs spanning the entire NRY, as an approach for physical mapping, has generated overlapping clone maps (Foote *et al.*, 1992; Jones *et al.*, 1994, Taylor *et al.*, 1996). Complete physical mapping, however, has been found to be difficult in regions rich in repetitive sequences. As a result of evolutionary processes, the human Y chromosome appears to be endowed with a particularly high density of long and short interspersed repetitive sequences as well as tandem repeats. It was demonstrated that sequence tagged sites (STSs) being used for assembling and orientating YACs/cosmids within the contig can correspond to Y-specific repetitive sequences dispersed on multiple sites on the Y chromosome (Foote *et al.*, 1992; Taylor *et al.*, 1996; Yen, 1998). STSs corresponding to repetitive sequences are particularly frequent in deletion interval 3 and 6 (Jones *et al.*, 1994) in which the male-specific expressed gene family members of *DAZ, RBMY, CDY* and *TSPY* are mainly clustered. A further problem is represented by the structural instability of repeat containing inserts cloned in YACs: YACs and also cosmids containing Y-specific tandem repeats had sustained large-scale deletion (Neil *et al.*, 1990; Taylor *et al.*, 1996). These complications in YAC contig assembly can give rise both to gaps and to incorrect assignment of YACs within the contig. Thus, it might well be the case that the YAC contig mapping data of gene families are incomplete or incorrect. In certain NRY regions.

In a recent article (Tilford *et al.*, 2001) reported a high-resolution physical map of the euchromatic, centromeric and heterochromatic regions of the human NRY providing the basis for sequencing the human Y chromosome. To achieve the construction of this physical map the authors included "genomic (YAC) clone subtraction" (Reijo *et al.*, 1995) to generate additional STSs and "dissection of sequence family variants" (Saxena *et al.*, 2000) to distinguish among NRY-specific amplified regions that could not be distinguished by STS content or restriction fingerprinting. Obviously, these methods led to a higher resolution and refinement of the physical map of the human Y chromosome and thus provide a big step forward for sequencing this unusual human chromosome. Nevertheless, it should be noted that these experimental procedures neither solve the problem of structural instability of repeat containing inserts of YACs nor do they generate additional YAC clones to close the gaps in the published Y contig. Not to forget that the high concentration of repeats, indeed, may cause the total failure of efficient cloning of certain repeat containing Y-chromosomal segments.

"Fiber-FISH" as a mapping strategy

The "fiber-FISH" technique can be used as a physical mapping strategy. It allows the direct visualization of cloned DNA sequences along chromatin "fibers" released from interphase nuclei (Heng *et al.*, 1992; Parra and Windle, 1993; Fidlerova *et al.*, 1994; Senger *et al.*, 1994). There are several articles published that are concerned with physical mapping of genes and gene families on the human Y chromosome applying the "fiber-FISH" technique (Conrad *et al.*, 1996; Gläser *et al.*, 1997b, 1998b; Saxena *et al.*, 2000; Röttger *et al.*, 2000). The problems that can arise when applying this "fiber-FISH" technique (e.g. mechanically stretching of chromatin, incomplete and broken "fibers", chromatin "fibers" are from different stages of the cell cycle, hybridization efficiency of FISH) are described in more detail in Gläser *et al.*, (1997b, 1998b). This should make clear that the proper interpretation and evaluation of the results of this procedure needs a lot of experience.

Although the "molecular Y chromosome community" is convinced that the human Y chromosome is not far from being completely sequenced (Tilford *et al.*, 2001) we are pursuing the application of the "fiber-FISH" technique as a physical mapping strategy. We belief that this technique can provide an important contribution to overcome the still existing problems (see above) of physical gene mapping occurring when building up a Y contig.

In our ongoing studies we physically map the multi-copy gene families *CDY, DAZ, RBMY* and *TSPY* on released chromatin of the human Y chromosome. We use the Yq heterochromatin and the Y centromere-specific alphoid satellite sequences, as well as already mapped single copy genes as markers for orientation along the "fibers". Comparison of these "fiber-FISH" results with the physical map of the human Y chromosome recently published by Tilford *et al.* (2001) will help to get a more complete picture with respect to number, position and arrangement of the members of the specific gene families.

References

Conrad C, Hierl T, Gläser B, Taylor K, Zeitler S, Chandley A, Schempp W (1996). High-resolution fluorescence *in situ* hybridization of *RBM-* and *TSPY*-related cosmids on released Y chromatin in humans and pygmy chimpanzees. *Chromosome Res* 5: 23–30.

Fidlerovà H, Senger G, Kost M, Sanseau P, Sheer D (1994). Two simple procedures for releasing chromatin from routinely fixed cells for fluorescence *in situ* hybridization. *Cytogenet Cell Genet* 65: 203–205.

Foote S, Vollrath D, Hilton A, Page DC (1992). The human Y chromosome: overlapping DNA clones spanning the euchromatic region. *Science* 258: 60–66.

Gläser B, Hierl T, Taylor K, Schiebel K, Zeitler S, Papadopoullos K, Rappold G, Schempp W (1997a). High-resolution fluorescence *in situ* hybridization of human Y-linked genes on released chromatin. *Chromosome Res* 5: 23–30.

Gläser B, Grützner F, Taylor K, Schiebel K, Meroni G, Tsioupra K, Pasantes J, Rietschel W, Toder R, Willmann U, Zeitler S, Yen P, Ballabio A, Rappold G, Schempp W (1997b). Comparative mapping of human Xp22 genes in hominoids – evolutionary linear instability of their Y homologues. *Chromosome Res* 5: 167–176.

Gläser B, Grützner F, Willman U, Stanyon R, Arnold N, Taylor K, Rietschel W, Zeitler S, Toder R, Schempp W (1998a). Simian Y chromosomes: species-specific rearrangements of *DAZ, RBM*, and *TSPY* versus contiguity of *PAR* and *SRY. Mamm Genome* 9: 226–231.

Gläser B, Yen P, Schempp W (1998b). Fibre-fluorescence *in situ* hybridization unravels apparently seven *DAZ* genes or pseudogenes clustered within a Y-chromosome region frequently deleted in azoospermic males. *Chromosome Res* 6: 481–486.

Gläser B, Myrtek D, Rumpler Y, Schiebel K, Hauwy M, Rappold GA, Schempp W (1999). Transposition of SRY into the ancestral pseudoautosomal region creates a new pseudoautosomal boundary in a progenitor of simian primates. *Hum Mol Genet* 8: 2,071–2,078.

Heng HHQ, Squire J, Tsui L-C (1992). High-resolution mapping of mammalian genes by *in situ* hybridization to free chromatin. *Proc Natl Acad Sci USA* 89: 9,509–9,513.

Jones MH, Khwaja OSA, Briggs H, *et al.* (1994). A set of ninety-seven overlapping yeast artificial chromosome clones spanning the human Y chromosome euchromatin. *Genomics* 24: 266–275.

Lahn BT, Page DC (1999). Four evolutionary strata on the human X chromosome. *Science* 286: 964–967.

Morton NE (1991). Parameters of the human genome. *Proc Natl Acad Sci USA* 88: 7,474–7,476.

Neil DL, Villsante A, Fisher RB, Vetrie D, Cox B, Tyler-Smith C (1990). Structural instability of human tandemly repeated DNA sequences cloned in yeast artificial chromosome vectors. *Nucleic Acids Res* 18: 1,421–1,428.

Ohno S (1967). *Sex Chromosomes and Sex Linked Genes*. Berlin: Springer.
Parra I, Windle B (1993). High resolution visual mapping of stretched DNA by fluorescent hybridization. *Nat Genet* 5: 17–21.
Reijo R, Lee T-Y, Salo P, Alagappan R, Brown LG, Rosenberg M, Rozen S, Jaffe T, Straus D, Hovatta O, de la Chapelle A, Silber S, Page DC (1995). Diverse spermatogenic defects in humans caused by Y chromosome deletions encompassing a novel RNA-binding protein gene. *Nat Genet* 10: 383–392.
Röttger S, Pasantes JJ, Baldermann C, Reichl E, Yen PH, Hansmann I, Schempp W (2000). Familial mosaicism of del(Y) and inv del(Y). *Cytogenet Cell Genet* 91: 208–211.
Saxena R, de Vries JWA, Repping S, Alagappan RK, Skaletzky H, Brown LG, Ma P, Chen E, Hoovers JMN, Page DC (2000). Four *DAZ* genes in two clusters found in the *AZFc* region of the human Y chromosome. *Genomics* 67: 256–267.
Senger G, Jones T, Fidlerovà H (1994). Released chromatin: linearized DNA for high resolution fluorescence *in situ* hybridization. *Hum Mol Genet* 3: 1,275–1,280.
Taylor K, Hornigold N, Conway D, *et al.* (1996). Mapping the human Y chromosome by fingerprinting cosmid clones. *Genome Res* 6: 235–248.
Tilford CA, Kuroda-Kawaguchi T, Skaletzky H, Rozen S, Brown LG, Rosenberg M, McPherson JD, Wylie K, Sekhon M, Kucaba TA, Waterston RH, Page DC (2001). A physical map of the human Y chromosome. *Nature* 409: 943–945.
Wimmer R, Kirsch S, Rappold GA, Schempp W (2001). Direct evidence for a *Homo-Pan* clade. *Chromosome Res* (in press).
Yen P (1998). A long-range restriction map of deletion interval 6 of the human Y chromosome: a region frequently deleted in azoospermic males. *Genomics* 54: 5–12.

1.5. The Male Hypermethylation (MHM) Region on the Chicken Z Chromosome: Female-Specific Transcription and its Biological Implication

Mika Teranishi[1] and Shigeki Mizuno[2]

[1]*Division of Genetic Ecology in Critical Environment, Department of Environmental Life Science, Graduate School of Life Science, Tohoku University, Sendai 980-8577, Japan;*
[2]*Department of Agricultural and Biological Chemistry, College of Bioresource Sciences, Nihon University, Fujisawa 252-8510, Japan*

Introduction

The human DM-related transcript 1 (*DMRT1*) and *DMRT2* genes are located at the distal region of chromosome 9p24.3 and suggested to be involved in the differentiation of testis in a dosage dependent manner, because XY individuals hemizygous for these genes exhibit a high frequency of XY feminization (Raymond *et al.*, 1999a). In mouse embryos, the expression of *DMRT1* gene is first detected by RT–PCR at E9.5, the earliest stage of genital ridge formation, and its expression in genital ridges and in early gonads, as detected by RT–PCR and *in situ* hybridization, continues until E14.5 in both male and female embryos. At E15.5, its expression in the female declines significantly and becomes testis-specific in adults (Raymond *et al.*, 1999b). In chickens, a single *DMRT1* gene is located on the short arm of the Z chromosome, expressed in genital ridges of both male and female embryos at as early as stage 19 (~3-day) (Raymond *et al.*, 1999b) but later its expression becomes male (testis)-specific as in mammals (this study; Figure 2A,B). In early chicken embryos at stages 25–31 (4.5- to 7-day), the level of *DMRT1* mRNA is about 2-fold higher in males than in females (Raymond *et al.*, 1999b and this study), which might have been attained by the absence of a mechanism to shut off the transcription on one of the Z chromosomes in males (Kuroda *et al.*, 2001). It has been speculated that the apparently dosage-dependent expression of the *DMRT1* gene in early chicken embryos may have a role in the sex determination (Raymond *et al.*, 1999b). The exclusive expression of the *DMRT1* gene

M. Schmid and I. Nanda (eds.), Chromosomes Today, Volume 14, 45–53.

in adult testis both in mammals and birds may also suggest that its gene expression is required for some essential functions in the differentiated testis.

We fortuitously found an about 460-kb region which lies very close to the *DMRT1* gene on the chicken Z chromosome. This region was designated the male hypermethylation or male hypermethylated (MHM) region, because the entire region is highly methylated and its transcription is silenced in males but is hypomethylated and transcribed into non-coding RNA species (MHM-RNA) which accumulate in the nucleus adjacent to the *DMRT1* locus in females (Teranashi *et al.*, 2001). Here we like to summarize the results described in the previous paper (Teranishi *et al.*, 2001), show some additional results and discuss unique features of MHM-RNA among several non-coding RNA species found in various systems.

Properties of the MHM region and its female-specific transcripts

The MHM region was first noted with a Z-chromosome-linked cDNA clone, pCO10-6 containing an about 0.5-kb insert, isolated from the cDNA library of lampbrush-stage chicken oocytes. A genomic clone, pCOC10618, containing a 16-kb sequence, was subsequently isolated using pCO10-6 as a probe. FISH to the lampbrush ZW bivalent with the biotinylated pCOC10618 revealed that the probe hybridized to a particular pair of loops on the short arm of the Z chromosome and suggested that the region was transcribed from a particular strand (sense strand). A cDNA clone pCC1-11, containing about 3-kb sequence of the MHM region was later isolated from the cDNA library of 8-day female chicken embryos.

Southern blot and quantitative slot-blot hybridization to the genomic DNA revealed that the region consisted of approximately 210 tandem repeats of *Bam*HI 2.2-kb unit. Comparison of digestion patterns obtained with a combination of *Bam*HI and a methylation-insensitive *Msp*I or a methylation-sensitive *Hpa*II, *Hae*II, *Hha*I or *Bst*UI indicated that the entire region is highly methylated in the male but hypomethylated in the female. A partial sequence of the genomic clone pCOC10618 and the whole sequence of the cDNA clone pCC1-11 were determined and subjected to the homology plot analysis (Figure 1A) and the search for open reading frames (ORFs) (Figure 1B). The homology plot suggested that the 2.2-kb unit was tandemly repeated in both genomic and cDNA sequences and that the two sequences were nearly identical. ORFs in either strand or in any reading frame of the cDNA sequence were relatively short; the longest one was 426 bp. These results supported the notion that the MHM region was transcribed into intron-less, non-coding RNA molecules. Northern blot hybridization revealed that transcription of the MHM region took place not only in oocytes but also in embryos and various tissues of female chickens. The transcripts were around 9.5-kb long, restricted in nuclei and did not have poly(A) tails. *In situ* hybridization to the lampbrush Z chromosome with a riboprobe transcribed from either strand indicated that the MHM region was transcribed from one particular strand. The transcriptional silencing of MHM region in males is most likely caused by hypermethylation, because the MHM region in male embryonic fibroblasts became hypomethylated and the 9.5-kb transcripts were produced after the treatment with 5-azacytidine for about three cell generations.

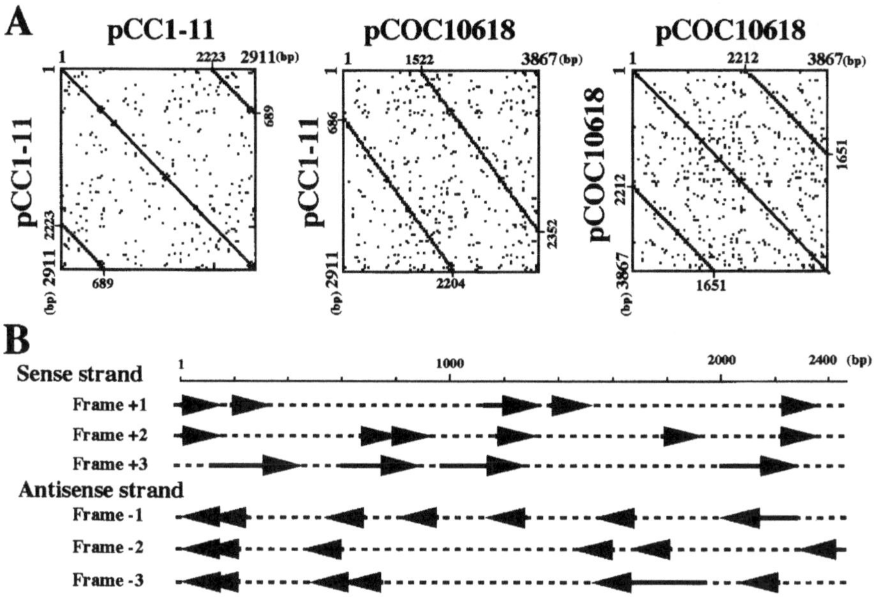

Figure 1. *Comparison of genomic and cDNA sequences of the MHM region (A) and search for ORFs (B). A partial sequence (3,867 bp) of the genomic clone pCOC10618 and the entire sequence of the cDNA clone pCC1-11 (2,911 bp) were determined and deposited with DDBJ/EMBL/GenBank nucleotide sequence databases under accession numbers AB046698 and AB046699, respectively. Homology plots [with DNASIS ver. 3 (Hitachi Software Engineering, Tokyo, Japan): check size, 10; matching bases, 8] suggest that about 2.2-kb unit is arranged tandemly both in genomic and cDNA sequences and that the cDNA sequence is nearly identical with the genomic sequence. ORFs (starting with the initiation codon and ending with a termination codon, shown with thick arrows) are all relatively short on both DNA strands and on each reading frame.*

Interesting aspects of the MHM phenomenon is its timing of establishment and a possible role of W chromosome. The MHM region in a sperm is hypermethylated. In a female ZW zygote, Z chromosome was derived from a sperm but its MHM region becomes largely hypomethylated during the early development to about the 1-day embryonic stage. Conversely, in a male ZZ zygote, one of Z chromosomes was derived from an ovum, but MHM regions on both Z chromosomes are hypermethylated in 1-day male embryos. It is thus likely that the sex-dependent hyper- or hypo-methylation of the MHM region is established after fertilization and during the early development up to the 1-day embryonic stage. Furthermore, in the male triploid chickens (ZZZ + 3 sets of autosomes), MHM regions on all the three Z chromosomes were hypermethylated and not transcribed, whereas in the intersex triploid chickens (ZZW + 3 sets of autosomes), MHM regions on the two Z chromosomes were hypomethylated and actively transcribed. It is thus conceivable that the presence of W chromosome positively regulates the establishment of hypomethylation and transcriptionarily active state of the MHM region by some unknown mechanisms.

Inverse relationship between transcriptional activities of the MHM region and the adjacent *DMRT1* locus in female chickens

FISH probed with the genomic clone pCOC10618 for the MHM region and a BAC clone for the *DMRT1* gene has proved that these two loci at the middle of the short arm of the mitotic Z chromosome are located indistinguishably close. On the lampbrush Z chromosome, the *DMRT1* gene was located at each base of a particular pair of loops on which DNA sequences and transcripts of the MHM region are located. The *DMRT1* locus was not found on the axial chromomere. These results imply that transcripts from the MHM region accumulate adjacent to the *DMRT1* gene locus on the loop of the lampbrush Z chromosome at the diplotene stage of the female meiosis.

The major 1.6-kb transcripts of the *DMRT1* gene were detectable by Northern blot hybridization in gonads of both male and female chicken embryos with the intensity ratio of approximately 2 (male) to 1 (female) (Figure 2A). However, soon after hatching, the level of *DMRT1* transcripts in the ovary declined significantly and became undetectable in ovaries of adult chickens, whereas the level of transcripts in testes remained high (Figure 2A). In the male chicken, *DMRT1* transcripts were detected only in the testis (Figure 2B). Thus, in post-hatched female chickens, transcription of the *DMRT1* gene is silenced, whereas the adjacent MHM region is hypomethylated and transcribed actively and its transcripts accumulate adjacent to the *DMRT1* locus in the nucleus.

Implications of biological roles for non-coding RNA species

As summarized in Table 1, some non-coding RNA species have been implicated to function in the regulation of transcription, translation or genomic imprinting. If the accumulation of MHM transcripts in a nucleus is involved in the transcriptional silencing of the adjacent *DMRT1* gene, the role of MHM-RNA somewhat resembles that of *Xist* RNA. However, a significant difference is that *Xist* RNA spreads in cis over a wide region of the X chromosome and is suggested to participate in the X chromosome inactivation, whereas the MHM-RNA does not spread significantly and may participate in local inactivation in a cis-acting manner. The *roX1* and *roX2* RNAs of *Drosophila* also spread over the single X chromosome in the male forming compensasome complexes with several protein components. The complex acetylates histone H4 and causes hyperactivation of the transcription of X-linked genes. It has been suggested that *roX1* and *roX2* genes on the X chromosome act as entry points of the compensasome complexes. In the case of MHM-RNA, it stays at or in the close vicinity of the site of transcription on the single Z chromosome in the female. There may be no other homologous sequence regions on the Z chromosome which might act as receptors of MHM-RNA or its complex with proteins. The cis-limited local silencing effect of

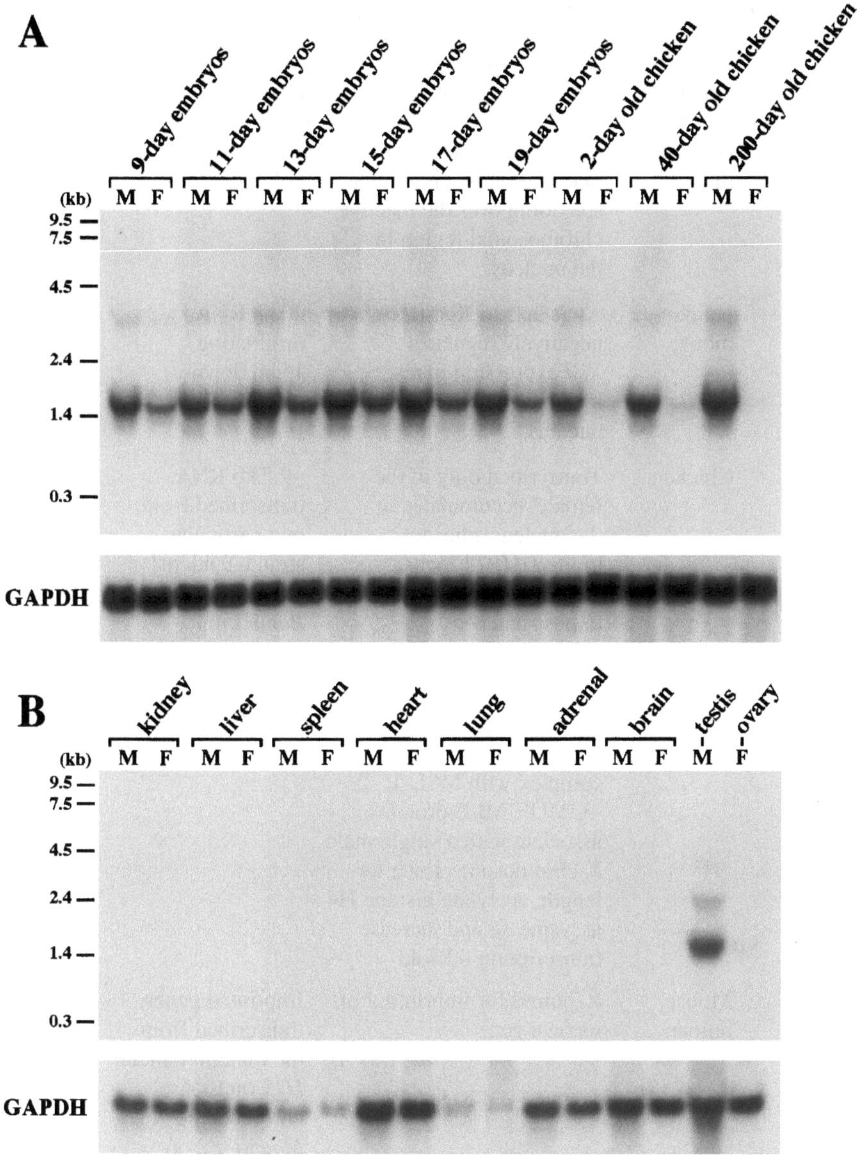

Figure 2. *Northern blot hybridization with* ^{32}*P-labelled chicken* DMRT1 *gene probe or chicken* GAPDH *cDNA probe against total RNA preparations from (A) gonads of male (M) or female (F) chicken embryos or testis or ovary of post-hatched chickens as indicated or (B) different tissues as indicated from 40-day male or female chickens. The* DMRT1 *probe was the 293-bp sequence of the 3′-side region of the chicken* DMRT1 *gene (1,172–1,464 bp of the sequence AF123456 in DDBJ/EMBL/GenBank nucleotide sequence databases).*

Table 1. *Some non-coding RNA species*

RNA	Species	Putative functions	Comments	References
Xist (X inactive specific transcript)	Human, mouse	Mediates X chromosome inactivation by cis-limited spreading over the inactive chromosomal region in the nucleus	16.5–19.3 kb poly(A)$^+$-RNA	(1) (2)
Tsix	Human, mouse	Antisense to Xist RNA, negatively regulates *Xist* expression in cis, localized at *Xic* in the nucleus	40 kb RNA, originating 15 kb downstream of *Xist*	(3) (4)
MHM	Chicken	Transcribed only in the female, accumulates in the nucleus adjacent to the *DMRT1* locus, affecting its silencing in post-hatched chickens (?)	~9.5 kb RNA, transcribed from one particular strand, tandem repeats of a 2.2-kb *Bam*HI unit	(5)
roX1, roX2 (RNA on X chromosome)	*Drosophila*	X-linked genes transcribed into male-specific RNAs, form compensasome complex with MSL-1, -2, -3, MOF, MLE proteins, associate with a single male X chromosome along its length, acetylate histone H4 at lysine 16 and increase transcription ~2-fold		(6) (7) (8) (9)
H19, Meg3/Gtl2	Mouse, human	Required for imprinting of its own gene	Imprinted genes, transcribed from the maternal allele; *H19* on mouse chromosome 7, *Meg3/Gtl2* on mouse chromosome 12, human chromosome 14	(10) (11) (12) (13)
Alternative transcript, BD RNA, from exon-1 region of *SNRPN* gene	Human	Acts in cis on imprinting of *SNRPN* gene by altering the chromatin structure?	Imprinted gene, transcribed from the paternal allele of chromosome 15	(11) (14)

Table 1. *(continued)*

RNA	Species	Putative functions	Comments	References
Bsr (brain-specific RNA)	Rat	Intron encodes a brain-specific box C/D snoRNA	Tandem repeats of ~0.9 kb unit, preferentially expressed in central nervous system	(15) (16)
Five transcripts *from meu* (meiotic expression up-regulated) genes	Fission yeast		Induced expression in meiosis/sporulation-specific manner, form hairpin structures	(17)
lin-4, let-7	Nematode (*C. elegans*)	Appear at specific stages during the larval development, base pair with the 3′-UTRs of target mRNAs and affects their translation (?)	lin-4 RNA, 22 nucleotides let-7-RNA, 21 nucleotides	(18) (19) (20) (21)

(1) Clemson *et al.* (1996), (2) Hong *et al.* (2000), (3) Lee *et al.* (1999), (4) Lee and Lu (1999), (5) Teranishi *et al.* (2001), (6) Franke and Baker (1999), (7) Kelley *et al.* (1999), (8) Meller *et al.* (2000), (9) Smith *et al.* (2000), (10) Pfeifer *et al.* (1996), (11) Latham (1999), (12) Miyoshi *et al.* (2000), (13) Takada *et al.* (2000), (14) Dittrich *et al.* (1996), (15) Komine *et al.* (1999), (16) Cavaille *et al.* (2001), (17) Watanabe *et al.* (2001), (18) Lee *et al.* (1993), (19) Wightman *et al.* (1993), (20) Reinhart *et al.* (2000), (21) Moss (2000).

Other non-coding RNAs are listed in Erdmann *et al.* (2001).

MHM-RNA, if any, seems to be advantageous, because it exists on the otherwise active, single Z chromosome in the female. The reason why the *DMRT1* gene is transcribed in female embryos even in the presence of MHM transcripts is unsolved. Perhaps, transcriptional regulation of the *DMRT1* gene is complex and the negative role of MHM-RNA is overcome by positive factors in embryos, while the state of permanent repression of the *DMRT1* gene in post-hatched female chickens may be facilitated with the local accumulation of MHM-RNA.

References

Cavaille J, Vitali P, Basyuk E, Huttenhofer A, Bachellerie JP (2001). A novel brain-specific box C/D small nucleolar RNA processed from tandemly repeated introns of a noncoding RNA gene in rats. *J Biol Chem* 276: 2,6374–2,6383.

Clemson CM, McNeil JA, Willard HF, Lawrence JB (1996). XIST RNA paints the inactive X chromosome at interphase: evidence for a novel RNA involved in nuclear/chromosome structure. *J Cell Biol* 132: 259–275.

Dittrich B, Buiting K, Korn B, Rickard S, Buxton J, Saitoh S, Nicholls RD, Poustka A, Winterpacht A, Zabel B, Horsthemke B (1996). Imprint switching on human chromosome 15 may involve alternative transcripts of the *SNRPN* gene. *Nat Genet* 14: 163–170.

Erdmann VA, Barciszewska MZ, Szymanski M, Hochberg A, de Groot N, Barciszewski J (2001). The non-coding RNAs as riboregulators. *Nucleic Acids Res* 29: 189–193.

Franke A, Baker BS (1999). The rox1 and rox2 RNAs are essential components of the compensasome, which mediates dosage compensation in *Drosophila. Mol Cell* 4: 117–122.

Hong YK, Ontiveros SD, Strauss WM (2000). A revision of the human *XIST* gene organization and structural comparison with mouse *Xist. Mamm Genome* 11: 220–224.

Kelley RL, Meller VH, Gordadze PR, Roman G, Davis RL, Kuroda MI (1999). Epigenetic spreading of the *Drosophila* dosage compensation complex from *roX* RNA genes into flanking chromatin. *Cell* 98: 513–522.

Komine Y, Tanaka NK, Yano R, Takai S, Yuasa S, Shiroishi T, Tsuchiya K, Yamamori T (1999). A novel type of non-coding RNA expressed in the rat brain. *Brain Res Mol Brain Res* 66: 1–13.

Kuroda Y, Arai N, Arita M, Teranishi M, Hori T, Harata M, Mizuno S (2001). Absence of Z-chromosome inactivation for five genes in male chickens. *Chromosome Res* 9 (in press).

Latham KE (1999). Epigenetic modification and imprinting of the mammalian genome during development. *Curr Topics Dev Biol* 43: 1–49.

Lee JT, Lu N (1999). Targeted mutagenesis of *Tsix* leads to nonrandom X inactivation. *Cell* 99: 47–57.

Lee JT, Davidow LS, Warshawsky D (1999). *Tsix*, a gene antisense to *Xist* at the X-inactivation centre. *Nat Genet* 21: 400–404.

Lee RC, Feinbaum RL, Ambros V (1993). The *C. elegans* heterochronic gene *lin-4* encodes small RNAs with antisense complementarity to *lin-14. Cell* 75: 843–854.

Meller VH, Gordadze PR, Park Y, Chu X, Stuckenholz C, Kelley RL, Kuroda MI (2000). Ordered assembly of roX RNAs into MSL complexes on the dosage-compensated X chromosome in *Drosophila. Curr Biol* 10: 136–143.

Miyoshi N, Wagatsuma H, Wakana S, Shiroishi T, Nomura M, Aisaka K, Kohda T, Surani MA, Kaneko-Ishino T, Ishino F (2000). Identification of an imprinted gene, *Meg3/Gtl2* and its human homologue *MEG3*, first mapped on mouse distal chromosome 12 and human chromosome 14q. *Genes Cells* 5: 211–220.

Moss EG (2000). Non-coding RNAs: lightning strikes twice. *Curr Biol* 10: R436–R439.

Pfeifer K, Leighton PA, Tilghman SM (1996). The structural H19 gene is required for transgene imprinting. *Proc Natl Acad Sci USA* 93: 13,876–13,883.

Raymond CS, Parker ED, Kettlewell JR, Brown LG, Page DC, Kusz K, Jaruzelska J, Reinberg Y, Flejter WL, Bardwell V, Hirsch B, Zarkower D (1999a). A region of human chromosome 9p required for testis development contains two genes related to known sexual regulators. *Hum Mol Genet* 8: 989–996.

Raymond CS, Kettlewell JR, Hirsch B, Bardwell VJ, Zarkower D (1999b). Expression of *Dmrt1* in the genital ridge of mouse and chicken embryos suggests a role in vertebrate sexual development. *Dev Biol* 215: 208–220.

Reinhart BJ, Slack FJ, Basson M, Pasquinelli AE, Bettinger JC, Rougvie AE, Horvitz HR, Ruvkun G (2000). The 21-nucleotide *let-7* RNA regulates developmental timing in *Caenorhabditis elegans. Nature* 403: 901–906.

Smith ER, Pannuti A, Gu W, Steurnagel A, Cook RG, Allis CD, Lucchesi JC (2000). The *drosophila* MSL complex acetylates histone H4 at lysine 16, a chromatin modification linked to dosage compensation. *Mol Cell Biol* 20: 312–318.

Takada S, Tevendale M, Baker J, Georgiades P, Campbell E, Freeman T, Johnson MH, Paulsen M, Ferguson-Smith AC (2000). *Delta-like* and *gtl2* are reciprocally expressed, differentially methylated linked imprinted genes on mouse chromosome 12. *Curr Biol* 10: 1,135–1,138.

Teranishi M, Shimada Y, Hori T, Nakabayashi O, Kikuchi T, Macleod T, Pym R, Sheldon B, Solovei I, Macgregor H, Mizuno S (2001). Transcripts of the MHM region on the chicken Z chromosome accumulate as non-coding RNA in the nucleus of female cells adjacent to the *DMRT1* locus. *Chromosome Res* 9: 147–165.

Watanabe T, Miyashita K, Saito TT, Yoneki T, Kakihara Y, Nabeshima K, Kishi YA, Shimoda C, Nojima, H (2001). Comprehensive isolation of meiosis-specific genes identifies novel proteins and unusual non-coding transcripts in *Schizosaccharomyces pombe. Nucleic Acids Res* 29: 2,327–2,337.

Wightman B, Ha I, Ruvkun G (1993). Posttranscriptional regulation of the heterochronic gene *lin-14* by *lin-4* mediates temporal pattern formation in *C. elegans. Cell* 75: 855–862.

1.6. Neo-X and Neo-Y Chromosomes in *Drosophila miranda*

S. Steinemann and M. Steinemann
Institut für Zoologie, TU Darmstadt, 64287 Darmstadt, Germany

Introduction

Many eukaryotic taxa inherit a heteromorphic sex chromosome pair. It is a generally accepted hypothesis that originally homologous chromosomes develop after the occurrence of a sex differentiator into two structurally and functionally different partners (Muller, 1932; Bull, 1983; Charlesworth, 1991; Steinemann and Steinemann, 1998; Lucchesi, 1999), X and Y chromosome (or Z and W). Often the Y chromosome evolves into a heterochromatic and genetically inert chromosome (chromosome degeneration). This evolutionary process has the final consequence that the male lineage becomes hemizygous for the heterosomal genes, a situation which is compensated by a special mechanism, the gene dosage compensation. The assumption, originally made for *Drosophila*, was extended by Ohno (1967) for the development of mammalian sex chromosomes. The constraints evolving the structurally and functionally different X and Y chromosomes have been the object of speculation since the discovery of sex chromosomes by McClung (1901). For the evolution of sex chromosomes it is assumed that starting from the differential locus in the heterogametic sex the state of permanent heterozygosity is spread further into the flanking regions (for review Lucchesi, 1994; Rice 1996; Charlesworth, 1996).

The degeneration of the Y is characterized by two phenomena: (1) the change from a euchromatic into a heterochromatic chromatin structure and (2) the erosion of genetic activity on the Y. The switch in chromatin structure can not be explained by an accumulation of point mutations. So the question arises how to explain the heterochromatinization of the Y chromosome? To what extend is the gene expression on the Y influenced by the heterochromatinization? On the other hand, the decrease of

M. Schmid and I. Nanda (eds.), Chromosomes Today, Volume 14, 55–63.

active genes on the Y is paralleled by the evolution of the dosage compensation mechanism on X-chromosomal genes. How is dosage compensation established on an evolving X chromosome? What types of molecular changes are involved in the evolutionary transformation of autosomes into dosage compensated X chromosomes? To answer these questions the *Drosophila miranda* model system is especially informative.

Results and discussion

According to Muller (1940) the different karyotypes of *Drosophila* could be derived from a basic karyotype consisting of five acrocentric rod shaped chromosome elements A, B, C, D, E and a dot shaped F element, not including the Y chromosome. In *Drosophila melanogaster* the X chromosome is a rod shaped, acrocentric chromosome (element A in Muller's terminology), while the Y represents a j-shaped submetacentric, completely heterochromatic chromosome. In contrast, all members of the *pseudo-obscura* subgroup have a metacentric X chromosome derived by fusion of the original X with an autosome (Mullers's element D, corresponding to chromosome arm *3L* in *D. melanogaster*). In *D. miranda*, which diverged from *D. pseudoobscura* only about 2 Myr (Bario *et al.*, 1992), a second autosome (Muller's element C, corresponding to chromosome arm *2R* in *D. melanogaster*) has fused with the Y chromosome (MacKnight, 1939; Steinemann, 1982), generating a neo-Y (i.e., Y–C) and a neo-X (i.e., the remaining, unfused C element) designated X2, Figure 1. The rearrangement results in the exceptional situation of genes that are normally located on autosomes now being found on a developing secondary sex chromosome pair, i.e., the X2 and neo-Y chromosome in *D. miranda*, Figure 2. As the Y–C fusion has occurred recently on an evolutionary time scale, both chromosomes are within an evolutionary process in remodelling the chromosome architecture of both chromosomes. The neo-Y is on its way to degenerate and the X2 to become dosage compensated (Bone and Kuroda, 1996; Marin *et al.*, 1996 Steinemann *et al.*, 1996;). In polytene nuclei from 3rd instar male larvae the neo-X chromosome (X2) is somatically unpaired with the neo-Y (cf. Figure 2). The X2 is early replicating (Das *et al.*, 1982) and shows partial dosage compensation (Strobel *et al.*, 1978; Steinemann *et al.*, 1996). Neo-X and the neo-Y chromosome correspond in the two sibling species *D. pseudoobscura* and *Drosophila persimilis* to autosome pair 3 (Steinemann *et al.*, 1984).

We obtained first indications that the cytologically observable structural differences between the X2 and the neo-Y of *D. miranda* could be the result from an invasion of middle repetitive DNA-sequences (transposable elements) into the fused chromosome arm of the neo-Y (Steinemann, 1982). In order to compare defined regions on the X2 with homologous regions on the neo-Y and from autosomes, e.g., from *D. melanogaster* and the two sibling species of *D. miranda, D. persimilis* and *D. pseudoobscura*, we used chromosome homologies as experimental strategy (Steinemann, 1982; Steinemann *et al.*, 1984). We could confirm that the neo-Y and X2 chromosome from *D. miranda* is homologous with chromosome element C. As a consequence the X2 and neo-Y from *D. miranda* is homologous with *2R* of *D. melanogaster* and chromosome 3 in the two sibling species, cf. Figure 1.

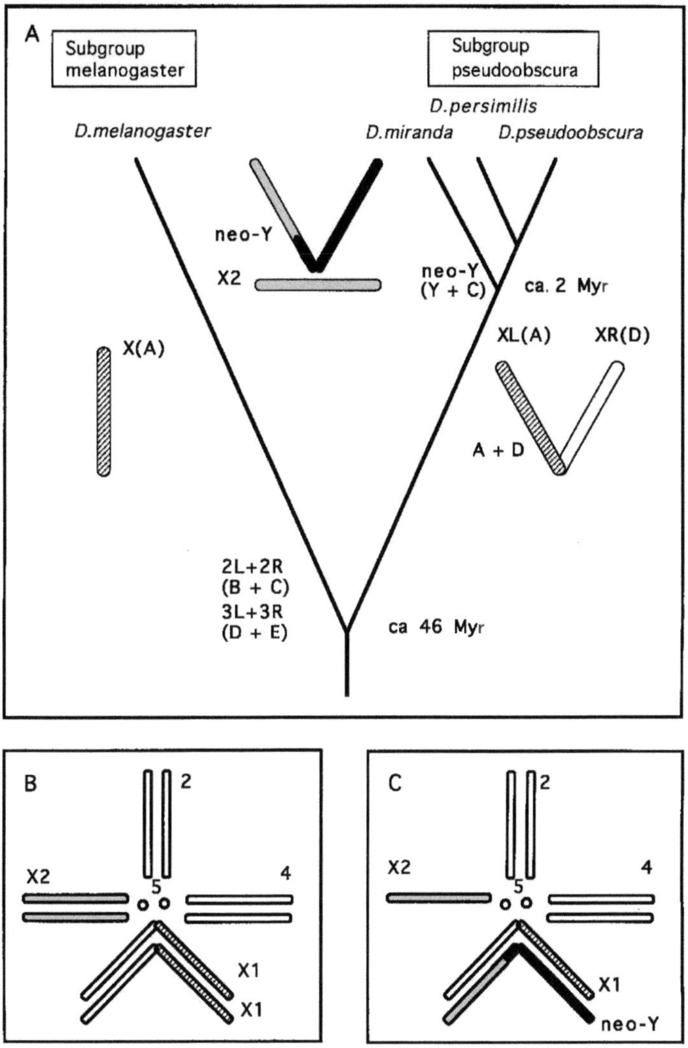

Figure 1. *Relationships and chromosome fusions generating new sex chromosomes in selected species of* Drosophila. *(A) Cartoon of chromosome fusions. Chromosomes are labelled according to the basic* Drosophila *karyotype of five rods, designated A–E, and one dot, designated F (Muller, 1940; Patterson and Stone, 1952). Times of divergence are taken from Barrio et al. (1992) and Russo et al. (1995). (B) and (C) Representations of the* D. miranda *female (B),* 2n = *10 and male (C),* 2n = *9 karyotypes. In all parts chromosome element A is striped; element C is stippled, and the Y chromosome is black.*

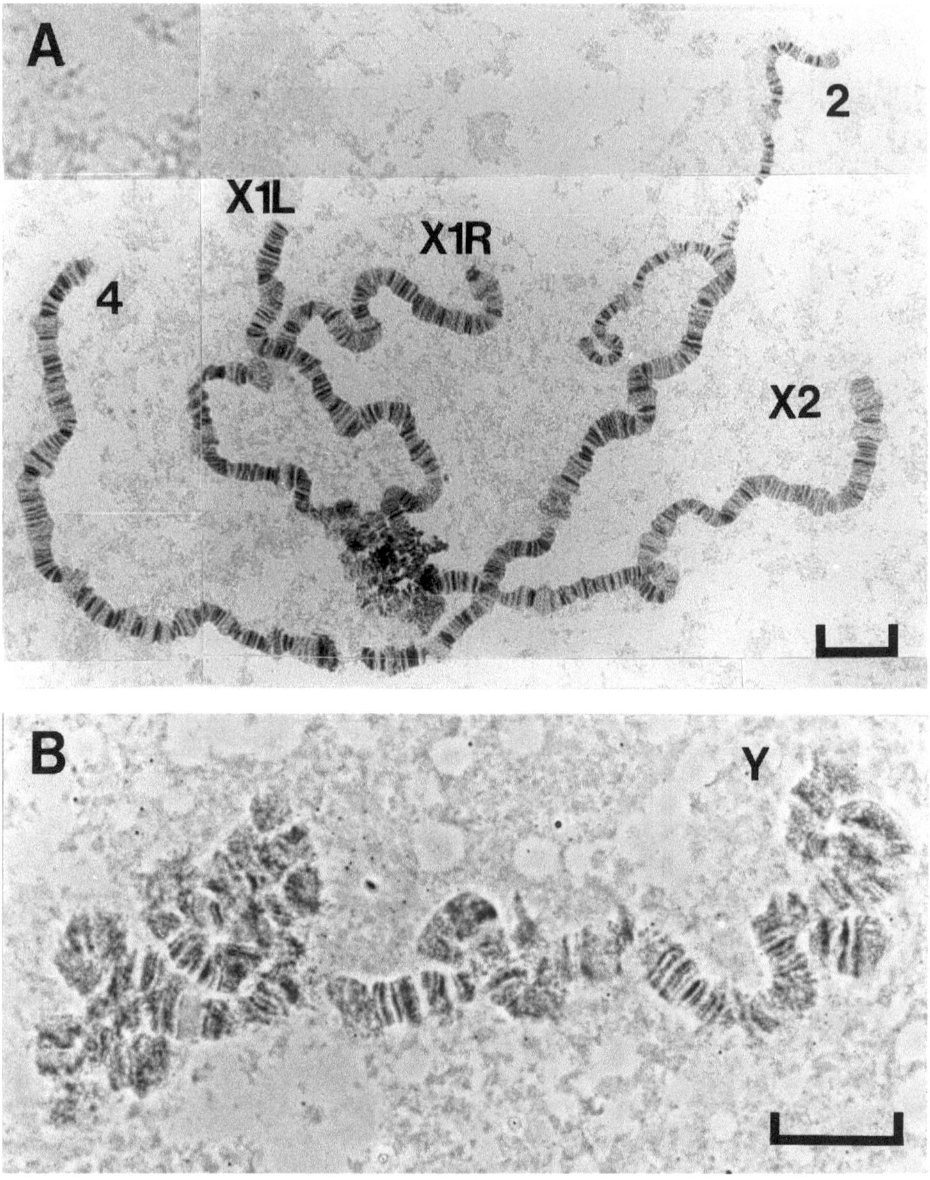

Figure 2. Salivary gland squash preparations of a polytene nucleus of D. miranda *females (A) and the neo-Y chromosome (B) of males, stained with carmine-acetic acid and photographed with phase contrast. The chromosome arms are designated according to Dobzhansky (1935). The bar represents 20 μm. (A) The short chromosome 5 is missing in the nucleus shown. (B) Neo-Y chromosome of a* D. miranda *male. The neo-Y chromosome is still polytenized. The structural difference between the former homologues, X2 and neo-Y, is obvious.*

We have selected three test genes on the X2, *larval cuticle protein* genes (*Lcp1-4*), *α-Amylase* genes (*Amy1-3*) and the *Krüppel* gene (*Kr*). The three loci are distributed along the X2 with the proximal *Lcp1-4*, the more central positioned *Amy1-3* and the distally located *Kr*. We use these genes as entry gates to analyse the mechanisms of the evolutionary processes in progress on the neo-Y/neo-X chromosome pair at the molecular level, Figure 3.

For the *α-Amylase* (*Amy1-3*) we could show genetically that there is no active enzyme on the *neo-Y* chromosome. In addition we obtained biochemical evidence that in *D. miranda* the *Amy* locus is dosage compensated (Steinemann *et al.*, 1986). We were able to show that the *Amy1* gene on the X2 chromosome is intact and expressed. The *Amy2* gene contains a complete coding sequence, however it is not expressed. This was proven by germline transformation experiments. *Amy3* is eroded and not functional. Therefore the *Amy1* is the dosage compensated gene (Steinemann and

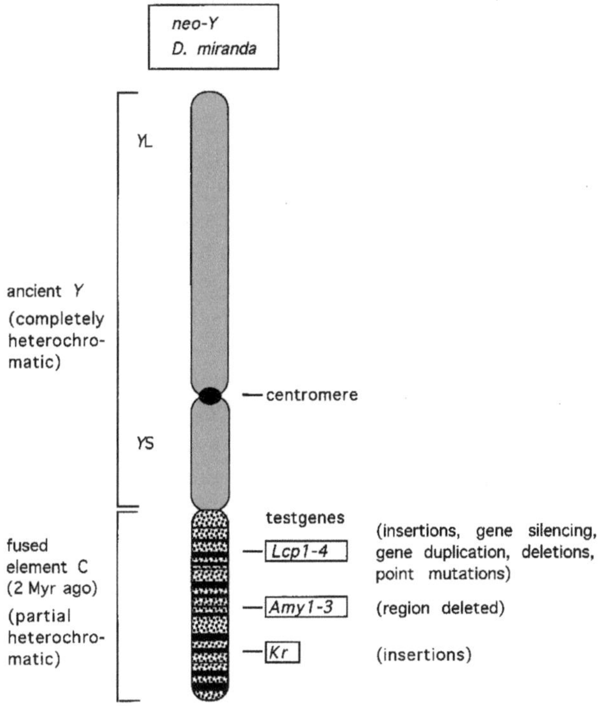

Figure 3. *Schematic drawing of the neo-Y chromosome of* D. miranda. *The neo-Y is the result of a fusion of chromosome element C (Muller's terminology), represented in* D. melanogaster *by the right arm of chromosome 2 (2R), with the ancient Y, about 2 Myr ago. The ancient Y is shown in a fine grey stippled texture. The partial heterochromatic structure of the fused element C is indicated by a coarse stippled texture, interrupted by banded regions. YS: Short arm of the Y; YL: Large arm of the Y. Test genes:* Larval cuticle protein *genes* (Lcp1-4); α-Amylase *genes* (Amy1-3) *and the* Krüppel *gene* (Kr).

Steinemann, 1999). Several evidences indicate that the *Amy* cluster on the neo-Y is already deleted (Steinemann and Steinemann, 1999).

The detailed molecular analysis of the complete X2- and neo-Y-chromosomal *Lcp1-4* region revealed a massive accumulation of insertions, a few deletions and a large duplication (Steinemann and Steinemann, 1992). The duplication enclosing the *neo-Y* allele of *Lcp2* and the *TRIM* retrotransposon, a non-LTR retrotransposon, was analysed in detail and possible mechanisms generating the duplication are discussed (Steinemann and Steinemann, 1993). A second insertion, designated *TRAM*, revealed to be as well a new retrotransposon. We had sequenced the element and analysed its structure. *TRAM* revealed to be a retrotransposon of the LTR-type (Steinemann and Steinemann, 1997). *TRIM* (Steinemann and Steinemann, 1991), *TRAM* and the *IS*Y2/3 insertions (Steinemann and Steinemann, 1992), detected in the neo-Y-chromosomal *Lcp* region, reveales a biased genomic distribution, e.g., *IS*Y3, Figure 4. The X2 chromosome derived *Lcp1-4* clone labels exclusively the *Lcp1-4* locus (Figure 4A). Using the *IS*Y3 probe from the neo-Y-chromosomal *Lcp* region we detect about 20 sites labelled on the X2 chromosome (Figure 4B). In contrast the neo-Y shows a dense labelling over the whole chromosome (Figure 4C). In comparison with the former homologue X2 the elements revealed a massive accumulation on the neo-Y. These findings are corroborated by results from the KR region (Steinemann and Steinemann, in press). We concluded from these findings that the neo-Y functions as a trap for retrotransposons (Steinemann and Steinemann, 1992).

Beside the unravelling of structural changes on the DNA level, we are interested in the influence of the observed chromatin changes on the resident gene expression. Due to several point mutations in the X2- and neo-Y-chromosomal *Lcp* alleles, we revealed amino acid substitutions in the derived amino acid sequences. We used the amino acid exchanges as genetic markers to identify the neo-Y- and X2-chromosomal encoded LCPs, respectively. We were able, applying the protein microsequencing technique on the separated LCP bands, to distinguish the products from *neo-Y* and *X2* alleles. The analysis of the neo-Y-chromosomal alleles revealed a strongly reduced activity of *Lcp3* and nonexpression of *Lcp1, Lcp2* and *Lcp4* (Steinemann *et al.*, 1993). To test the activity of the neo-Y-chromosomal *Lcp* alleles and the possible influence from flanking insertions, we designed constructs. DNA fragments from the neo-Y containing the complete *Lcp* genes were subcloned, either flanked by the *TRIM* retrotransposon and *IS*Y1-3 sequences or without. After germline transformation of *D. melanogaster* embryos we analysed transgenic lines for ectopic expression. We were able to show that the neo-Y-chromosomal *Lcp1* and *Lcp3* alleles are normally expressed in the heterologous system when the flanking transposon and insertion sequences are deleted. Together with the flanking insertion sequences the *Lcp1* and *Lcp3* genes are not (*Lcp1*) or strongly reduced (*Lcp3*) expressed. We concluded from these results that the analysed neo-Y-chromosomal *Lcp1* and *Lcp3* genes are silenced to different degrees at their resident locus (Steinemann and Steinemann, 1992). With the demonstration of the silencing of neo-Y-chromosomal genes we have introduced a new aspect in the discussion of the evolutionary process of Y chromosome degeneration.

For a working hypothesis we assume that the degeneration of the Y chromosome is paralleled by the development of dosage compensation on the X. As a start for

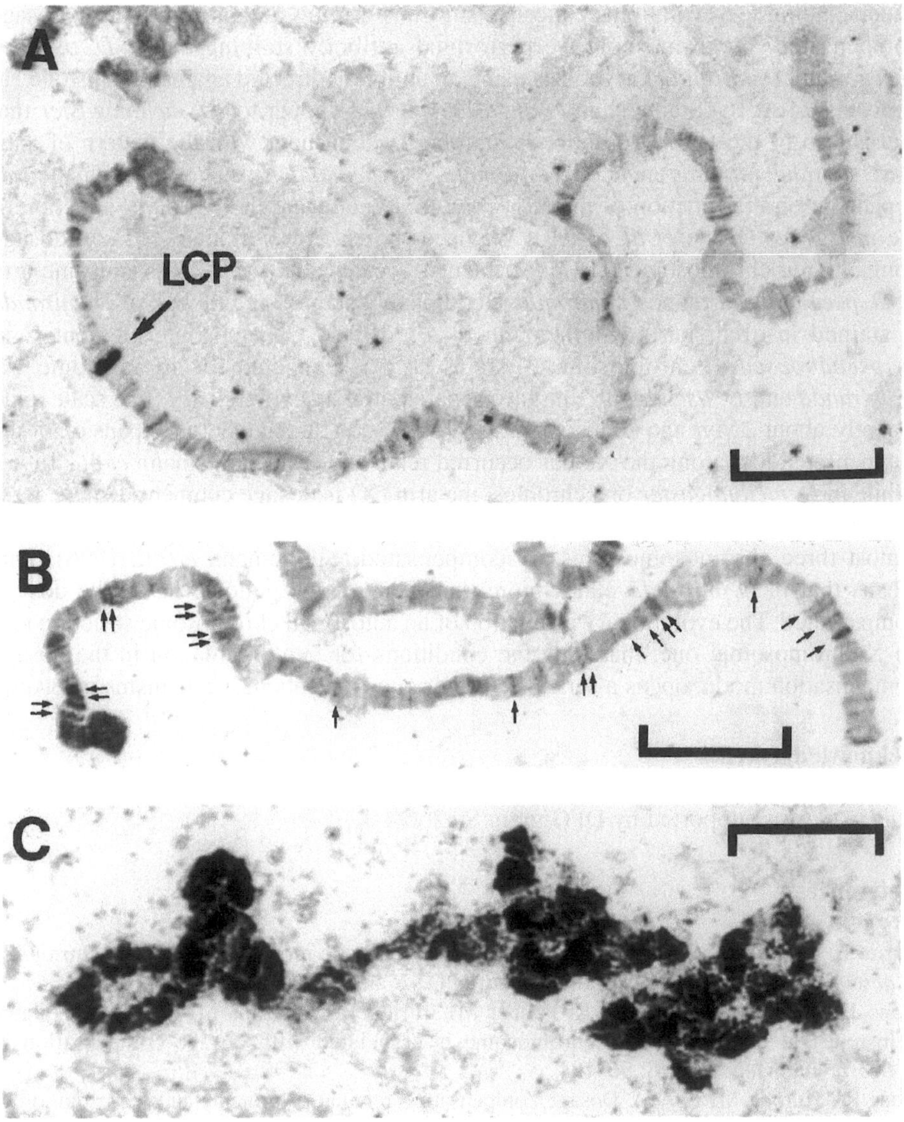

Figure 4. In situ *hybridization of biotinylated probes on the neo-X (X2) and neo-Y chromosome of* D. miranda. *The probes had been biotinylated by nick translation and hybridized as detailed in Steinemann and Steinemann (1991). For signal detection the protocol for immunoperoxidase staining supplied with the DETEK I-hrp kit (Enzo Biochem) was followed. The bar represents 20μm. (A)* In situ *hybridization of a 0.7 kb HindIII fragment (cf. Figure 2 in Steinemann and Steinemann, 1991) from a X2-chromosomal* Lcp *clone. The X2-chromosomal* Lcp *locus is labelled (arrow). (B)* In situ *hybridization with an* IS*Y3-homologous probe derived from the neo-Y-chromosomal* Lcp *region (cf. Figure 4 in Steinemann and Steinemann, 1992). About 20 sites are labelled along the neo-X chromosome (arrows). (C) The probe used in (B) hybridized on the neo-Y. The neo-Y chromosome shows a strong labelling distributed over the whole chromosome. Comparing the former homologues neo-X and neo-Y a biased distribution of the* IS*Y3 transposon is obvious.*

understanding the evolutionary mechanism transforming an autosome into a dosage compensated X chromosome we performed antibody staining of the *D. pseudoobscura* and *D. miranda* karyotypes using an antibody directed against the histone H4 isoform, H4.Ac16 (Steinemann *et al.*, 1996). It was shown for *D. melanogaster* that the pattern of the H4.Ac16 antibody staining is congruent with the pattern of anti-MSL staining, proteins encoded by the *male-specific lethal* (*msl*) genes. The pattern is dependent on the function of the *Sex lethal* (*Sxl*) gene and the wild-type function of the *msl* genes (Bone *et al.*, 1994). We used the H4.Ac16 as marker for a dosage compensated chromosome. The experiments revealed that the *XR* chromosome arm of *D. pseudoobscura* and *D. miranda* is completely stained and the X2 of *D. miranda* is stained in great parts (Steinemann *et al.*, 1996). Chromosome element C in *D. pseudoobscura*, chromosome 3, stains as it is expected for an autosome. As *D. miranda* and *D. pseudoobscura* have separated on an evolutionary time scale quite recently, about 2 Myr ago (Barrio *et al.*, 1992), we concluded that the expansion of the compensated loci along the X2 has occurred relatively fast (Steinemann *et al.*, 1996). While in *D. melanogaster* one chromosome arm (X) is dosage compensated, we were able to show that in *D. pseudoobscura* two chromosome arms and in *D. miranda* almost three chromosome arms are compensated (Steinemann *et al.*, 1996). This means that in *D. miranda* males more than 50% of the genome has to be dosage compensated. The evolutionary adaptation of an autosomal chromosome structure into an X-chromosomal one, matching the conditions for gene regulation in the dosage compensation mode, opens a series of further questions about mechanisms involved.

Acknowledgements

This work was supported by DFG grant Ste266/4-1.

References

Barrio E, Latorre A, Moya A, Ayala FJ (1992). Phylogenetic reconstruction of the *Drosophila obscura* group, on the basis of mitochondrial DNA. *Mol Biol Evol* 9: 621–635.
Bone JR, Lavender J, Richman R, Palmer MJ, Turner BM, Kuroda MI (1994). Acetylated histone H4 on the male X chromosome is associated with dosage compensation in *Drosophila. Genes Dev* 8: 96–104.
Bone RJ, Kuroda MI (1996). Dosage compensation regulatory proteins and the evolution of sex chromosomes in *Drosophila. Genetics* 144: 705–713.
Bull JJ (1983). *Evolution of Sex Determining Mechanisms.* Menlo Park, CA: Benjamin/Cummings, pp. 248–269.
Charlesworth B (1991). The evolution of sex chromosomes. *Science* 251: 1,030–1,033.
Charlesworth B (1996). The evolution of chromosomal sex determination and dosage compensation. *Curr Biol* 6: 149–162.
Das M, Mutsuddi D, Duttagupta AK, Mukherjee AS (1982). Segmental heterogeneity in replication and transcription of the *X2* chromosome of *Drosophila miranda* and conservativeness in the evolution of dosage compensation. *Chromosoma* 87: 373–388.
Lucchesi JC (1994). The evolution of heteromorphic sex chromosomes. *BioEssays* 16: 81–83.
Lucchesi JC (1999). On the origin of sex chromosomes. *BioEssays* 21: 188–190.

MacKnight RH (1939). The sex-determining mechanism of *Drosophila miranda. Genetics* 24: 180–201.
Marin I, Franke A, Bashaw GJ, Baker BS (1996). The dosage compensation system of *Drosophila* is co-opted by newly evolved X chromosomes. *Nature* 383: 160–163.
McClung CE (1901). Notes on the accessory chromosome. *Anat Anz* 20: 220–226.
Muller HJ (1932). Some genetic aspects of sex. *Am Nat* 66: 118–138.
Muller HJ (1940). Bearings of the *Drosophila* work on systematics. In: Huxley I, Hrsg. eds. *The New Systematics.* Oxford: Oxford University Press, pp. 185–268.
Ohno S (1967). *Sex Chromosomes and Sex-Linked Genes.* Springer-Verlag, New York.
Patterson JT, Stone WS (1952). *Evolution in the Genus Drosophila.* New York: The Macmillan Company.
Rice WR (1996). Evolution of the *Y* Sex Chromosome in Animals. *BioScience* 46: 331–343.
Russo CAM, Takezaki N, Nei M (1995). Molecular phylogeny and divergence times of Drosophilid species. *Mol Biol Evol* 12: 391–404.
Steinemann M (1982). Multiple sex chromosomes in *Drosophila miranda*: A system to study the degeneration of a chromosome. *Chromosoma* 86: 59–76.
Steinemann M, Pinsker W, Sperlich D (1984). Chromosome homologies within the Drosophila obscura group probed by *in situ* hybridization. *Chromosoma* 91: 46–53.
Steinemann M, Steinemann S, Fakler J (1986). Differential expression and dosage compensation of the α-Amylase gene in *Drosophila miranda. Experientia* 42: 577–579.
Steinemann M, Steinemann S (1991). Preferential *Y* chromosomal location of TRIM, a novel transposable element of *Drosophila miranda, obscura* group. *Chromosoma* 101: 169–179.
Steinemann M, Steinemann S (1992). Degenerating *Y* chromosome of *Drosophila miranda*: A trap for retrotransposons. *Proc Natl Acad Sci USA* 89: 7,591–7,595.
Steinemann M, Steinemann S, Lottspeich F (1993). How *Y* chromosomes become genetically inert. *Proc Natl Acad Sci USA* 90: 5,737–5,741.
Steinemann M, Steinemann S (1993). A duplication including the *Y* allele of *Lcp2* and the TRIM retrotransposon at the *Lcp* locus on the degenerating *neo-Y* chromosome of *Drosophila miranda*: Molecular structure and mechanisms by which it may have arisen. *Genetics* 134: 497–505.
Steinemann M, Steinemann S, Turner BM (1996). Evolution of dosage compensation. *Chromosome Res* 4: 185–190.
Steinemann M, Steinemann S (1997). The enigma of *Y* chromosome degeneration: *TRAM*, a novel retrotransposon is preferentially located on the *neo-Y* chromosome of *Drosophila miranda. Genetics* 145: 261–266.
Steinemann M, Steinemann S (1998). Enigma of *Y* chromosome degeneration: *Neo-Y* and *Neo-X* chromosomes of *Drosophila miranda* a model for sex chromosome evolution. *Genetica* 102/103: 409–420.
Steinemann S, Steinemann M (1999). The *Amylase* gene cluster on the evolving sex chromosomes of *Drosophila miranda. Genetics* 151: 151–161.
Steinemann S, Steinemann M (2001). Biased distribution of repetitive elements: a landmark for neo-Y chromosome evolution in *Drosophila miranda. Cytogenet Cell Genet*, in press.
Strobel E, Pelling C, Arnheim N (1978). Incomplete dosage compensation in an evolving *Drosophila* sex chromosome. *Proc Natl Acad Sci USA* 75: 931–935.

1.7. The X Chromosome Plays a Special Role During Speciation

Hildegard Kehrer-Sawatzki and Horst Hameister
Department of Human Genetics, University Ulm, Albert-Einstein Allee 11
D-89069 Ulm, Germany

According to the concept of Mayr, 1942, a species is defined as an interbreeding population. Speciation itself is a long lasting continuous process by which a formerly unique species gets separated into two different species by reproductive isolation. Darwin viewed speciation as "the mystery of mysteries". According to our present view speciation describes the last steps during the evolutionary process which gave rise to the respective species. e.g. human speciation lasts back 5–6 million years when the common ancestor of human and chimpanzee lived. We view evolution as a discontinuous process of single speciation events.

The analysis of genetic factors influencing speciation processes revealed that it is not a change in a single gene or in a single chromosome by which two species become separated, but multiple changes in a large number of genes end up in a specific trait development (Davis and Wu, 1996; Phillips, 1999). Simultaneously to specific trait development new species become more and more reproductively isolated. In interspecies hybrids these reproductive isolating effects can be studied directly. These isolating effects are subdivided into premating (premeiotic or prezygotic) and post-mating (postmeiotic or postzygotic) factors (for review see Coyne, 1992). Postmating factors interfere with fertility of the mating partners and viability of the interspecies offspring. According to Haldane's rule, 1922, postmating effects are mostly observed in the heterogametix sex (XY, X0, ZW, Z0). This biologically well founded rule is synonymous with "the large X chromosome effect", which *Drosophila* geneticist use to describe the disproportional large frequency of sterility genes on the X chromosome (Turelli and Orr, 1995). Both, Haldane's rule and "the large X chromosome effect" are due to the fact that genes which influence the number and viability of offspring are preferentially localized on the sex chromosomes.

M. Schmid and I. Nanda (eds.), Chromosomes Today, Volume 14, 65–71.

Most probably, genetic influences on postmating effects are active in todays human beings too. In agreement with Haldane's rule they especially affect the heterogametic male sex. This assumption is corroborated by the observation that 5–10% of all males suffer from sterility or subfertility. Further, the male : female ratio is calculated to be at fertilization 1.4 (A. Gropp, pers. comm.), at birth 1.06, and at reproductional age the male : female ratio is nearly 1.0. Recently, Saifi and Chandra (1999) reported a three times higher rate of sex- and reproduction-related genes on the human X chromosome. This is completely in line with Haldane's rule.

Our knowledge about the genetic factors influencing premating effects relevant to the speciation process is poor. It is well established that premating effects are exerted by specific habitat requirements, by the day-times being awake, and different behaviour. Common to many species and very efficient during speciation is mate choice, which nearly always is a female mate choice. This is simply due to the fact that in most plant and animal species the number of female gametes is limited and the female investment to raise offspring exceeds the contribution of the male. When we now will turn to the premating characters which work in human one can simply apply Haldane's rule: i.e. which character shows a large X chromosome effect – as has been shown for sex- and reproduction-related genes? For clinical geneticists the answer is clear: it is mental performance. It is long known that there is an excess of males among the mentally retarded patients (Penrose, 1938; Lehrke, 1974; Herbst and Miller, 1980). There is some controversy as to use mental retardation as being an indicator of a mental performance gene. In our context this is done in analogy to sterility genes which in interspecies hybrid studies are taken as genes which interfere with fertility. In a QTL model of general intelligence, or "g" according to Plomin (1999) the same genes which cause mental retardation contribute to the normal range of individual intelligence. Mapping studies have repeatedly shown a high number of X-linked mental retardation genes. Since 1984 biannual workshops take place registering all newly identified X chromosomal mental retardation syndromes (Opitz and Sutherland, 1984; Chiurazzi *et al.*, 2001). In Table 1 we have listed separately for each chromosome the number of identified genes and the number of identified genes, which in the mutated state cause mental retardation. Evaluation along this line counts the different allelic forms of one and the same gene only once.

As is obvious from Table 1 there is indeed an overrepresentation of genes affecting mental retardation on the X chromosome. The frequency of genes affecting mental performance is 3.5 times higher on the X chromosome than on autosomes. When evaluating the association of a trait with the X chromosome, it is obvious that due to the expression of recessive traits in the hemizygous male one has to take into account an ascertainment bias. This relatively stable ascertainment bias becomes visible when a database like OMIM is queried for common defect phenotypes as is shown in Table 2.

To our surprise this ascertainment bias is with 2.35 more X chromosomal entries relatively constant. But even when this correction factor is taken into consideration, it is evident that mental retardation genes are still 3.1 times more frequent on the X chromosome than on autosomes.

Table 1. *Comparison of the percentage of identified genes per chromosome which in the mutated state cause mental retardation. The data are based on OMIM statistics from 28 January 2001*

Chromosome	Identified genes	Mental retardation as a symptom	% of genes with mental retardation as a symptom
1	658	20	3.04
2	406	8	1.97
3	336	9	2.69
4	258	10	3.88
5	283	6	2.12
6	397	12	3.02
7	308	8	2.60
8	222	7	3.15
9	255	8	3.14
10	230	6	2.61
11	418	8	1.91
12	351	11	3.13
13	109	5	4.59
14	206	8	3.88
15	175	5	2.86
16	251	11	4.38
17	403	9	2.23
18	96	2	2.08
19	438	9	2.05
20	147	7	4.76
21	90	4	4.44
22	146	7	4.79
All autosomes	6182	180	2.91
X	406	41	10.1

If we follow this line of interpretation, general intelligence in human has to be seen as a character sexually selected for. Already Darwin (1874) realized relatively fantastic phenotypes in the living world which for him made clear that these phenotypes could not have developed by natural selection. He noted the existence of a second, in some way counter-active selection process: sexual selection. Sexually selected traits mostly contrast traits, which are selected for naturally. The traits sexually selected for are often luxurious traits which may become a burden for the individual, as e.g. peacock's tail and elk's antlers. It was R.A. Fisher who revived the discussion on sexual selection as a different mechanism from natural selection. In his famous treatise "The genetical theory of natural selection", published 1930, he considered sexual selection

Table 2. *Ascertainment bias for X chromosomal linkage of frequent clinical phenotypes compared to mental retardation as symptom. The data are based on OMIM statistics from 28 January 2001. The frequeny in column 2 and 3 was calculated relative to the total number of X chromosomal (*n = *412) and autosomal (*n = *6,367) OMIM entries (mode of inheritance proven)*

	Number of respective clinical phenotypes on the			
Clinical phenotype	X chromosome (%)	Autosomes (%)	% X chromosomal loci/ % autosomal loci	
Polydactyly	5 (1.21)	35 (0.55)	2.20	2.35
Cleft palate	10 (2.34)	56 (0.88)	2.76	
Facial dysplasia	7 (1.70)	45 (0.71)	2.39	
Skeletal dysplasia	11 (2.67)	76 (1.19)	2.24	
Growth retardation	13 (3.16)	94 (1.48)	2.14	
Mental retardation	103 (25.0)	222 (3.49)	7.16	

necessary to explain the rapid evolution of such traits as colours in fish and phantastic plumage in birds. He described the evolution of a sexually selected for character as a run-away process. The genes for sexual character development in the male and sexual preference for that character in the female get genetically correlated. In such case there is positive feed-back in both systems. In the next years sexual selection again was not very well recognized until in the mid 1980s scientists from the the famous Chicago school (Lande and Arnold, 1985) showed that this run-away process of sexual character development receives a rocket-like speed when both systems, the male sexual character development and the female perception system for that character, develop jointly in the same system. This is given where the male sexual character is a specific behaviour, like courtship dancing in *Drosophila* species. Now, there is positive feed-back in both sexes on the same genes.

As mentioned before, usually the development of a sexual character is luxurious and thus contradicts natural selection. For instance, when peacocks develop even larger tails they cannot escape predators. Therefore after some time the run-away process of sexual ornament development gets balanced by natural selection. It is the uniqueness of human evolution that the trait sexually selected for is cognitive ability, which at the same time is selected for in humans by the struggle for survival. In humans the development of the mating character is not balanced by natural selection but even reinforced. This is an ongoing process with exponential acceleration.

If we accept this hypothesis one could conclude that only males should be carrier of this special sexual character. But as pointed out before, for the development of this character the perception of the specific cognitive abilities by the female is as important as the presentation of these abilities by the male. Speaking and listening to someone use mainly the same neural functions. Insofar selectee and selector apply the same genetic system. The discussion who performs better simply boils down to the controversion between an artist and his criticizers, or between the scientist and his reviewer. We can learn from evolution: if both, artist and criticizer, agree on the same aim, development becomes fast and faster.

Insofar we realize mate choice as the driving force for the development of general intelligence. Mate choice is a real universal behavioural trait in nature. Even in yeast there are two mating types (Herskowitz *et al.*, 1992). For sexual conjugation one partner does choose and the other partner has to change its mating type by switching the mating type cassette. This behavioural trait works in its simplest form the same way an organism reacts to extern stimuli. It is easy for us to explain the answer of an organism towards changes of light or temperature. It is completely unknown and random in which way an organism reacts to the opposite sex. For the first time there runs a decision making neuronal process which is unpredictable. According to the discussion brought forward here we see mate choice not only as the driving force for the development of general intelligence in human but also as an initiating stimulus at the roots of the development of a higher functioning central nervous system in nature.

It is worthwhile to consider which different characters were developed in nature thanks to the driving force of sexual selection.

If only natural selection were in power one could imagine that the world would be populated by grey, fat rats. But, as G.F. Miller (2000), in his intriguing monograph "The mating mind" has outlined both selection forces working at the same time made possible: peacock's tail, irish elk's antlers, butterfly's wings, nightingale's song, bowerbird's nests, orchid's blossoms, van Gogh's lilies, and Mozart's Requiem.

At the end we will come back to genes and chromosomes. As human beings we are convinced that human evolution and especially evolution of the specific human mental performance must bear something special. In Table 3 we have listed genes which in humans cause non-syndromal mental retardation, i.e. genes which show in the mutated state only influence on mental performance.

A is indicated in Table 3 these genes are in universal use in the animal kingdom. They are mostly engaged in basic cellular signalling cascades and there is no new human or primate specific gene. There are no new genes, but recruitment of old genes for new functions drives evolution to higher complexity (Duboule and Wilkins, 1998). When we consider the leading role of the X chromosome during all speciation processes in the last 300 million years it is inescabale to realize that the genes of the most conserved gene arrangement (Glas *et al.*, 1999) are engaged in a dominant manner in morphological and functional diversification. Further, comparative mapping data have shown that the human karyotype resembles the ancient mammalian founder karyotype very much. This ancient karyotype was already a "wohlbalanziertes Genom". Evolution to higher complexity took place not on the genome level but by establishment of complex epistatic interactions between these genes. This bears witness for the leading role of the parsimony principle during evolution.

Table 3. *Identified human MRX genes*

Gene symbol	Gene	Position	Expression	Function	Orthologous EST sequence
FMR2	FRAGILE X MENTAL RETARDATION 2	Xq27.3	Neocortex, Purkinje cells, hippocampus, testis, thymus, placenta	Nuclear protein, putative transcription activator	*Drosophila melanogaster*
GDI1	GDP DISSOCIATION INHIBITOR 1	Xq28	Differentiated neuronal cells of the central and peripheral nervous system	Helps to maintain a soluble pool of Rab-GDP, participates in vesicle fusion	*Saccharomyces cerevisiae*
OPHN1	OLIGOPHRENIN 1	Xq12	Fetal brain, lung, kidney, placenta	Rho-GTPase activating protein	*Onchocerca volvulus*
PAK3	p21 (CDKN1A)-ACTIVATED KINASE 3	Xq22	Brain, spinal cord, lung, testis	Downstream effector of Rho-GTPases	*Saccharomyces cerevisiae*
RPS6KA3	RIBOSOMAL PROTEIN S6 KINASE, 90 KD, POLYPEPTIDE 3	Xp22.2	Brain, kidney, pituitary and thyroid gland	MAPK-activated signalling pathway	*Danio rerio*
IL1RAPL1	INTERLEUKIN 1 RECEPTOR ACCESSORY PROTEIN-LIKE 1	Xq21	Not known	IL1 signalling during inflammation	*Danio rerio*
TM4SF2	TRANSMEMBRANE 4 SUPERFAMILY MEMBER 2	Xp11.4	Cerebral cortex, hippocampus, adult heart, brain, liver, kidney, pancreas	Organisation of actin-cytoskelett via β-integrins	*Torpedo marmorata*
ARHGEF6	RHO GUANINE NUCLEOTIDE EXCHANGE FACTOR 6	Xq26	Ubiquitously	Rho-GTPase activating	*Danio rerio*
RPS6KA6	RIBOSOMAL PROTEIN S6 KINASE, 90 KD, POLYPEPTIDE 6	Xp21.3–p22.1	Brain, kidney, placenta, pancreas	MAPK-activated signalling pathway	*Danio rerio*

References

Chiurazzi, Pietro, Ben CJ Hamel, Giovanni Neri (2001). XLMR genes: update 2000. *Eur J Hum Genet.* 9: 71–81.

Coyne JA (1992). Genetics and speciation. *Nature* 355: 511–515.

Coyne JA, Charlesworth B (1989). Genetic analysis of X-linked sterility in hybrids between three sibling species of drosophila. *Heredity* 62: 97–106.

Davis, Andrey W, Chung-I Wu (1996). The broom of the sorcerer's apprentice: the fine structure of a chromosomal region causing reproductive isolation between two sibling species of *Drosophila. Genetics* 143: 1287–1298.

Duboule D, Wilkins AS (1998). The evolution of "bricolage". *Trends Genet* 14: 54–59.

Fisher RA (1930) *The Genetical Theory of Natural Selection.* Oxford: Clarendon Press.

Glas, Rob, Jennifer A Marshall-Graves, Roland Toder, Malcom Ferguson-Smith, Patricia C O'Brien (1999). Cross-species chromosome painting between human und marsupial directly demonstrates the ancient region of the mammalian X. *Mamm Genome* 10: 1115–1116.

Haldane JBS (1922). Sex-ratio und unisexual sterility in hybrid animals. *J Genet* 21: 101–109.

Herskowitz I, Rine J, Strathern J (1992). Mating-type determination and mating type interconversion in Saccharomyces cerevisiae. In: *The Molecular und Cellular Biology of the Yeast Saccharomyces cerevisiae. Gene Expression*, vol. II. Cold Spring Harbor: Cold Spring Harbor Laboratory Press, pp. 583–656.

Herbst DS, Miller JR (1980). Non-specific X-linked mental retardation. II The frequency in British Columbia. *Am J Med Genet* 28: 378–382.

Lande R Arnold SJ (1985). Evolution of mating preference and sexual dimorphism. *J Theor Biol* 117: 651–664.

Lehrke R (1974). X-linked mental retardation and verbal disabilities. *Birth Defects* 10: 1–7.

Mayr, Ernst (1942). *Systematics and the Origin of Species.* New York: Columbia University Press.

Miller, Geoffrey F (2000). *The mating mind: how sexual choice shaped the evolution of human nature.* Heinemann/Doubleday 2000.

Opitz JM, Sutherland GR (1984). Conference report: International Workshop on the fragile X and X-linked mental retardation. *Am J Med Genet* 17: 5–94.

Penrose LS (1938). A clinical und genetic study of 1280 cases of mental defects (The Colchester Survey). *MRC Special Report* 229 (Her Majesty's Stationery Office, London 1938).

Phillips PC (1999). From complex traits to complex alleles. *Trends Genet* 15: 6–8.

Plomin R (1999). Genetic and general cognitive ability. *Nature* 402: C25–C29.

Saifi GM, Chandra HS (1999). An apparent excess of sex- and reproduction-related genes on the human X-chromosome. *Proc R Soc Lond B* 266: 203–209.

Turelli M, Orr HA (1995). The dominance theory of Haldane's rule. *Genetics* 140: 389–402.

1.8. Sex Chromosome Evolution: Evidence from Fish, Fly and Moth Species

Walther Traut

Institut für Biologie, Medizinische Universität Lübeck, Ratzeburger Allee 160, D23538 Lübeck, Germany

Introduction

In several plant taxa and in many animal taxa, one chromosome pair controls the development to a female or a male. This is commonly referred to as "chromosomal sex determination". One or more genes for the primary sex determining signal are located on these chromosomes but even if multiple genes are involved, they are transmitted as a single unit through meiosis. The sex determining mechanism, therefore, is formally "monofactorial" in terms of transmission genetics, it follows the Mendelian backcross scheme and produces a primary 1 : 1 ratio of females : males among offspring. Chromosomal sex determination is of two types, XX/XY or WZ/ZZ, depending on whether females or males are the heterogametic sex, but otherwise there is no basic difference between the two. There are variants of both types e.g. XX/X, $X_1X_1X_2X_2/X_1X_2Y$, XX/XY_1Y_2, Z/ZZ, W_1W_2Z/ZZ and even more complex systems. But even multiple sex chromosomes segregate as a single unit in meiosis of the heterogametic sex.

Sex chromosomes are thought to have evolved independently from autosomes many times. Yet they show surprisingly similar properties besides determining the sex of the carrier. The Y or W chromosomes lose genetic functions and show a tendency towards heterochromatinization. Often sex chromosomes form a heteromorphic pair in the heterogametic sex: they are visibly different in size and/or centromere location. The evolution of a heteromorphic sex chromosome pair from an autosome pair requires three kinds of genetic changes (for a review see Bull, 1983):

(1) acquisition of the sex determining function (i.e., these chromosomes acquire a dominating role in the generation of the primary sex determining signal)

M. Schmid and I. Nanda (eds.), Chromosomes Today, Volume 14, 73–82.

(2) establishment and extension of a non-recombining chromosome region
(3) molecular and morphological differentiation.

Grosso modo, these changes may be seen as steps in the evolution of a differentiated sex chromosome pair. Acquisition of the sex determining function is necessarily first step, the other two may be interwoven. In the following, I comment on these changes and present relevant cases from fish, fly and moth species.

Acquisition of the sex determining function

The necessity to establish a new sex determining mechanism arises when hermaphroditic or monoecious organisms evolve into gonochoristic or dioecious ones. Chromosomal sex determination has arisen in conjunction with such changes from monoecy to dioecy in plants (e.g., *Melandrium, Rumex*, see Stebbins, 1971) or from hermaphroditism to gonochorism in animals (*Schistosoma*, see Grossman *et al.*, 1981).

But establishment of a new sex determining mechanism does not require transformation from hermaphroditism to gonochorism. This can be learned from animal groups like reptiles, amphibians or insects where various sex determining mechanisms are found although these groups are purely or almost purely gonochoristic. Conceivably, a new chromosomal sex determining mechanism may arise from other types of sex determination, from environmental sex determination, from polygenic sex determination or from mechanisms of maternal determination of all-female and all-male broods. Perhaps the most common way is the replacement of an already existing sex chromosome system by a new one. One can observe that even in some recent species. In the platyfish, *Xiphophorus maculatus*, XY and WZ systems coexist, whereby Z and Y are identical (Kallman, 1984). In the housefly, *Musca domestica*, we witness the substitution of an old sex determining system with differentiated sex chromosomes by new ones derived from former autosomes (Dübendorfer *et al.*, 1992; Rubini *et al.*, 1983). In another fly species, *Megaselia scalaris*, new Y chromosomes come into being under our own eyes in the lab. The *Maleness* factor is mobile. It changes its position in the genome at a low but perceptible rate. It can insert transposon-like into any of the three chromosomes (Traut, 1994). The chromosome, it inserts into, then is the new Y chromosome, its homologues becomes the new X chromosomes (Figure 1). All strains isolated from the field had Y chromosomes derived from chromosome #2. Two of them, "Wien" and "Tennessee", must have had their Y chromosomes for quite a while since they go back to the same transposition event and yet the strains come from independent sources. From these original strains, we isolated one line with a Y chromosome derived from chromosome #1, two lines with a Y from the former X, and two lines with a Y from chromosome 3. Thus we have a collection of strains with different sex chromosomes and of different age for the study of the early steps of sex chromosome evolution.

Rather often autosomes become sex chromosomes in a more trivial way: because they have been pulled into an existing sex chromosome system by fusion or translocation with one or the other of the two original sex chromosomes. The result is a more

a

Strain	Y chromosome
Except45	#1
Florida	#2
Tennessee	#2
Wien	#2
Except11	#2 (former X)
Except46	#2 (former X)
Except1	#3
Except10	#3

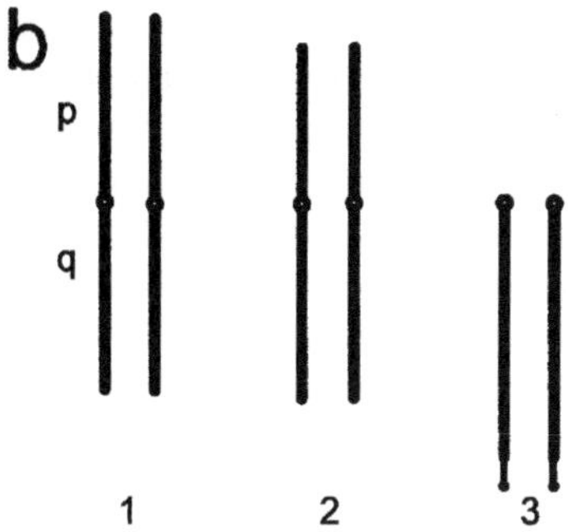

Figure 1. *The fly* M. scalaris. *(a) Origin of the Y chromosome in different strains; (b) karyogram of a Wien male, #2 is the homomorphic XY pair in this strain.*

complex sex chromosome system like $X_1X_1X_2X_2/X_1X_2Y$ (e.g., in *Drosophila miranda*, see Steinemann and Steinemann, 1997). The involvement of an autosome may occur repeatedly. The record for a multiple sex chromosome system is held by a race of the termite species *Kalotermes approximatus* with 10 X and nine Y chromosomes (Syren and Luykx, 1981).

Establishment and extension of a non-recombining region

The establishment and extension of a non-recombining region in the Y or the W chromosome is thought to be due to the occurrence of sexually antagonistic alleles on a primitive sex chromosome, alleles benefiting one sex and/or being detrimental to the other sex (Rice, 1987a). In this scenario, selection favours the establishment of tight linkage of the good allele with the respective sex locus. It is mostly unclear how the tight linkage is brought about, but inversions are known crossover suppressors and, therefore, prime candidates for this genetic change. Lepidoptera and many flies have found another way. They abolished meiotic recombination altogether in the heterogametic sex. Thus, when a new Y chromosome arises in such a fly genome or a new W in a lepidopteran genome, the whole sex chromosome forms a non-recombining compartment from the start.

Such a non-recombining compartment in an otherwise recombining genome has uncommon properties. In contrast to the X and the autosomes which recombine, segregate and associate in Mendelian fashion, its transmission through the generations is clonal. It occurs only in the heterogametic sex and is constantly heterozygous. This region will evolve separately from its originally homologous region, molecular differentiation and decay of genetic functions will occur. But the region is not totally shut off from the rest of the genome: gene conversion can take place and mobile elements have access to it and even accumulate in this region.

In present-day sex chromosomes, the region can be small, large or even cover the entire chromosome. A very small non-recombining region is found in the Y chromosome of the midge *Chironomus thummi* (Kraemer and Schmidt, 1993). In mammals, on the other hand, the non-recombining region covers most of the Y chromosome. In the human

Y chromosome, there is only a small recombining region of roughly 2 Mb left which is one of the two pseudoautosomal regions of the Y chromosome. Enlargement of the non-recombining region appears to have happened in at least four steps in the evolutionary past of the Y chromosome, presumably by inversions (Lahn and Page, 1999).

Enlargement of a non-recombining region appears to be no problem when the sex determining region is at the distal end of the Y chromosome. This can be seen in the guppy fish, *Poecilia reticulata.* All known genes from the X recombine with alleles on the Y chromosome, and there is another set of genes presumably including the sex determiner that is confined to the Y chromosome (review: Kirpičnikov, 1987). The cytogenetic equivalent can be seen when the chromosomes are subjected to comparative genomic hybridization with differently labelled genomic DNAs from females and males (Traut and Winking, 2001). X and Y chromosome share the same structure over most of their length, but the Y chromosome has a large additional segment of male-specific sequence at its distal end (Figure 2a,b). In synaptonemal complex spreads, we see that the distal region of the Y chromosome does not pair with the X in early pachytene while it does so in late pachytene (Figure 2c,d). This meiotic behaviour indicates

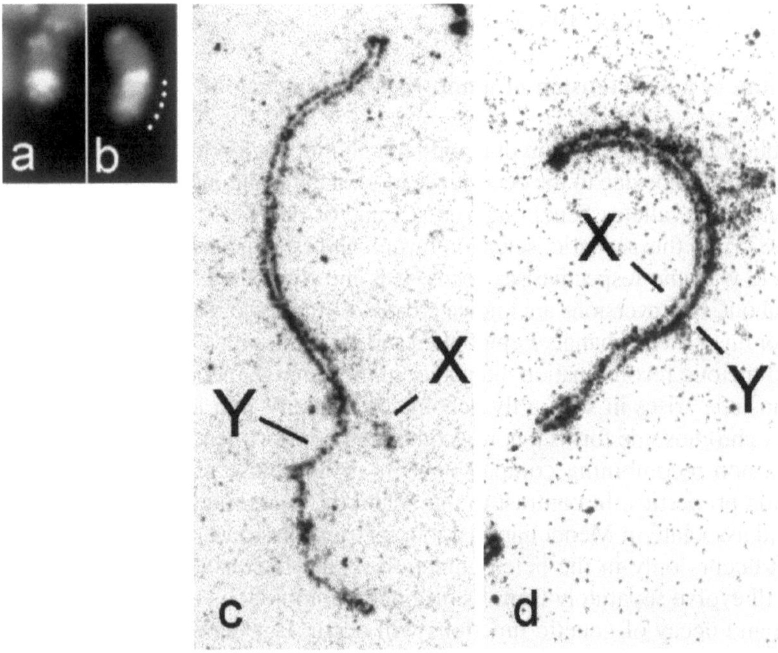

Figure 2. *The guppyfish* P. reticulata. *(a) X chromosome; (b) Y chromosome, the dots mark a segment of the Y chromosome with male-specific DNA; (a,b) stained by comparative genomic hybridization (CGH); (c) synaptonemal complex of an XY bivalent at early pachytene with unsynapsed distal ends; (d) XY bivalent at late pachytene with synapsed ends and a thick Y chromosomal lateral element; (c,d) silverstained, EM micrographs.*

non-recombination of the terminal region. It is conceivable that a sex-determining region at the distal end can easily accumulate additional sequence and extend at the free end. X and Y become a heteromorphic chromosome pair even without such complications as loss of recombination between other X/Y loci, degradation of Y chromosomal genes and establishment of a dosage compensation mechanism.

Molecular and morphological differentiation

Considerable attention has been paid to the decay of genetic functions in the Y chromosome (e.g. Charlesworth, 1994; Muller, 1964; Rice, 1987b; Steinemann *et al.*, 1993). But this is only one aspect of the molecular and morphological differentiation of sex chromosomes.

In the fly species *M. scalaris*, new and older Y chromosome strains are kept in the lab (see Figure 1). The best we could see of XY differentiation in one of the new Y strains was a length difference between X and Y for an amplified DNA fragment (Traut, 1994). But this difference had been an amplified fragment length polymorphism (AFLP) in the autosome of the source strain before this chromosome pair was singled out to become X and Y by a transposition event. Thus an AFLP has become fixed as the first X/Y difference in this strain. From one of the older strains, the "Wien" strain, we isolated 24 DNA probes that revealed length differences of restriction fragments between X and Y when used in Southern hybridization. The type of molecular differences is apparent in a short stretch of isolated Y DNA and its homologous region of the X chromosome (Traut and Wollert, 1998). The comparison shows considerable sequence identity between X and Y, interrupted by numerous single basepair substitutions and a couple of larger rearrangements: insertion of a transposon (*MIM*), a duplication and an undefined indel (Figure 3a). These few rearrangements break up homology between the two chromosomes and contribute more rapidly to the differentiation than the many single basepair substitutions. Another stretch of Y DNA recently isolated by Suck (2000) proved to be crammed with mobile elements (Figure 3b).

One can obtain a general idea about the degree of molecular differentiation by applying comparative genomic hybridization (CGH) to chromosome spreads. We label routinely whole genome DNA of females with one fluorochrome, mostly FluorX (emitting green fluorescence), and DNA of males with another fluorochrome, Cy3 (emitting red fluorescence as a contrast), and use both probes together *in situ* hybridization. The two probes compete for target sites in the chromosomes. From karyotypes with molecularly differentiated sex chromosomes, e.g. those of the human, the house mouse or *Drosophila melanogaster*, we expect and observe an intermediate fluorescence, i.e. brown–yellow, in autosomes, a shift to green fluorescence in the X because the female-derived green probe contains DNA from two X chromosomes while the male-derived red probe contains only DNA from one X, and a predominance of red signal in the Y chromosome because there is no Y DNA in the green female-derived probe (Traut *et al.*, 1999). In *M. scalaris*, however, CGH did not differentiate the X or Y chromosomes, not even in the old Y chromosome strains. Thus, molecular differentiation is not far enough advanced yet in these strains to show up in CGH.

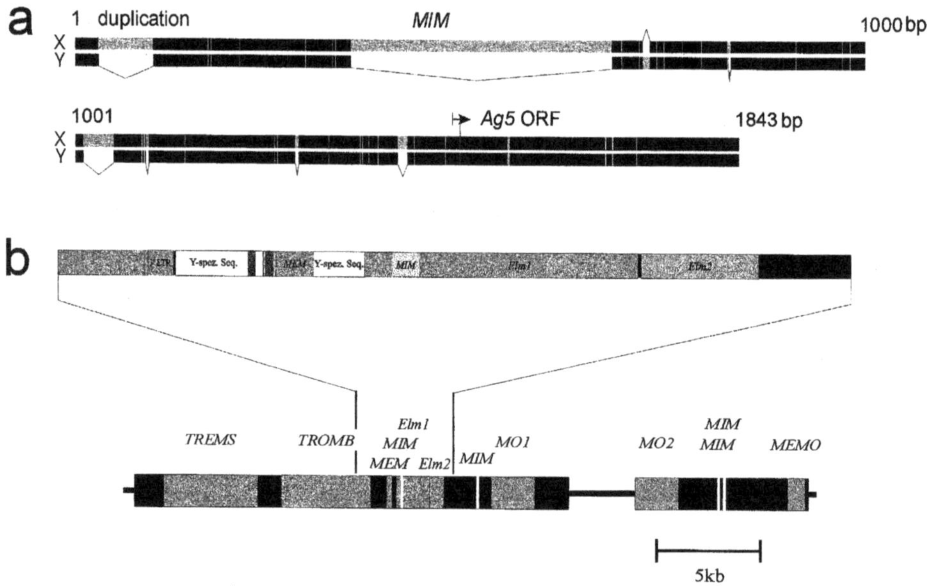

Figure 3. Megaselia scalaris. *(a) An X chromosome segment from the "Wien" strain aligned with the homologous sequence from the Y chromosome, black indicates identity, light grey sequence difference, Ag5 ORF is the coding sequence for a vespid antigen5-like protein, light grey vertical lines mark single nucleotide exchanges (data from Traut and Wollert, 1998); (b) a segment of the Y chromosome from the "Wien" strain, grey blocks indicate mobile element sequence, MEMO, MIM, MO1-2, Elm1-2, TREMS and TROMB are mobile elements (modified from Suck, 2000).*

This is different in W chromosomes of the insect order Lepidoptera. The Lepidopteran W chromosome is old, it can be traced back not in absolute time but in terms of phylogeny to the common root of the Ditrysia and Tischeriidae which comprise about 98% of all Lepidoptera species (Lukhtanov, 2000; Traut and Marec, 1996). Moths have usually small mitotic chromosomes which do not show much structural detail. The pachytene stage of meiosis, however, especially that of female meiosis, affords structural details at a fine resolution (Figure 4a). Autosomal bivalents at this stage display a fairly homologous chromomere pattern in the synapsed chromosomes. W and Z are completely paired. This facilitates comparison. The Z chromosome of the meal moth, *Ephestia kuehniella*, has a fairly normal sequence of chromomeres and interchromomeres while the W forms a condensed heterochromatin strand over nearly the full length. With CGH, this strand is highlighted with greenish fluorescence indicating the accumulation of repetitive, mostly female-specific DNA (Figure 4c).

This is also visible in mitotic complements where CGH highlights the otherwise unrecognizable W chromosome. The W chromosome shows up not only in the meal moth, *E. kuehniella*, but also in the wax moth, *Galleria mellonella*, and the silkworm, *Bombyx mori* (Traut *et al.*, 1999). The strong fluorescence indicates an accumulation

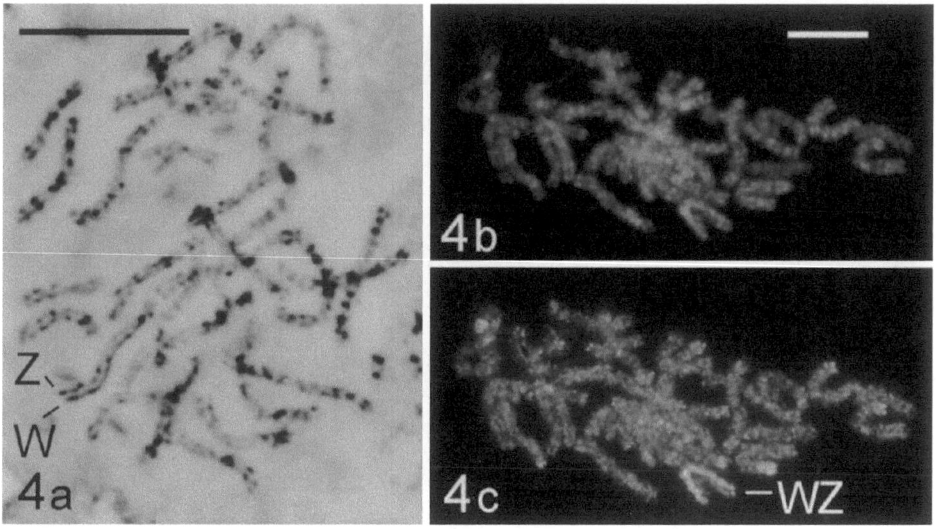

Figure 4. The meal moth E. kuehniella. *(a) Pachytene of female meiosis, orcein-stained; (b) pachytene of female meiosis, DAPI-stained; (c) the same complement, CGH with FluorX-labelled DNA from females (green) and Cy3-labelled DNA from males (red), the W chromosome emits bright greenish fluorescence, the Z has reddish fluorescence while fluorescence from the other chromosomes is yellow–brown. Bars 10 μm.*

of repetitive sequences in the W chromosomes of all three species. The W chromosome is obviously a "trap" (Steinemann *et al.*, 1993) or "graveyard" (Kjellman *et al.*, 1995) for transposable elements similar to the *Drosphila* and the hominid Y chromosomes. A comparison of signal ratios from the female- and the male-derived probes, however, reveals a difference among the W chromosomes (Sahara *et al.*, in preparation). The meal moth and wax moth W chromosomes bind significantly more female-derived than male-derived probe. This indicates the presence of female-specific repetitive sequences. In contrast in the silkworm, the signal ratio from the W is about the same as that from autosomes. This means female- and male-derived probes bind equally well to the W chromosome: the W chromosome has accumulated repetitive sequences which are present in both, female and male genomes. This explains the difficulties to isolate sequences from the silkworm W chromosome; the few that have been isolated were either not W-specific or contained retroposon sequences (Abe *et al.*, 1996; 1998).

The W chromosome of another moth, *Orgyia antiqua*, is exceptional in this series. Only half of the W chromosome is highlighted in mitotic (Traut *et al.*, 2001) and pachytene complements (Figure 5c). The chromosome complement of *O. antiqua* is known to have recently undergone a series of chomosome fusions including the W chromosome (Traut and Clarke, 1997). Thus, the W chromosome has an old part which has accumulated repetitive sequences and a new part which has not. In conventionally stained pachytene spreads, about half of the WZ bivalent displays

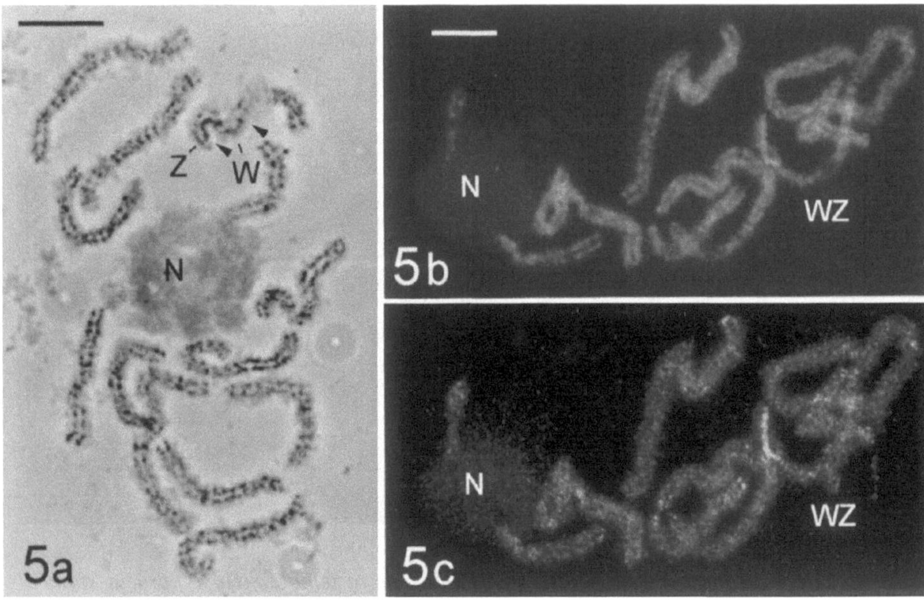

Figure 5. The moth O. antiqua. *(a) Pachytene of female meiosis, orcein-stained; (b) pachytene of female meiosis, DAPI-stained; (c) the same complement, CGH with FluorX-labelled DNA from females (green) and Cy3-labelled DNA from males (red), a segment of the W chromosome fluoresces brightly greenish while the remaining chromosomes emit yellow–brown fluorescence. Arrowheads mark the borders of the differentiated W chromosome segment, N nucleolus, Bars 10 μm.*

a homologous chromomere pattern – this is the recent acquisition from an autosome – and half of it, the old part, displays structural differentiation, it has a non-homologous pattern (Figure 5a). This W chromosome, therefore, presents two stages of W chromosome differentiation in one chromosome, rather similar to the neo-Y chromosome of *D. miranda*.

Summary

The creation of new sex chromosomes is not confined to the distant past, we can observe the rise of new sex chromosome systems in the field and in the lab. The event depends on changes of the primary sex determining signal or on changes of the location of the primary sex determining factor. In the guppyfish, we see a heteromorphic sex chromosome pair although the genes from the X recombine with the Y. Heteromorphism is due to the location of a Y-specific region including the putative sex determiner at the distal end of the Y chromosome. Accumulation of repetitive sequences, sex-specific or common ones, is a general phenomenon in old W as well as Y chromosomes.

Acknowledgement

My thanks for fruitful discussions and joint efforts go to Ulrike Eickhoff (Lübeck, Germany), František Marec (České Budějovice, Czech Republic), Ken Sahara (Sapporo, Japan) and Heinz Winking (Lübeck, Germany).

References

Abe H, Kanehara M, Terada T, Ohbayashi F, Shimada T, Kawai S, Suzuki M, Sugasaki T, Oshiki T (1998). Identification of novel random amplified polymorphic DNAs (RAPDs) on the W chromosome of the silkworm, *Bombyx mori*, and the wild silkworm, *B. mandarina*, and their retrotransposable element-related nucleotide sequences. *Genes Genet Syst* 73: 243–254.

Abe H, Shimada T, Kawai S, Ohbayashi F, Harada T, Yokoyama T, Oshiki T, Kobayashi M (1996). Nucleotide sequence of the random amplified polymorphic DNA (RAPD) on the W chromosome of the silkworm, *Bombyx mori* (Lepidoptera: Bombycidae). *Appl Entomol Zool* 31: 633–637.

Bull JJ (1983). *Evolution of Sex Determining Mechanisms*. Menlo Park: The Benjamin/ Cummings Publishing Company.

Charlesworth B (1994). The effect of background selection against deleterious alleles on weakly selected, linked variants. *Genet Res* 63: 213–228.

Dübendorfer A, Hilfiker-Kleiner D, Nöthiger R (1992). Sex determination mechanisms in dipteran insects: the case of *Musca domestica*. *Develop Biol* 3: 349–356.

Grossman AI, Short RB, Cain GD (1981). Karyotype evolution and sex chromosome differentiation in Schistosomes (Trematoda, Schistosomatidae). *Chromosoma* 84: 413–430.

Kallman KD (1984). A new look at sex determination in poeciliid fishes. In: Turner BJ, ed. *Evolutionary Genetics of Fishes*. New York: Plenum Press, pp. 95–171.

Kirpičnikov VS (1987). Genetische Grundlagen der Fischzüchtung. Berlin: VEB Deutscher Landwirtschaftsverlag.

Kjellman C, Sjögren H-O, Widegren B (1995). The Y chromosome: a graveyard for endogenous retroviruses. *Gene* 161: 163–170.

Kraemer C, Schmidt ER (1993). The sex determining region of *Chironomus thummi* is associated with highly repetitive DNA and transposable elements. *Chromosoma* 102: 553–562.

Lahn BT, Page DC (1999). Four evolutionary strata on the human X chromosome. *Science* 286: 964–967.

Lukhtanov VA (2000). Sex chromatin and sex chromosome systems in nonditrysian Lepidoptera (Insecta). *J Zool Syst Evol Res* 38: 73–79.

Muller HJ (1964). The relation of recombination to mutational advance. *Mutation Res* 1: 2–9.

Rice WR (1987a). The accumulation of sexually antagonistic genes as a selective agent promoting the evolution of reduced recombination between primitive sex chromosomes. *Evolution* 4: 911–914.

Rice WR (1987b). Genetic hitchhiking and the evolution of reduced genetic activity of the Y sex chromosome. *Genetics* 116: 161–167.

Rubini PG, Vecchi M, Franco MG (1983). Recenti progressi nello studio delle variazioni dei determinanti sessuali in popolazioni naturali di *Musca domestica* L. Atti 13 Congr. Nat. It. Ent. Sestriere – Torino, pp. 527–534.

Stebbins GL (1971). *Chromosomal Evolution in Higher Plants*. London: Arnold.

Steinemann M, Steinemann S (1997). The enigma of Y chromosome degeneratiom: TRAM, a novel retroposon is preferentially located on the neo-Y chromosome of *Drosophila miranda*. *Genetics* 145: 261–266.
Steinemann M, Steinemann S, Lottspeich F (1993). How Y chromosomes become genetically inert. *Proc Natl Acad Sci USA* 90: 5,737–5,741.
Suck G (2000). Geschlechtsspezifische und mobile Sequenzen bei der Fliege *Megaselia scalaris* (Diptera), Medizinische Universität Lübeck.
Syren RM, Luykx P (1981). Geographic variation of sex-linked translocation heterozygosity in the termite *Kalotermes approximatus* Snyder (Insecta: Isoptera). *Chromosoma* 82: 62–88.
Traut W (1994). Sex determination in the fly *Megaselia scalaris*, a model system for primary steps of sex chromosome evolution. *Genetics* 136: 1,097–1,104.
Traut W, Clarke CA (1997). Karyotype evolution by chromosome fusion in the moth genus Orgyia. *Hereditas* 126: 77–84.
Traut W, Eickhoff U, Schorch J-C (2001). Identification and analysis of sex chromosomes by comparative genomic hybridization (CGH). In: Sharma AK, Sharma A, eds. *Methods in Cell Science*. Kluwer Academic.
Traut W, Marec F (1996). Sex chromatin in Lepidoptera. *Quart Rev Biol* 71: 239–256.
Traut W, Sahara K, Otto TD, Marec F (1999). Molecular differentiation of sex chromosomes probed by comparative genomic hybridization. *Chromosoma* 108: 173–180.
Traut W, Winking H (2001). Meiotic chromosomes and stages of sex chromosome evolution in fish: zebrafish, platyfish and guppy. *Chromosome Res* (in press).
Traut W, Wollert B (1998). An X/Y DNA segment from an early stage of sex chromosome differentiation in the fly *Megaselia scalaris*. *Genome* 41: 289–294.

2. Meiosis

2.1. Expression and Chromatin Binding Specificity of Ki-67 in Male and Female Meiosis of the Mouse

Heinz Winking[1], Johannes Gerdes[2] and Walter Traut[1]
[1] *Institut für Biologie, Medizinische Universität Lübeck, D-23538 Lübeck, Germany;*
[2] *Department of Immunology and Cell Biology, Division of Molecular Immunology, Research Center Borstel, D-23845 Borstel, Germany*

Introduction

Eukaryotic chromosomes consist of a single long DNA molecule tightly packaged by histone and other proteins. Some proteins like the core histone proteins are permanent components of chromosomes, whereas others might only be temporarily linked to chromosomes or chromosomal segments. Such a protein is the Ki-67 protein (pKi-67). The pKi-67 is exclusively expressed in proliferating cells in all stages of the cell cycle but absent in G_0 cells. (Gerdes *et al.*, 1984). This expression pattern have made antibodies raised against pKi-67 a valuable tool for diagnosis and prognosis of cancers (Gerdes, 1990). In mitotically proliferating cells the location of pKi-67 is within the nucleus but changes of the location during the cell cycle are obvious. Depending on the stage of interphase a preferential colocalization of pKi-67 with heterochromatin and nucleoli are reported (Bridger *et al.*, 1998; Starborg *et al.*, 1996; Verheijen *et al.*, 1989b). During mitosis pKi-67 seems to be a constituent of a perichromosomal layer (Gautier *et al.*, 1992; Verheijen *et al.*, 1989a). Data on expression and location of pKi-67 during the meiotic divisions are scarce. Wrobel *et al.* (1996) reported on the expression of pKi-67 in the seminiferous epithelium of the bovine testis. In all stages of meiosis except leptotene/zygotene, the presence of pKi-67 was found indicating a discontinuous expression of pKi-67 during meiosis.

The knowledge of an intriguing stage-dependent association to nuclear and chromosomal domains of pKi-67 during the mitotic cell cycle and the lack of corresponding data on meiotic cells prompted us to investigate the expression and location of pKi-67 during male and female meiosis of the mouse. Two different mouse

M. Schmid and I. Nanda (eds.), Chromosomes Today, Volume 14, 85–91.

strains were used, one with an all acrocentric karyotype (strain NMRI; $2n = 40$) and one with nine pairs of Robertsonian translocations (strain CD; $2n = 22$). Presence and location was analysed by immunostaining using a polyclonal rabbit antiserum against the murine Ki-67 protein (designated S24; (Kubbutat *et al.*, 1994)). Detection of S24 was done either with goat anti rabbit IgG conjugated with Alexa Fluor 568 or by the peroxidase reaction in the presence of 3,3'diaminobenzidine. Meiocytes were analysed either *in situ* in gonadal sections after fixation with 4% paraformaldehyde or in cytogenetic air dried preparation after fixation with methanol : acetic acid (3 : 1). In air dried, immunostained preparations the cells were counterstained by 4'6-diamidino-2-phenylindole (DAPI) for visualization of chromosomes. In the mouse, DAPI strongly stains pericentromeric heterochromatin and can thus be used as a marker for pericentromeric heterochromatin domains in interphase and metaphase nuclei.

Expression of pKi-67 in female meiosis

Ovarian histology
To study the expression of pKi-67 during female meiosis, immunostaining of pKi-67 in histological sections of ovaries of 14–18 day fetuses (plug day = day 0 of pregnancy), of ovaries of young born females (day 3 post-partum) and of ovaries of adult females was performed. In the mouse female germ cells enter meiosis during fetal life around day 14 of pregnancy (Dietrich and Mulder, 1983), which was confirmed by this study. In ovaries of 14–18 day fetuses, the meiotic prophase stages from leptotene to diplotene were present. The nuclei of these stages were labelled after immunostaining (Figure 1 (a) and (b)). Labelling was found over the whole nucleus with some dark foci. These foci were less prominent or even absent in oocytes of day 14 and day 18 fetuses. In ovaries of young females (day 3 post-partum), the oocytes have reached the dictyotene stage. Oocytes in this stage of meiotic development were found free of label. In ovaries of adult females, oocytes within small primary follicles were found without label. However, in growing primary follicles, labelling reappeares and a nucleus with a prominent nucleolus is found pKi-67 positive. This type of labelling is maintained during the maturation process of follicles until the tertiary follicle stage. Summerizing the results obtained from the analysis of immunostained sections it can be concluded that from leptotene to diplotene pKi67 is present, it is absent in dictyotene oocytes of small small follicles and present again in oocytes of growing primary follicles up to mature Graafian or tertiary follicles. Thus, the expression of pKi-67 in female meiosis is discontinuous.

Analysis of air dried preparations
To investigate further the location of pKi-67, air dried preparations of ovaries of day 14–18 fetuses were immunostained and subsequently counterstained with DAPI. Identification of specific meiotic stages in cytogenetic preparations is notoriously difficult, especially early prophase I stages. However, the arrangement of centromeres and size of the cells allowed a fairly precise identification. In the cytogenetic preparation of ovaries of day 14 and 15 fetuses, prophase cells were present which showed strong label of randomly distributed to aligned centromeric areas after immunostaining.

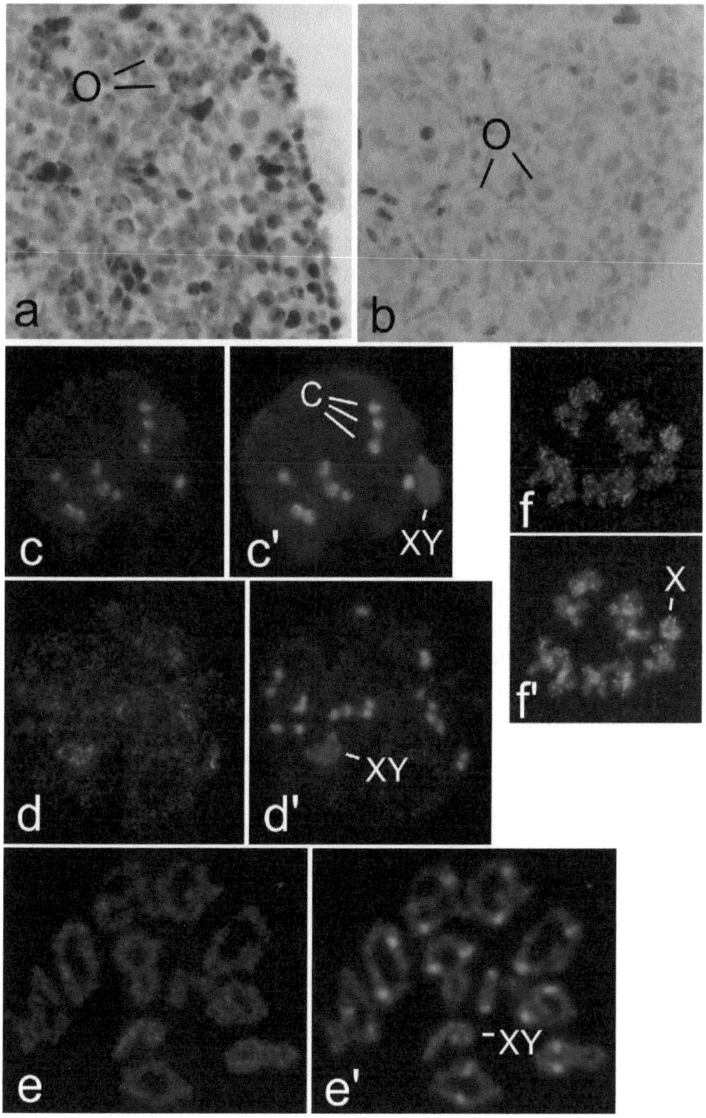

Figure 1. Distribution of pKi-67 in sections of ovaries and in meiocyte nuclei of male meiosis. (a) Ovary of a 15 day NMRI fetus, oocytes are labelled, immunostaining with the peroxidase reaction; (b) Ovary of a three day newborn NMRI female, oocytes are free of label, immunostaining with the perxidase reaction; (c–f) male meiotic cells of a CD mouse with nine pairs of Robertsonian chromosomes, immunostained for pKi-67 (c' – f') pKi-67 (red) and DAPI (blue) images merged; (c–c') pachytene, note colocalisation of DAPI and pKi-67 positive spots, lack of pKi-67 in the sex vesicle; (d–d') diplotene, note splitting of homologous centromeres, pKi-67 label on the sex vesicle; (e–e') diakinesis, note labelled chromosome arms and diminished affinity of pKi-67 to centromeric heterochromatin; (f–f') metaphase II, note labelled chromosome arms and the strong label on the X. O – oocyte; XY – sex chromosomes: XX – chromosome; C – centromere.

These cells were classified as meiotic cells in leptotene to zygotene. In ovaries of day 16–17 fetuses, centromeres fuse into a few aereas and show strong pKi-67 labelling. According to chromosome condensation, these cells were classified as late zygotene or pachytene. In ovaries of day 18 fetuses separation of bivalents can be seen typical for diplotene. In these cells the centromeres don't show label anymore. Taken together most probably the following sequential order of events is occuring during fetal female meiosis. The centromeric aereas accumulate label from leptotene to zygotene, in pachytene centromeres show the strongest label and from late pachytene to diplotene label vanishes from the centromeres. The cytogenetic data suggest that the foci seen in sectioned oocytes most probably represent centromeric areas.

Expression of pKi-67 in male meiosis

Testis histology
In histological sections the cycle of the seminiferous epithelium of a testis of an adult mouse can be subdivided into 12 stages (stages I–XII; (Oakberg, 1956)). Each stage shows an association of specific cells of the spermatogenic cycle, facilitating the identification of testicular cells. Staging of immunostained sections was done on subsequent sections stained with the periodic acid and Schiff reagent (PAS). In stages VI–XII all meiotic stages from resting spermatocytes I to metaphase II (MII) are occurring. The analysis of immunostained sections revealed a continuous expression of pKi-67 from resting spermatocytes I up to MII. Cells in leptotene/zygotene and diplotene/diakinesis showed a somewhat weaker signal intensity and less pronounced chromocenters than pachytene and resting spermatocytes II. The chromatin in MI and MII appeared as an intensely stained lump without subchromosomal differentiation. Young spermatids in stage I were also found pKi-67 positive and showed one to four intense foci.

Analysis of air dried preparations
Resting spermatocytes I, cells in premeiotic S-phase and early leptotene cells could not be identified with certainty. Leptotene/zygotene cells in male meiosis showed a similar pattern as in female meiosis, namely strongly labelled centromeres. In pachytene centromeric heterochromatin of autosomes was found strongly labelled. However, the sex chromosomes forming a sex vesicle showed no label at all including the centromeric heterochromatin of the X chromosome (Figure 1(c) and (c')). With progression of meiosis the labelling of the autosomal centromeres diminihes, whereas the sex chromosomes show the opposite staining behaviour, namely accumulation of label with the exception of the centromere of the X chromosome (Figure 1(d) and (d')). At MI and MII chromosomes are labelled but the centromeres of autosomes and of the X chromosome are free of label (Figure 1(e), (e′), (f) and (f′)). In some MII the X- and Y-chromsomal arms were heavier stained than those of autosomes (Figure 1(f) and (f′)). Resting spermatocytes II could not be identified with certainty. However, round spermatids can be identified by their size and concentration of heterochromatin into one or a few spots. Most of these cells did not show label after immunostaining, but in some cells centromeric areas and or areas adjacent to the heterochromatin spots show label.

These cells most probably represent young spermatids of stage I of the spermatogenic epithelium in which pKi-67 has preferential affinity to centromeric heterochromatin and the nucleolus.

As shown, pKi-67 changes during meiosis its affinity to different chromosomal or nuclear regions; to centromeric heterochromatin, to the nucleolus and to the periphery of chromosomes in MI and MII. It has previously been shown that during the mitotic cell cycle, pKi-67 similarly changes affinity to nuclear domains (Bridger *et al.*, 1998; Starborg *et al.*, 1996; Verheijen *et al.*, 1989a,b). Even more, redristribution of pKi-67 from centromeric heterochromatin to the periphery of chromosomal arms occurs in meiosis and mitosis at a similar stage. This remarkable event takes place in prophase of mitosis prior to prometaphase (Starborg *et al.*, 1996) and in meiosis I in late pachytene to diplotene (this chapter). In immunostained sections of the testis, resting spermatocytes II exhibit a speckled nucleus indicating redistribution of pKi-67 from perichromosomal location in MI to specific nuclear domains in resting spermatocytes II, most probably the aereas of the centromeric heterochromatin. If this is the case, a further redistribution takes place prior to MII where pKi-67 again shows perichromosomal location excluding centromeric aereas. In the next stage of spermatogenesis, young spermatids, heterochromatin again is associated with pKi-67, thus, indicating cycling redistribution processes of pKi-67 in mitotic and the two meiotic divisions.

A remarkable finding is the lack of affinity of pKi-67 to the inactivated sex chromosomes forming the sex vesicle during pachytene. At the same time, the autosomal centromeric heterochromatin shows strong affinity to pKi-67, indicating that pKi-67 is able to discriminate between the two types of heterochromatin facultative heterochromatin of the sex chromosomal arms and constitutive heterochromatin around the centromeres of the autosomes. Interestingly the centromere of the X chromosome shows the same lack of affinity to pKi-67 than the sex chromosomal arms. The lack of affinity of pKi-67 to the sex chromosomes during pachytene does not simply reflects inactivation processes, since during further meiotic progress the still inactivated sex chromosomes get into intimate contact to pKi-67 as visualized by immunostaining.

For the bull a discontinuous expression of pKi-67 during meiosis has been shown (Wrobel *et al.*, 1996). Cell in leptotene/zygotene were found pKi-67 negative. In the mouse, however, a continuous expression during meiosis including leptotene/zygotene was found. Most probably these differences reflect species specificities.

It is a well established fact that Ki-67 is expressed in proliferating cells and is absent in G_0 cells (Gerdes *et al.*, 1983, 1984). In this chapter, it has been shown that part of a specific stage of oocyte maturation, the dictyotene stage, lack pKi-67. Dictyotene oocytes are arrested, depending on the species, for month or years, thus resembling properties of G_0 cells. With progression of meiosis the expression of pKi-67 reappears. In growing oocytes, which are still in diplotene, the nucleolus has strong affinity to pKi-67. In G_0 cells which were stimulated to re-enter the cell cycle again, expression of pKi-67 takes place simultaneously with DNA replication (Gerdes *et al.*, 1984). This shows that expression or reactivation of pKi-67 may take place at different stages of the cell cycle, in stimulated G_0 cells at the end of G_1 and in dictyotene oocytes after DNA replication in prophase of meiosis.

Many details about nuclear locations of pKi-67 are known. However, the most interesting question about the function of pKi-67 remains unsolved yet (Endl and Gerdes, 2000a). But there are at least some hints on the role of pKi-67. There is evidence that pKi-67 plays a role in proliferation-dependent processes and in regulation of structural and functional reorganization during mitosis and meiosis (Endl and Gerdes, 2000b; Starborg *et al.*, 1996). In a very recent paper it is shown that pKi-67 interacts with members of the heterochromatin protein 1 (HP1) family indicating that pKi-67 may play a role in the regulation of higher-order chromatin structure (Scholzen *et al.*, in press). There is evidence that HP1 proteins may play a role in transcriptional expression and sex chromosome inactivation during spermatogenesis (Hoyer-Fender *et al.*, 2000; Lachner *et al.*, 2001; Motzkus *et al.*, 1999). In this context it is interesting to note the peculiar lack of affinity of pKi-67 to the sex vesicle and its strong affinity to centromeric heterochromatin during interphase and part of the meiotic prophase.

Conclusions

During meiosis pKi-67 shows affinity to several nuclear domains: centromeric heterochromatin, nucleolus, and chromosome arms. During progression of meiosis pKi-67 shows a cycling affinity to pericentromeric heterochromatin. Strong affinity exists during interphase and early prophase and diminished affinity during metaphase to telophase. Strong affinity to the nucleolus of growing oocytes was seen as well. Expression of pKi-67 was continuous in male meiosis but discontinuous in female meiosis, where arrested oocytes in dictyotene are pKi-67 negative. The sex vesicle in pachytene of male meiosis was found pKi-67 negative as well.

Acknowledgement

The skillful technical assistence of Ulrike Eickhoff, Constanze Reuter, Bettina Baron-Lühr Margret Kernbach, and Patricia Prilla is gratefully acknowledged.

References

Bridger JM, Kill IR, Lichter P (1998). Association of pKi-67 with satellite DNA of the human genome in early G1 cells. *Chromosome Res* 6: 13–24.

Dietrich AJ, Mulder RJ (1983). A light- and electron microscopic analysis of meiotic prophase in female mice. *Chromosoma* 88: 377–385.

Endl E, Gerdes J (2000a). The Ki-67 protein: fascinating forms and an unknown function. *Exp Cell Res* 257: 231–237.

Endl E, Gerdes J (2000b). Posttranslational modifications of the Ki-67 protein coincide with two major checkpoints during mitosis. *J Cell Physiol* 182: 371–380.

Gautier T, Robert-Nicoud M, Guilly MN, Hernandez-Verdun D (1992). Relocation of nuclear proteins around chromosomes at mitosis. A study by confocal laser scanning microscopy. *J Cell Sci* 102: 729 – 737.

Gerdes J (1990). Ki-67 and other proliferation markers useful for immunohistological diagnostic and prognostic evaluations in human malignancies. *Sem Cancer Biol* 1: 199–206.

Gerdes J, Lemke H, Baisch H, Wacker H-H, Schwab U, Stein H (1984). Cell cycle analysis of a cell proliferation-associated human nuclear antigen defined by the monoclonal antibody Ki-67. *J Immunol* 133: 1,710–1,715.

Gerdes J, Schwab U, Lemke H, Stein H (1983). Production of a mouse monoclonal antibody reactive with a human nuclear antigen associated with cell proliferation. *Int J Cancer* 31: 13–20.

Hoyer-Fender S, Singh PB, Motzkus D (2000). The murine heterochromatin protein M31 is associated with the chromocenter in round spermatids and is a componenet of mature spermatozoa. *Exp Cell Res* 254: 72–79.

Kubbutat MHG, Key G, Duchrow M, Schlüter C, Flad HD, Gerdes J (1994). Epitope analysis of antibodies recognizing the cell proliferation associated nuclear antigen previously defined by the antibody Ki-67. *J Clin Pathol* 47: 524–528.

Lachner M, O'Caroll D, Rea S, Mechtler K, Jenuwein T (2001). Methylation of histone H3 lysine 9 creates a binding site for HP1 proteins. *Nature* 410: 116–120.

Motzkus D, Singh PB, Hoyer-Fender S (1999). M31, a murine homolog of Drosophila HP1, is concentrated in the XY body during spermatogenesis. *Cytogenet Cell Genet* 86: 83–88.

Oakberg E (1956). A desription of spermatogenesis in the mouse and its use in analysis of the cell cycle of the seminiferous epithelium and germ cell renewal. *Am J Anat* 99: 391–413.

Scholzen T, Endl E, Wohlenberg C, van der Sar S, Cowell IG, Gerdes J, Singh PB (in press). The Ki-67 protein interacts with members of the heterochromatin protein 1 (HP1) family: A potential role in the regulation of higher-order chromatin structure. *J Pathol.*

Starborg M, Gell K, Brundell E, Höög C (1996). The murine Ki-67 proliferation antigen accumulates in the nucleolar and heterochromatic regions of interphase cells and at the periphery of the mitotic chromosomes in a process essential for cell cycle progression. *J Cell Sci* 109: 143–15.

Wrobel KH, Bickel D, Kujat R (1996). Immunohistochemical study of seminiferous epithelium in adult bovine testis using monoclonal antibodies against Ki-67 protein and proliferating cell nuclear antigen (PCNA). *Cell Tissue Res* 283: 191–201.

Verheijen R, Kuijpers HJH, Beck JLM, van Dierendonck JH, van Driel R, Brakenhoff GJ, Ramaekers FCS (1989a). Ki-67 detects a nuclear matrix-associated proliferation-related antigen. II. Localization in mitotic cells and association with chromosomes. *J Cell Sci* 92: 531–540.

Verheijen R, Kuijpers HJH, Schlinglmann RO, Boehmer ALM, van Driel R, Brakenhoff GJ, Ramaekers FCS (1989b). Ki-67 detects a nuclear matrix-associated proliferation-related antigen. I. Intracellular localization during interphase. *J Cell Sci* 92: 123–130.

2.2. Conservation and Variability of Meiotic Processes – Lessons from the Unconventional Meiosis of *C. elegans*

Josef Loidl[1], Pawel Pasierbek[1] and Ann M. Rose[2]
[1]*Institute of Botany, University of Vienna, Rennweg 14, A-1030 Vienna, Austria;*
[2]*Department of Medical Genetics, University of British Columbia, Vancouver, Canada V6T 1Z3*

Introduction

The nematode *Caenorhabditis elegans* is one of the genetically best-studied organisms. Its genome was completely sequenced, and, having a determinate cell lineage, the fate of every single cell during embryonic development was uncovered (for review see Riddle *et al.*, 1997). *C. elegans* is a hermaphrodite with $2n = 12$ holocentric chromosomes. Its meiosis shows several unusual features whose investigation may provide clues to how evolution came up with different solutions to fundamental requirements as pairing and segregation.

Most meiotic recombination is likely to depend on recombinational repair of programmed meiotic DNA double-stranded breaks (DSBs) which are induced by Spo 11p (Keeney *et al.*, 1997; Bergerat *et al.*, 1997). In the budding yeast, the fungus *Coprinus cinereus*, the mouse and *Arabidopsis* it was shown that this or an additional function of Spo11p is also required for the formation of the synaptonemal complex (SC), the proteinaceous structure which intimately links homologous chromosomes (Alani *et al.*, 1990). *C. elegans* and *Drosophila* are different in that Spo 11p is not required to initiate SC formation (Dernburg *et al.*, 1998; McKim and Hayashi-Hagihara, 1998). On the other hand, each *C. elegans* chromosome has a single locus (the HRR) whose presence allows it to pair and recombine with its homologous partner. It is not known what specific role the HRR plays in promoting recombination. Here, we provide evidence, that homologous prealignment, which takes place at a stage prior to that of synapsis and recombination, does not dependend on the presence of the HRR.

M. Schmid and I. Nanda (eds.), Chromosomes Today, Volume 14, 93–101.

Another peculiarity of *C. elegans* meiosis is that, despite the fact that the mitotic chromosomes acquire multiple spindle attachments, chromosomes are functionally monocentric in both meiotic divisions (Zetka and Rose, 1995). In meiosis I, centromeric activity is restricted to a region somewhere between the single chiasma and the chiasma-distal (polar) chromosome end. Here we will discuss several possibilities how the kinetic region of meiotic chromosomes may be determined and organized.

Results and discussion

The HRR

Genetic studies have revealed that each of the six *C. elegans* chromosomes has a specialized domain near one of its ends, which is crucial for the pairing of homologous chromosomes. Partial duplications of chromosomes which lack this region are unable to recombine with and segregate from the normal full-length chromosome. In reciprocal translocation heterozygotes, only the portions of chromosomes adjacent to this region form crossovers, whereas recombination is suppressed in translocated parts without it. The consequence is that such translocation heterozygotes do not form quadrivalents (see Albertson *et al.*, 1997). The simplest explanation for crossover suppression, put forward by Rosenbluth and Baillie (1981) was that each chromosome contained a pairing center (similar to those hypothesized by Sandler (1956) and Hawley (1980)); a region without such a domain could not find its pairing partner. The term "homolog recognition regions" (HRRs) was adopted by McKim *et al.* (1988) to emphasize the role of these regions in the homolog-specific functions of recombination and segregation during meiosis I. Villeneuve (1994) restated the Rosenbluth and Baillie hypothesis using the term "meiotic pairing centers".

A first hint that HRRs are not exclusive sites of homology recognition came from the observation of presynaptic alignment (Pasierbek *et al.*, 2001). There, homologous chromosomes are associated in parallel at a distance before their intimate pairing via the SC is initiated. This distance alignment can be explained by the assumption that many sites along a pair of chromosomes are capable of homology recognition and formation of a physical link with their counterparts. Here we show that SC-independent homologous associations can occur between chromosomes which do not share an HRR.

We used a strain heterozygous for the reciprocal translocation *eT1* involving chromosomes *III* and *V*. In this strain the non-HRR arms *III* R and *V* L are exchanged (Figure 1). Thus, these two arms are disconnected from their respective HRRs on the other side of the chromosomes. We asked whether or not the translocated portions *III* R and *V* L were able to associate with the nontranslocated partner. To this aim we performed fluorescence *in situ* hybridization (FISH) as a means to differentially highlight HRR and non-HRR ends of the chromosomes in nuclei at various stages of meiotic pairing. The distinct morphology of cell stages makes it easy to discriminate them by DAPI staining of the tubular ovarium of adult hermaphrodites, which contains a succession of cells at all stages of meiotic development. At the distal end of the ovarium, the mitotic zone, mitotic precursor cells are generated. This is followed by the

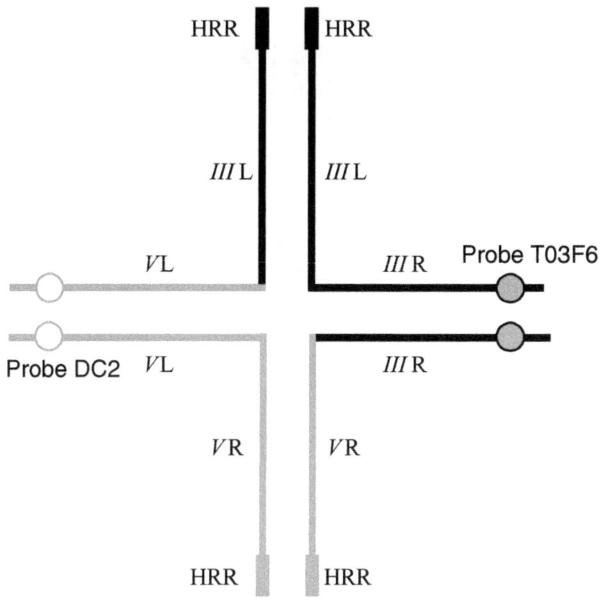

Figure 1. *The* eT1 *translocation (Rosenbluth and Baillie, 1981) and the FISH markers used.* eT1 *is a reciprocal translocation of the non-HRR-bearing chromosome regions* III *R (right) and* V *L (left). Distal parts of these regions were differentially labelled by FISH with the 34.5 kb cosmid T03F6 (grey dots) and the 38.1 kb cosmid DC2 (white dots), respectively. Recombination-competent meiotic pairing of homologous regions* III *L –* III *L and* V *R–*V *R is allowed by their respective HRRs, whereas regions* III *R –* III *R and* V *L –* V *L, belonging to different chromosomes, do not recombine.* III *R –* III *R and* V *L –* V *L associate readily in meiotic prophase (leptotene/zygotene) as can be seen from the fusion of the corresponding FISH signals (see Figure 2 and Table 1). Thus, their recombination defect is not due to the unability to recognize each other in the absence of HRR.*

transition zone, which is comprised of cells at leptotene and zygotene, the pachytene zone, and cells at diplotene. Oocytes become arrested at diakinesis and finish meiosis after fertilization. For the techniques of chromosome preparation and FISH see Pasierbek *et al.* (2001). We selected FISH probes for distal regions on *III* R and *V* L (Figure 1) and recorded the pairing frequencies of homologous regions (as seen by the association or fusion of like FISH signals) in cells before and after SC formation (Figure 2). As a control, we tested the association of the HRR near the right end of chromosome I. For this we used as a FISH probe PCR-amplified 5S rDNA (Pasierbek *et al.*, 2001).

Table 1 shows the association frequencies of homologous FISH signals in the translocation strain and in the wild type. In both strains, associations of HRR regions were rare in premeiotic nuclei of the mitotic zone, where they probably occur due to random contacts. They strongly increased during leptotene/zygotene and reached

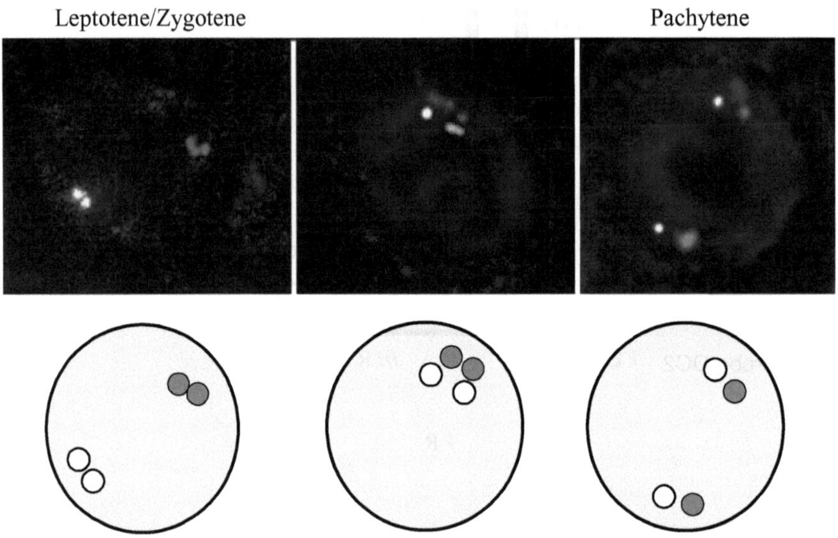

Figure 2. Association of homologous FISH signals in leptotene/zygotene nuclei and of nonhomologous FISH signals in pachytene nuclei. In the middle, an intermediate stage with all 4 labelled loci associated, is shown. FISH signals appear as white and grey dots and correspond to the loci shown in Figure 1. The lower row is a schematic interpretation of the microscopic images.

Table 1. *Association frequencies of FISH signals in HRR-distal chromosome regions (for the chromosomal location of FISH probes see Figure 1) and near a HRR*

	Both homol. signals associated (%)	One homol. signal associated (%)	All four signals separate (%)	One nonhom. signal pair associated (%)	Two nonhom. signal pairs associated (%)	HRR-near (5S rDNA) associated (%)
eT1 Transloc. strain						
Mitotic zone	2	10	83	3	2	10
Leptotene/Zygotene	33	40	27	0	7	80
Pachytene	2	0	1	14	83	100
Wild-type strain						
Mitotic zone	1	12	81	5	1	11
Leptotene/Zygotene	94	3	3	5	2	90
Pachytene	100	0	0	7	1	100

Both for the WT and the translocation strain, more than 50 nuclei were evaluated of each stage. The categories of signal associations add up to over 100% because homologous and nonhomologous signals can be associated at the same time. For an example see Figure 2 middle.

100% at pachytene. In the wild-type strain, the association of the HRR-distant homologous FISH markers showed the same stage-dependent increase. Also in the translocation strain, regions near the non-HRR chromosome ends were frequently homologously associated at leptotene/zygotene. This indicates that regions with identical DNA sequences, even if they do not share an HRR, recognize each other and associate prior to the formation of the SC (Pasierbek *et al.*, 2001).

At pachytene, associations between the HRR-distant homologous regions had disappeared in the translocation strain. Instead, nonhomologous associations of the translocated arm regions were now frequent (Figure 2). This suggests that synapsis promoted by the pairing of corresponding HRRs extends into the translocated nonhomologous segments (Figure 3). In fact, it had been shown in a translocation strain, that the translocated segments must have engaged in nonhomologous synapsis since 6 unconspicuous bivalents appeared in EM sections of SCs at pachytene (Goldstein, 1986b).

As the HRR is not required for the initial recognition of homologous loci, and yet it is indispensible for a crossover anywhere in a bivalent, what might then be its function? Could it be the site of SC initiation? There were two reports (Rosenbluth *et al.*, 1990; McKim *et al.*, 1993) which, on the basis of the behaviour of structural heterozygosities involving HRR-distant regions, clearly refute this explanation in its simplest form. For example, heterozygous terminal deletions at the non-HRR end suppress recombination for some distance into the adjacent proximal chromosome region, which they should not do if the HRR was sufficient for proper alignment and synapsis. Thus, the HRR is necessary but not always sufficient for crossover-competent pairing, and is unlikely to be the (sole) site of SC initiation.

Could the HRR then be necessary for establishing a meiosis-specific chromosome arrangement inside the nucleus? In a wide range of organisms, including fungi, plants and animals, it has been realized that chromosome ends attach to the nuclear envelope and cluster in a small sector prior to the initiation of synapsis. This "bouquet" -arrangement may be of importance to bring homologous chromosomes into register and to promote their pairing (see Scherthan, 2001). Synapsis initiates preferentially but not exclusively near the attachment site. A bouquet has not yet been shown in *C. elegans*, but Goldstein (1986a) observed that at least one end of each bivalent is associated with the nuclear periphery during pachytene. The identity of the HRR with the chromosomal attachment site to the nuclear envelope would explain many of its

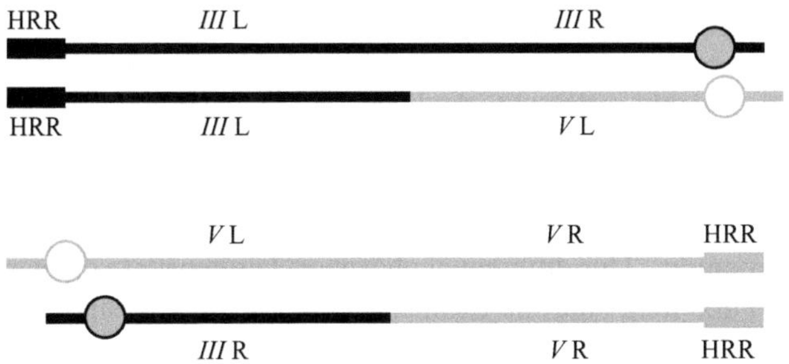

Figure 3. *At pachytene, when the SC is formed, translocated regions* III *R and* V *L become associated by virtue of the HRR-driven homologous synapsis of the regions* III *L –* III *L and* V *R –* V *R to which they are linked. This is indicated by the association of nonhomologous FISH signals (see Figure 2).*

properties, and this possibility has to be explored further. With regard to the independence of synapsis on Spo11p it is tempting to speculate that, whatever the HRR's precise role in the pairing process may be, it is part of a mechanism by which *C. elegans* substitutes for the pairing function of Spo 11p.

Different views on the nature of the meiotic centromere

In *C. elegans*, like in many other organisms with multiple spindle attachment sites during mitosis, chromosomes become functionally monocentric in both meiotic divisions. This is topologically necessary, as a chromosome with multiple spindle attachment sites, which has undergone crossing over, would risk having its recombined chromatids pulled to opposite poles. A possible mechanism to avoid this problem would be the existence of a defined locus, which is activated during meiosis and which functions to organize the meiotic kinetochore. An argument against such a constitutive meiotic centromere was put forward by Albertson and Thomson (1993). They observed that in *C. elegans* a FISH probe from near a chromosome end either assumed a position at the poleward bivalent ends or in the equatorial plane (Figure 4(a)). This led them to propose that either end of a chromosome can adopt centromeric function during the first meiotic division.

In *C. elegans*, bivalents have a single chiasma which is mostly close to one of the two ends, and which produces diakinesis/metaphase I rod bivalents with long and short axes of distinctly different lengths. Albertson and Thomson (1993) proposed that it is most likely the long axes of a metaphase bivalent which adopt kinetic function and polar orientation. In a heterozygous translocation bivalent it is always the nonexchanged portions of arms that point to opposite poles at metaphase I and thus are likely to harbour the centromeres (Albertson *et al.*, 1997).

The fact that the long portions of a bivalent are centromeric, suggests that centromere allocation is dictated by the position of the crossover. There are several possibilities how this could be achieved. One is that centromere sites are selected when a bivalent has assumed cross or rod shape at diakinesis/metaphase I. As a consequence of the surfaces exposed, the longer two of the four arms of a bivalent would be more likely to catch microtubules and assume centric activity. Alternatively, centromere position could be predetermined at pachytene when homologues are closely paired. This would ensure that corresponding regions (the same two ends) of the paired chromosomes would be assigned centric function. In this case, however, it is more difficult to explain why centromeres are invariably assigned to the longer of the two crossover-telomere intervals. A possible explanation may be offered by the observation that the region of the short crossover-telomere interval (i.e. the equatorial axis of the bivalent) is earmarked by the presence of histone H3 phosphorylated at Ser10 in diplotene/diakinesis (Pasierbek, unpublished observation). This differential phosphorylation pattern creates an asymmetry between the short and long chiasma-telomere intervals and may underlie their different behaviour with respect to centromere formation and other functions.

The alternative possibility, namely that meiotic chromosomes feature a unique constitutive, mappable centromere, is still valid. It can be imagined that condensed worm chromosomes are rigid and do not bend over easily at the centromere. In this case, the observation that chiasma-distant ends of bivalents always point to the poles

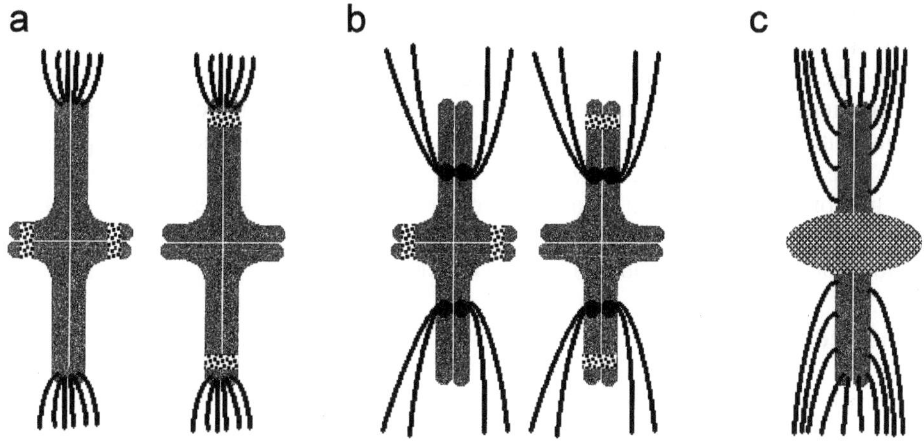

Figure 4 *Models of the meiotic centromere of* C. elegans. *Bivalents have always a single chiasma near one end and are therefore rod-shaped or cross-shaped with strongly unequal axes. It is assumed that the long arms organize the centromeres and are oriented towards the poles. (a) When the two ends of a bivalent were discriminated by one carrying a FISH marker (dotted region), it was found that either end of a chromosome can be poleward-directed and thus probably have adopted centric function (Albertson and Thomson, 1993). In this case the positions of the crossover and the centromeres are interdependent, but most likely the crossover is the primary event and dictates centromere site selection. (b) The observation that either end of a chromosome can be poleward-oriented is reconcilable with constitutive meiotic centromeres in a median region. (c) The meiotic centromere could be diffuse as in mitosis, provided that the short bivalent arms are protected from microtubule association. Because of their small size or because they are buried inside the mass of bivalents crowding at the equator of the cell they might have a reduced chance to catch microtubules. Alternatively, they may be masked by a substance or a chromatin modification (see text), as is shown here.*

(Albertson and Thomson, 1993), can be well reconciled with an invariable centromere position in a median chromosome region (Figure 4b). However, this concept of a constitutive meiotic centromere is challenged by the behaviour of chromatid fragments that form in a *rec-8* (RNAi) mutant (Pasierbek *et al.*, 2001). These fragments have a high chance to be moved to the poles at anaphase I; only in rare cases such fragments were lagging (Pasierbek and Loidl, unpubl. observ.). Thus, fragments irrespective of their place of origin in the chromosomes appear capable of capturing microtubules. On the other hand, in view of their small size, it cannot be excluded that they are carried along by a non-centromeric mechanism.

Finally, it is possible that *C. elegans* chromosomes in principle retain the capability of multiple spindle attachments during meiosis. It was shown by Howe *et al.* (2001) that HIM-10, a member of a conserved family of kinetochore proteins, localizes all along both mitotic and meiotic chromosomes, which could indicate that also meiotic chromosomes have a diffuse kinetochore. The conversion from holocentric to quasi-monocentric behaviour would then require prevention of the equatorial part of the bivalent (where kinetic activity would be fatal) from catching microtubules. Simply, their small size (mostly virtually absence) could make them a negligible target for microtubules. Dernburg (2001) proposed that the chiasma could terminalize, i.e., migrate toward the nearest end, which would eliminate the equatorial parts of metaphase bivalents. Alternatively, the equatorial parts could be actively protected from microtubule attachment by a substance or a chromatin modification (Fig. 4c). In this context, the observation that it is exactly this part of the bivalent which is enriched in phosphorylated H3 (see above) may be of relevance.

Acknowledgements

This work was supported by grants from the Austrian Science Fund (S8211 and P14642) and the Natural Sciences and Engineering Research Council of Canada (NSERC).

References

Alani E, Padmore R, Kleckner N (1990). Analysis of wild-type and rad50 mutants of yeast suggests an intimate relationship between meiotic chromosome synapsis and recombination. *Cell* 61: 419–436.

Albertson DG, Rose AM, Villeneuve AM (1997). Chromosome organization, mitosis, and meiosis. In: Riddle DL, Blumenthal T, Meyer BJ, *et al.*, eds. *C. elegans II.* Plainview, NY: Cold Spring Harbor Laboratory Press, pp. 47–78.

Albertson DG, Thomson JN (1993). Segregation of holocentric chromosomes at meiosis in the nematode, *Caenorhabditis elegans. Chromosome Res* 1: 15–26.

Bergerat A, de Massy B, Gadelle D, Varoutas P-C, Nicolas A, Forterre P (1997). An atypical topoisomerase II from archaea with implications for meiotic recombination. *Nature* 386: 414–417.

Dernburg AF (2001). Here, there, and everywhere: kinetochore function and holocentric chromosomes. *J Cell Biol* 153: F33–F38.

Dernburg AF, McDonald K, Moulder G, Barstead R, Dresser M, Villeneuve AM (1998). Meiotic recombination in *C. elegans* initiates by a conserved mechanism and is dispensable for homologous chromosome synapsis. *Cell* 94: 387–398.

Goldstein P (1986a). The synaptonemal complexes of *Caenorhabditis elegans*: pachytene karyotype analysis of hermaphrodites from the recessive *him-5* and *him-7* mutants. *J Cell Sci* 82: 119–127.

Goldstein P (1986b). The synaptonemal complexes of *Caenorhabditis elegans*: the dominant *him* mutant mnT6 and pachytene karyotype analysis of the X-autosome translocation. *Chromosoma* 93: 256–260.

Hawley RS (1980). Chromosomal sites necessary for normal levels of meiotic recombination in *Drosophila melanogaster*. I. Evidence for and mapping of the sites. *Genetics* 94: 625–646.

Howe M, McDonald KL, Albertson DG, Meyer BJ (2001). HIM-10 is required for kinetochore structure and function on *Caenorhabditis elegans* holocentric chromosomes. *J Cell Biol* 153: 1227–1238.

Keeney S, Giroux CN, Kleckner N (1997). Meiosis-specific DNA double-strand breaks are catalyzed by Spo11, a member of a widely conserved protein family. *Cell* 88: 375–384.

McKim KS, Hayashi-Hagihara A (1998). *mei-W68* in *Drosophila melanogaster* encodes a Spo11 homolog: evidence that the mechanism for initiating meiotic recombination is conserved. *Genes Dev* 12: 2,932–2,942.

McKim KS, Howell AM, Rose AM (1988). The effects of translocations on recombination frequency in *Caenorhabditis elegans. Genetics* 120: 987–1001.

Pasierbek P, Jantsch M, Melcher M, Schleiffer A, Schweizer D, Loidl J (2001). A *Caenorhabditis elegans* cohesion protein with functions in meiotic chromosome pairing and disjunction. *Genes Dev* 15: 1,349–1,360.

Riddle DL, Blumenthal T, Meyer BJ, *et al.* (1997). Introduction to *C. elegans.* In: Riddle DL, Blumenthal T, Meyer BJ, *et al.*, eds. *C. elegans II.* Cold Spring Harbor Laboratory Press, Plainview, NY: pp 1–22.

Rosenbluth RE, Baillie DL (1981). Analysis of a reciprocal translocation, *eT1(III;V)*, in *Caenorhabditis elegans. Genetics* 99: 415–428.

Rosenbluth RE, Johnsen RC, Baillie DL (1990). Pairing for recombination in LGV of *Caenorhabditis elegans*: a model based on recombination in deficiency heterozygotes. *Genetics* 124: 615–625

Sandler I (1956). Studies in $T(1;4)B^s$ in *Drosophila melanogaster.* Master's Thesis. University of Missouri,

Scherthan H (2001). A bouquet makes ends meet. *Nature Rev Mol Cell Biol* 2: 621–627.

Villeneuve AM (1994). A *cis*-acting locus that promotes crossing over between X chromosomes in *Caenorhabditis elegans. Genetics* 136: 887–902.

Zetka M, Rose A (1995). The genetics of meiosis in *Caenorhabditis elegans. Trends Genet* 11: 27–31.

2.3. Meiotic Chromosomes and Meiotic Mechanisms

Alberto J. Solari

Meiosis and sex: Ancient processes with scarce evidences of their evolutionary history

The fact that the common yeast (*Saccharomyces cerevisiae*) displays an elaborate, non-obligatory meiosis (reviewed in Roeder, 1997; Dresser, 2000) has been at the same time a bonus for geneticists, who have yeast as a very handy model, and a curse for evolutionists, who look for a primitive and perhaps simple kind of meiosis. In fact, few eukaryotes can be pointed out as absolute asexual, that is lacking sex reproduction and the vestiges of sexual ancestors. Perhaps South American trypanosomes (Tibayrenc and Ayala, 1999) and free-living euglenoids are some of the best known organisms lacking meiosis, but even these protists cannot be safely thought as coming from asexual ancestors. It may be remarked that the asexual stercorarian trypanosomes show large variations in nuclear DNA content and "electrokaryotype" (reviewed in Solari, 1995) while protists having regular sex cycles and meiosis, as *Plasmodium falciparum*, show a much more constant karyotype (Gardner, 1999).

Given that sex and meiosis are wide-embracing, ancient processes, it is remarkable that the fission yeast, *Schizosaccharomyces pombe*, has a simpler meiotic process compared to that of the budding yeast. In *S. pombe*, there are no synaptonemal complexes (SCs) (Kohli and Bahler, 1994; Egel, 2000), which are some of the best conserved structures of meiosis. Only a handful of lower eukaryots with sex cycles have been described as lacking SCs (Zickler and Kleckner, 1999), but only in *S. pombe* there are systematic observations along the whole meiosis, showing the presence of linear, segmented structures which are located parallel to chromosomes but never become paired (Kohli and Bahler, 1994). In fact, even meiotic axial structures may not be necessary in special cases during meiosis, as shown by the X and Y chromosomes in hemipterans (Solari, 1979) and in *Drosophila* spermatocytes (Meyer, 1961).

M. Schmid and I. Nanda (eds.), Chromosomes Today, Volume 14, 103–118

A large amount of information on the molecular mechanisms of meiosis has come from observations on *S. cerevisiae* in the last decade. However, meiosis was initially described in organisms somewhat more complex than yeast, and it shows a remarkable uniformity in basic features. This paper deals mainly with animal meiocytes, especially those from mammals, birds and insects.

Classical meiosis

This process shows much more dynamic changes in chromosome structure and function when compared with mitosis. Classically, three main processes occur in meiosis: chromosome synapsis, meiotic recombination and homologue segregation, and all occur in the first division. Chromosome synapsis is almost universally coupled to the assembly of a tape-like structure, the SC (Moses, 1956, 1968). Meiotic recombination is morphologically demonstrated by the appearance of chiasmata after pachytene, while at the molecular level results in the formation of recombinant molecules of DNA. Homologue segregation is shown at anaphase I, while the sister chromatid centromeres separate from each other only at anaphase II. These changes are being explained in recent years, especially the assembly and behaviour of the synaptonemal complex, and the structures and molecules participating in recombination.

The making of a synaptonemal complex

General

Synaptonemal complexes are proteinaceous structures of meiotic chromosomes at pachytene stage in a wide spectra of eukaryotes. The diversity of cell cycles, the particular, specific features of SCs, and the diversity of techniques used may give inconsistent data and controversial interpretations, which are mirrored in recent reviews (Zickler and Kleckner, 1999; Walker and Hawley, 2000). However, a basic structural pattern of the SC emerges from the consensus of these data (von Wettstein *et al.*, 1984; Dresser, 2000; for techniques, see Solari, 1998)

The lateral elements (LEs) or meiotic axes

There has been some debate on the singleness or multiple nature of LEs. The possible representation of the two chromatids in each LE underlies that discussion. Both established methods of fine structural study, serial thin sections and whole mounts after microspreading agree in the singleness of LEs (Moses, 1968; Westergaard and Wettstein, 1972; von Wettstein *et al.*, 1984; Solari, 1998). Apparent splitting into two or more fibers (Del Mazo and Gil-Alberdi, 1986; Dietrich *et al.*, 1992) may be the result of partial extraction after extensive detergent action (Triton X-100 or other), or the partial view (in a thin section) of the intrinsically multistranded structure of each LE. In fact, LEs are dynamic structures which show progressive changes in mass and length during pachytene (Moses *et al.*, 1977). The basic structure of each LE is microfibrillar (apparently 10 nm wide, Moses, 1968) although the view of single microfibrils is obscured by its highly packed condition with the usual techniques. The fibrillar substructure of axes is strikingly apparent in the axes of the mammalian sex chromosomes (X and Y). The differential region of the X chromosome axis in a number of mammals shows a clear multistranded structure (in mouse: Solari, 1970a; in the marsupial *Monodelphis dimidiata*; Solari and Bianchi, 1975; in Man: Solari, 1980; in the rodent *Galea musteloides*: Solari *et al.*, 1993) (Figure 1). This multistranded condition is not related to any visible duplex organization, but instead shows that some main components of the LEs pack themselves as fibrillar bundles.

The axes or LEs behave as units, not as doublets. This is observed in the switching of partners in the quadrivalents from autotetraploids in *Bombyx* (Rasmussen and Holm, 1979) and in autotetraploid chicken oocytes (Solari and Fechheimer, 1988). At the switching points there is no indication of doubleness in any of the axes (Figure 2).

The axes behave as units when triploid meiocytes begin synapsis in ZZW triplod chickens (Solari *et al.*, 1991). The three homologs form two SCs that share a LE (in the center of the trivalent). Thus, an axis is able to engage in two central spaces, showing lateral symmetry (Figure 3). A similar behaviour is seen in human germ cells trisomic for chromosome #21 (Speed, 1984). This fact seems to exclude any doubleness of LEs in the lateral plane. Doubleness of LEs has not been established either in cross sections of SCs. This fact is more clearly observed in the SCs of insects, that have thick LEs as in *Drosophila* females (Carpenter, 1975) or in hemipterans (Solari, 1979).

On the other hand, there are several indications of the presence of an *inner layer* flanking each LE towards the central space (Moses, 1968). This inner layer may differ in composition compared to the main LE (Moses, 1968). This inner layer stains much lighter than the main LE and is probably discontinuous, and forms an integral part of

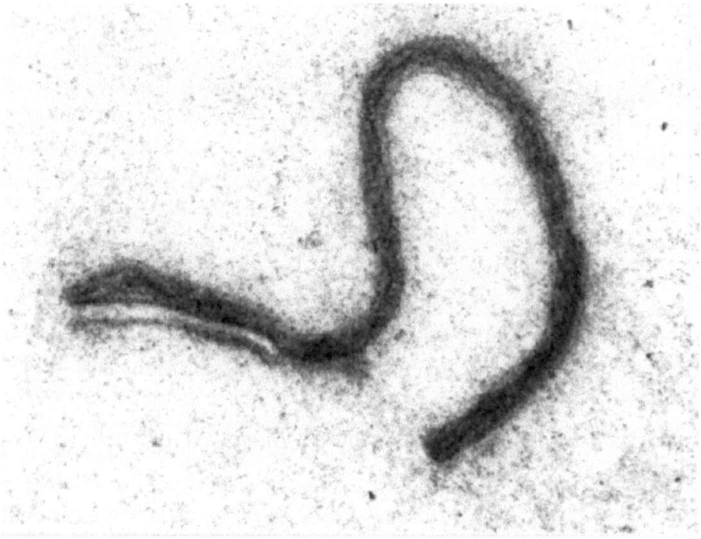

Figure 1. *XY pair of Galea musteloides. The X axis shows its multistranded structure (12,500×).*

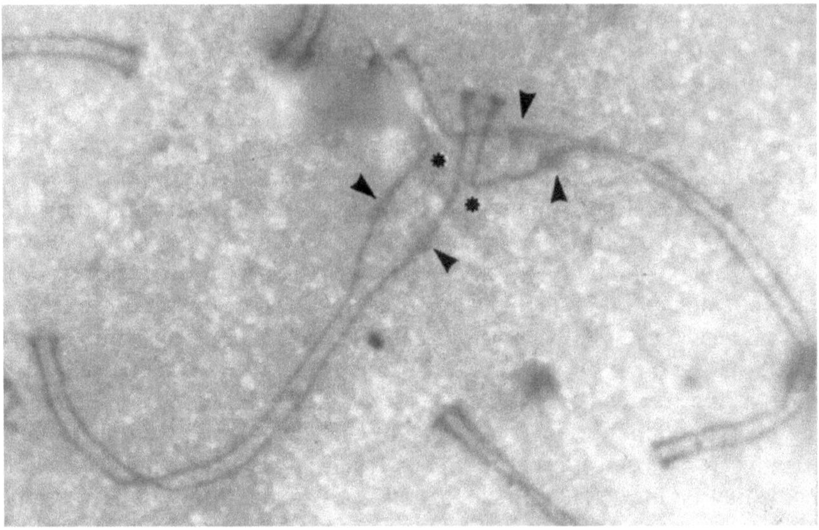

Figure 2. *Partner switchings (asterisks) in a quadrivalent of an autotetraploid chicken oocyte. The kinetochores are marked with arrowheads (12,000 ×).*

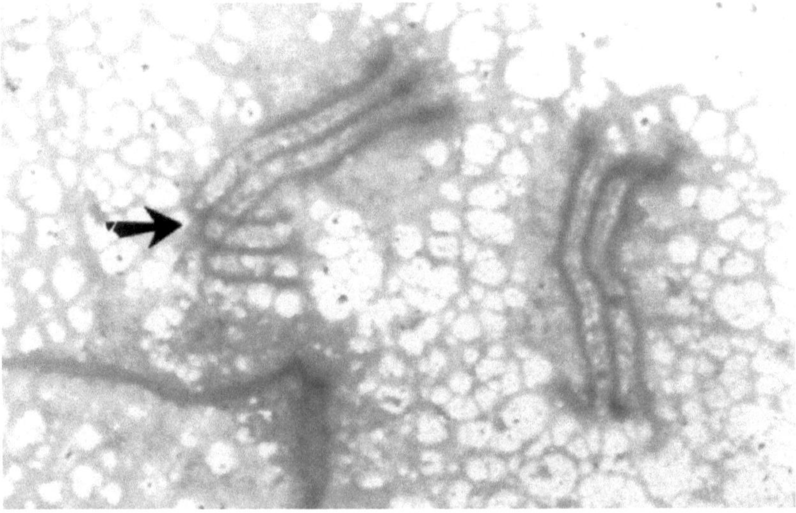

Figure 3. *Triple synapsis in a triploid ZZW chicken. The folded axes (arrow) show the planar structure of the triplet (20,000 ×).*

the central space, as the transverse fibrils end on this layer. In *G. musteloides* the inner layer and the main axis of the X chromosome separate cleanly from each other, but the inner layer is absent in the non-synapsed region. Thus, while a differential inner layer may exist in correspondence with transverse fibrils, there are no firm indications of doubleness in the main axis. Furthermore, no evidence of doubleness is found in LEs from hamster and rat spermatocytes after microspreading, formaldehyde fixation and negative staining (Solari, 1972).

Correspondingly, at least one of the proteins (SCP3/ Cor1) forming the longitudinal axes is immunolocalized as a continuous, single filament in the LEs of a number of mammalian (Dobson *et al.*, 1994; Schalk *et al.*, 1998) and avian species (Solari, unpubl. observ.).

The singleness of meiotic axes is especially outstanding at early diplotene, as seen by electron microscopy of thin serial sections (Solari, 1970b) and also in microspread diplotene nuclei from rodent spermatocytes (Solari, 1981). This fact is especially significant, as during diplotene the four chromatids of bivalents become visible by conventional light microscopy (Figure 4). However, the axes remain single, including at the sites corresponding to chiasmata as seen by electron microscopy (Solari, 1970b, 1981) and by immunofluorescence (Moens *et al.*, 1995).

In all the mammalian and avian species that have been reported, the meiotic axes attach themselves to the inner layer of the nuclear envelope from zygotene to the end of pachytene or early diplotene stages (von Wettstein *et al.*, 1984; Solari, 1993). The attachment of the LEs are cone-shaped expansions of each axis, that merge with a dense layer on the inner side of the nuclear envelope. These terminal expansions, or "terminal plaques" show an interesting departure in structure from the main axis: they are visibly frayed into multiple strands (Figure 5).

At early diplotene in rodent spermatocytes, the multistranded structure of the terminal plaques stands out compared to the single, thin axis (Solari, 1970b). Again, these observations show that the two chromatids of each homologue are not represented by separate units in meiotic axes, but that axial material can be distributed in multiple patterns according to the stage and the location in the nucleus. Neither the

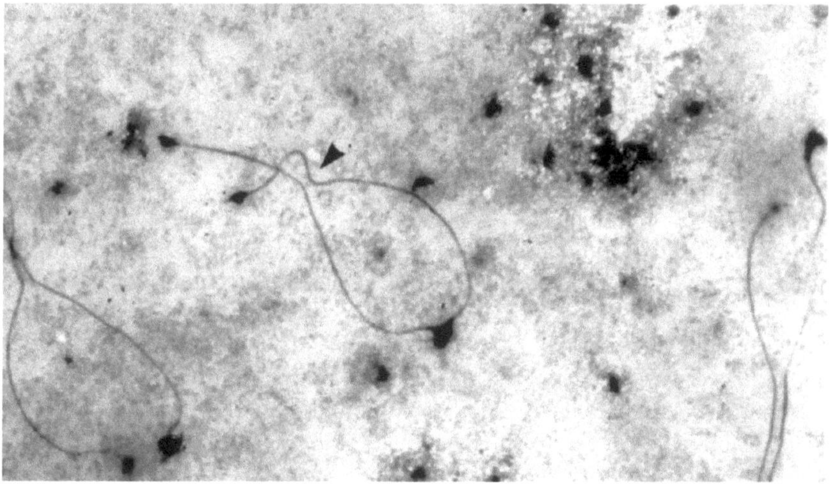

Figure 4. *Diplotene bivalents from spread mouse spermatocytes. The LEs of SCs are mostly separated as single axes. A convergence region is marked with an arrowhead (6,500 ×).*

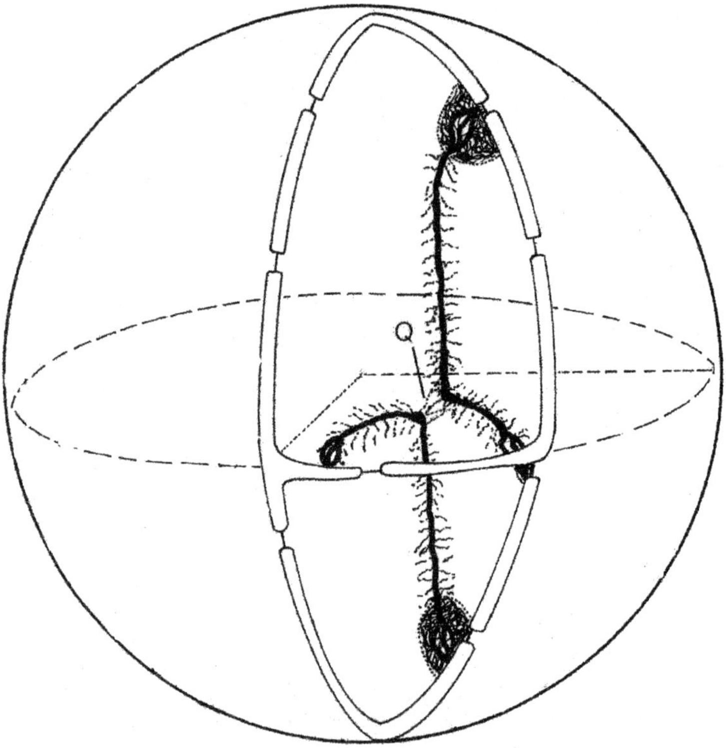

Figure 5. *Scheme of the shape and location of the two single axes from a diplotene bivalent in mouse spermatocytes. Two terminal plaques with axial splitting have dense, associated centromeric heterochromatin, and the other two are the distal ones and show the same splitting. Q: location of an interstitial chiasma.*

kinetochore knobs of LEs (Counce and Meyer, 1973) show any doubleness in the many insect, mamalian and avian species so far described (reviewed in von Wettstein *et al.*, 1984; Gillies, 1989).

The preceding observations lead to the conclusion that meiotic axes or LEs do not express the sister chromatin duality of each synaptic homolog, but instead, that these axial elements are possibly *joining* both sister chromatids, as first stated on EM grounds (Solari, 1970b, 1981). This conclusion holds on for diplotene stages, in which chiasmata establish a functional individualization of chromatids.

There are, however, some observations suggesting that *single* chromatids may form a meiotic axis in some mutants of fungi (*spo 76* mutant of *Sordaria* and *spo 22* mutant of *Coprinus*, cited by Zickler and Kleckner, 1999). Thus, it is possible that the interface between sister chromatids, that is usually the site for axial assembly may be substituted by specific, axially related regions of single chromatids in this assembly process under special conditions.

Whenever the duality of sister chromatids is compared to these structural data, we have stated the hypothesis that each chromatid has special regions, axially located, that are joined to each other by axial components (Solari, 1981). Now it can be added that not only the known axial proteins (SCP3/Cor1 and SCP2) are at those locations, but also cohesins (SMC1 and SMC3) as shown by immunolocalization in rat spermatocytes (Eijpe *et al.*, 2000). Axes may probably contain many other minor proteins, for instance the RecA-like protein Rad51, as in mouse and chicken meiocytes (Ashley *et al.*, 1995) and other proteins (reviewed in Zickler and Kleckner, 1999).

The disposal of axial components and the structure of chiasmatic regions at diplotene

Diplotene nuclei are easily identified in rodents, and were first studied with three-dimensional recontructions of serial thin sections in mice (Solari, 1970b). The presence of long stretches of single axes is typical of this stage. The other feature is the approaching of axes, forming the "convergence regions" of the diplotene axes (Solari, 1970, 1981; Moens and Spyropoulos, 1995). Excluding the regions in which the LEs have not yet diverged, and which have the usual SC structure, the convergence regions are points of sharp departure of two axes from a small region in which they are separated by the same distance as LEs during pachytene (Figures 4 and 5). This space shows changes from early diplotene, in which a short stretch of central element of the SC is visible, to mid-diplotene, in which this central region is invaded by chromatin (Solari, 1970b, 1981). The central region of the pachytene SCs begins to disintegrate at the end regions of bivalents, allowing the LEs to become separate, single axes, that separate progressively from each other except at the "convergence regions". Immunofluorescence observations with anti-Cor1 on mouse spermatocytes have fully confirmed this previous EM pattern of behaviour (Moens and Spyropoulos, 1995). Remnants of axial components are visible near centric locations in a number of organisms, as far as metaphase I (Solari, 1981; Solari and Tandler, 1991) and Cor1/SCP1 is immunolocalized at these places (Moens and Spyropoulos, 1995).

The central region of the SC in whole mounts

The central region is crossed by large numbers of "transverse filaments" (TFs) (Moses, 1968; Wettstein and Sotelo, 1971). However, the features of single filaments are difficult to study in thin sections, and other techniques have been useful in this respect (Solari, 1998). Negative staining of nuclei spread on liquid surfaces has given detailed information (Solari and Moses, 1973; Solari, unpubl.). DNase-resistant, straight filaments cross at least half of the central space in the bivalents of both hamster and cricket spermatocytes. The average width of the transverse filaments is 1.6 nm (Solari and Moses, 1973; Solari, unpubl.) (Figure 6).

These filaments are protease (chymotrypsin and pronase) sensitive: they are completely removed after 15 min of treatment (Solari, 1972). These unit filaments are often grouped in bundles of 2–5 filaments which are often diverging from each other either at the LE side or at the Central Element (CE). Many of the filaments are seen to enter the wide, scalariform CE of the cricket (and of other insects) and apparently end at the other side of the CE (Figure 6) There are places in which filaments coming from

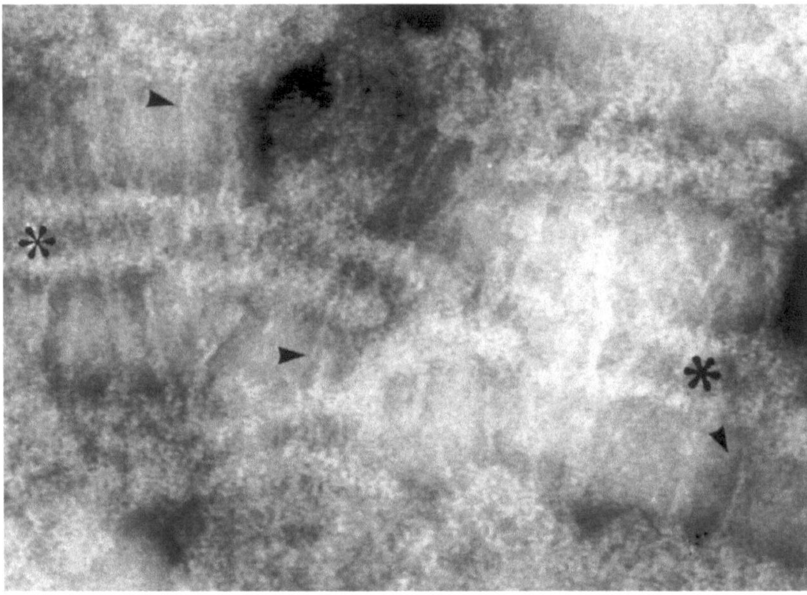

Figure 6. *Segment of a SC from a spread cricket spermatocyte. Individual TFs (arrowheads) cross the central space and enter the wide central element (asterisks). Negative stain with uranyl acetate (360,000 ×).*

opposite LEs overlap each other along the width of the full CE. This may be a more general pattern that is obscured by the amorphous material of the CE in cricket spermatocytes, as the overlapping of opposite filaments in the CE is clearly seen in grasshopper SC spreads (Solari, 1998). The observation of preserved SC regions and regions partly disorganized by the stretching forces at spreading shows that the filament length is the addition of half a central space plus the width of the CE. Thus, most, if not all transverse filaments of cricket and grasshopper SCs overlap at their axial ends all over the CE. The TFs from mammalian SCs are similar but have a much shorter overlap at a thin CE (Solari and Moses, 1973).

The CE of insects is largely different in width (25–28 nm) as compared with that of mammals (12–20 nm). Furthermore, the CE from insects is well organized, as two lateral densities that are traversed by the TFs and where some additional material of granular aspect is deposited, and a central region permeable to the negative stain, on which a more or less imperfect axial periodicity is observed, especially in grasshoppers.

The Central Region and the scalariform CE of insects in thin sections

As the stretching forces of microspreading distort the SC structure as a whole, a useful view can be obtained with thin, serial sections and 3-D reconstructions (Wettstein and Sotelo, 1967). A number of papers have reported the presence of a well-ordered array of fibrils in the CE of insects (reviewed in Westergaard and von Wettstein, 1972;

Schmekel *et al.*, 1993). A recent analysis of serial thin sections through the central region of insects from two groups, hemipterans and coleopterans, have shown that in sagittal sections a paracrystalline array is shown by dots with an average diameter of 3.2 nm, spaced with an axial periodicity of 10.3 nm and forming a definite series of rows: 12 in the coleopteran *Diloboderus abderus* (Figure 7), and six rows in the hemipterans *Triatoma infestans* and *T. pallidipennis*. In fact the higher number of rows seems to be restricted to the coleopteran order, as our own observations in a grasshopper (*Laplatacris dispar*) shows the same row number (6) as in hemipterans, and data on other insects suggest that six rows may be a widespread feature (illustrated in Westergaard and Wettstein, 1972). The dots on adjacent rows are not in register, but slightly shifted in axial direction, and thus the line joining corresponding dots in the 12 layers makes an angle of 20 degrees with the long axis (Figure 7). Cross-sections of SCs show that the 12 dot layers of sagittal sections are seen as 12 segmented lines or 12 rows of elongated, dot-like rows, according to the section plane. Recombination nodules (RNs) are often found sitting on the external row in these sections (Figure 8).

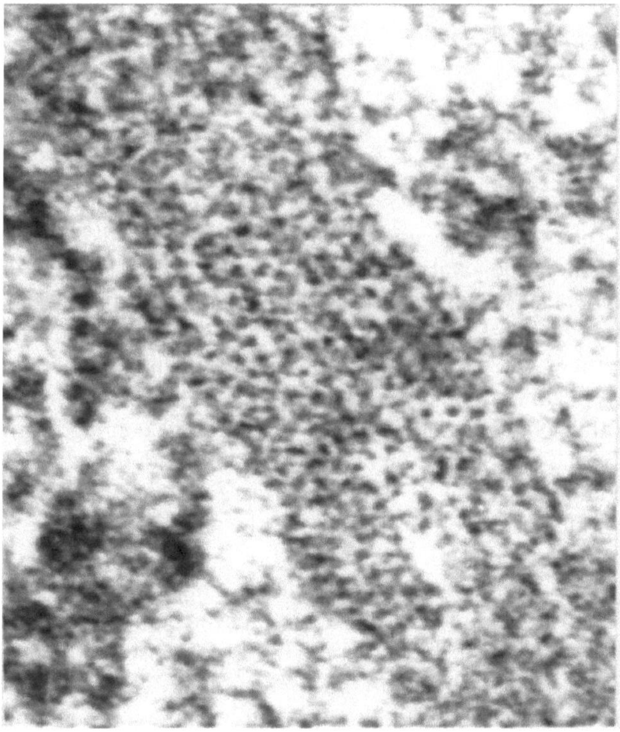

Figure 7. *Sagittal thin section through the CE of a coleopteran spermatocyte. The CE shows 12 rows of dot-like elements. The dots are not in register but they are shifted in each row to make an angle about 20° with the long axis (238,000 ×).*

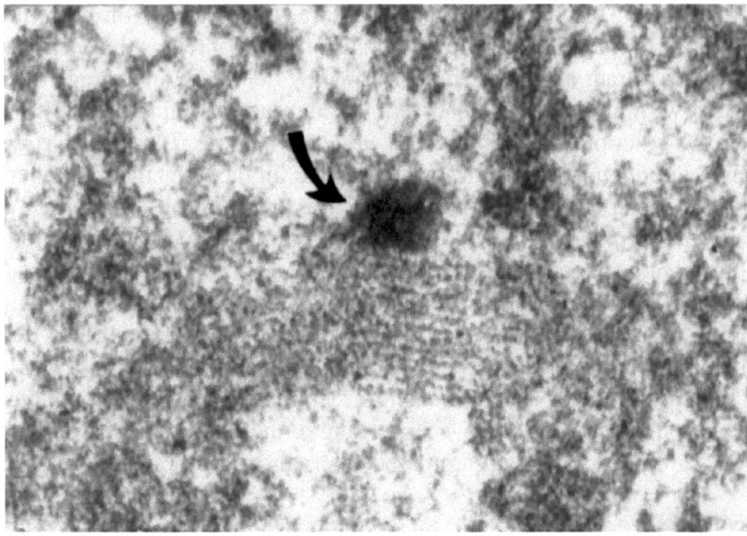

Figure 8. *Transverse section though a coleopteran SC at pachytene. The 12 rows of the CE are followed by a diffuse material associated to chromatin. A recombination nodule (arrow) sits on top of the CE (130,000 ×).*

In sections through a twist of the SC there is a transition between the scalariform structure seen in the frontal plane and the paracrystalline, dot-like array of sagittal sections. In this transition, the axial periodicities of the CE are substituted by bundles of transverse filaments that become denser at the center of the central region.

While both sections and negatively stained whole mounts cannot be thoroughly integrated in a perfect scheme, a model is proposed in which the bundles of TFs coming from the LEs become tightened at the entrance of the CE, perhaps with the addition of another (amorphous) component, overlap with TFs coming from the other side, transverse the CE and end on the other edge of this CE with the addition of other layer of amorphous component.

A different model has been proposed for the CE of the beetle *Blaps cribosa* (Schmekel *et al.*, 1993) in which every TF originates in one LE, crosses the CE through a transverse CE component and reaches the opposite LE. In a CE layer, short irregular "pillars" form the junctions between transverse and longitudinal CE components. The model of *Blaps* considers that the CE is composed of three or four distinct layers, while our observations show in the coleopteran 12 layers of dot-like elements, and 6 layers in other insect taxa. Both models agree in the more ordered arrangement of TF inside the CE, as compared to the outer regions. However, the TFs in the model of Schmekel *et al.* (1993) are usually 5 nm in diameter, while our observations on negatively stained specimens show a thinner (1.6 nm) unit which can be arranged in bundles (see above). The identity of the "pillar" (Schmekel *et al.*, 1993) or the "amorphous" (Solari and Moses, 1973) component, must wait for further chemical data on the CE of insects.

The TFs and CEs are able to self-assemble inside the nucleus and in the cytoplasm
Either planar or curved, single layered "mutiple complexes" (Moses, 1968) or three-dimensional, multilayered "policomplexes" (Moses, 1968) containing central regions identical to those of SCs, but completely independent from chromosomes, have been described in meiotic or post-meiotic germ cells of many organisms (Goldstein, 1987).

The most convincing hypothesis on their origin is the ability to autoassembly of their components, which are generally the remnants of the central regions of SCs after the establishment of diplotene. These multiple complexes may stay inside the nucleus, as in the spermatocytes of *Triatoma* species at the diffuse stage (Solari, 1979), or may be found in the cytoplasm of dividing spermatocytes as in humans (Solari and Vilar, 1978) or at the perinuclear cytoplasm of spermatids, as in many orthopterans (Wolf and Mesa, 1993). In most cases a clear distinction is made between these multiple structures and SCs: the multiple complexes lack truly organized LEs, which are substituted by large granular and dense material. The most convincing and complete study on the origin of these multiple structures has been made on yeast (*S. cerevisiae*) (Dong and Roeder, 2000). Pure Zip1 protein (the constituent of TFs in yeast) is able to self-assembly in two major types: polycomplexes, similar to those of other organisms, and "networks" which are a distinct type of organization. In polycomplexes, as in normal SCs, Zip1 assembles as bundles of dimers, and in each dimer two Zip1 proteins are joined in register, the NH_2-terminal domain lying towards the central element, while the carboxyl termini are oriented towards the sides (or towards the LEs in true SCs) (Dong and Roeder, 2000). A similar orientation is found in the SCs from mice, as regards the molecules of SCP1, the major component of TFs in mammals (Liu *et al.*, 1996) and other mammals (reviewed in Zickler and Kleckner, 1999).Thus, it is possible that these two proteins, although having unrelated aminoacid sequences, form essentially similar structures, by aggregation of in-register dimers into bundles: the TFs. The CE is most possibly formed by overlapping and interaction of the NH2 globular termini coming from bundles of TFs originated near each LE (Dong and Roeder, 2000).

Recombination nodules, recombination and chiasmata in mammals and birds

Recombination nodules (Carpenter, 1975) are round or elongated structures closely associated to the central region of SCs but clearly independent of the central structures of SCs. Avian meiotic chromosomes show well-distinguishable RNs of the so-called "late variety" and they have been studied in our lab in the past decade (Rahn and Solari, 1986; Solari, 1993; Pigozzi and Solari, 1998; 1999a). In avian meiocytes RNs are oval, strongly staining structures about 126 nm in average diameter, but with large variations according to the pachytene substage (Pigozzi and Solari, 1998). Each RN is organized as a cluster of subparticles, the "recombinomeres" (Pigozi and Solari, 1998), having high staining after phosphotungstic acid (PTA) and about a spherical shape (Figure 9). The number of recombinomeres in each RN varies from

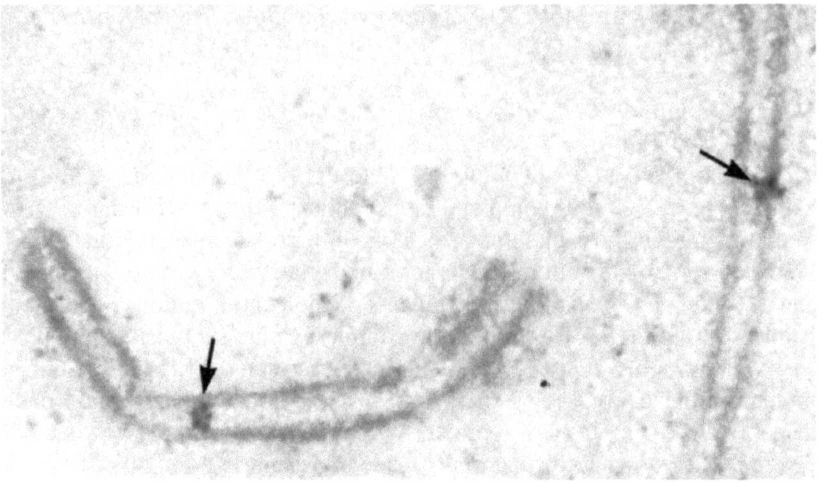

Figure 9. *Spread SCs from pigeon spermatocytes showing the "recombinomeres" in the RNs (arrows) (36,000 ×). From Pigozzi MI, Solari AJ (1998). Biocell 22: 177–186; by kind permission of the publishers.*

an average of 2.3 at late zygotene to 3.58 at mid-late pachytene, and the frequency of RN having high numbers of recombinomeres increases with the advance of pachytene (Pigozzi and Solari, 1998). Thus, RNs are also "dynamic" structures that change over time.

The relationship between these RNs and chiasmata is very well shown in birds. A study of both chiasma distribution and RN number and location in *Columba livia* (Pigozzi and Solari, 1999a) shows a very close agreement between the total number of chiasmata (60–66) and the mean number of RNs (64.7). Furthermore, specific bivalents have both very close numbers of chiasmata and RNs and their distribution along the bivalent is very similar. As expected, the number of RNs is directly proportional to the SC length, and the linear regression of RNs respect SC length is almost perfect (Pigozzi and Solari, 1999a). In fact, there are other features shared by RNs and chiasmata, as interference (Pigozzi and Solari, 1999b) and strict localization in sex chromosomes (Solari, 1993).

A good marker of RNs is the protein MLH1 that participates in the mismatch repair system (reviewed in Dresser, 2000; Zickler and Kleckner, 1999). As expected, anti-MLH1 decorates the sites of RNs in birds (chicken) (Pigozzi, unpubl.).

The sex chromosomes, partial synapsis and restricted recombination

One largely unmentioned fact is that homolog synapsis is *partial* in heteromorphic sex chromosomes of mammals and birds, as first proved by electron microscopy (in mouse, Solari, 1970; in Man, Solari and Tres, 1970; in chickens, Solari, 1977; Rahn

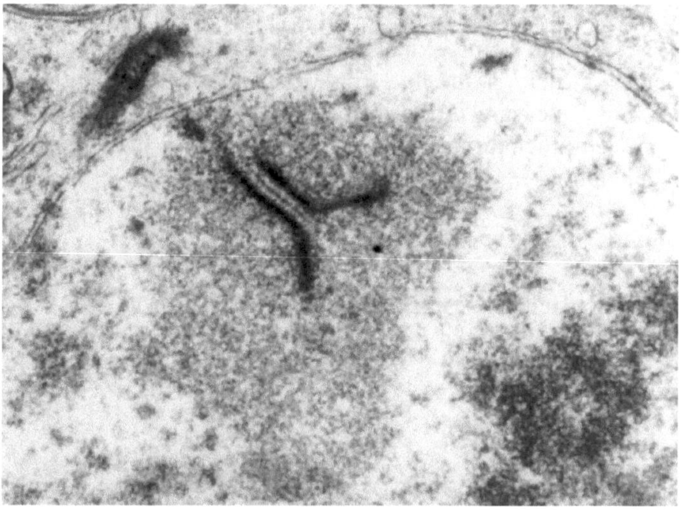

Figure 10. *Section through the XY pair of the mouse showing partial synapsis (24,000 ×). From Solari AJ (1973), by kind permission of the publishers.*

and Solari, 1986). SC extends only to a point (which is variable according to species and to stage) (Figure 10) and then the remaining part of the sex chromosomes have a single axis, which generally shows distinct differentiations (reviewed in Solari, 1993; Pigozzi, 1999). The fact that synapsis is interrupted cannot be exclusively ascribed to lack of homology, as SCs can form on non-homologous regions, generally in a phase of "synaptic adjustement" (Moses and Poorman, 1981). In fact, the avian ZW pairs show plainly this phenomenon (Solari, 1993).

Recombination is severely restricted in many heteromorphic X-Y and ZW pairs, as first demonstrated by RN studies in mammals and in birds (reviewed in Solari, 1993, and in Pigozzi, 1999). Recombination restriction has long been suspected to be a leading factor in the origin and evolution of heteromorphic sex chromosomes (Muller, 1964; reviewed in Solari, 1993). Recently, the finding of species forming a "missing link" between those having homomorphic sex chromosomes and those having highly heteromorphic gonosomes has allowed an integral scheme of the evolutionary changes in the avian ZW pair (Pigozzi, 1999).

Conclusion

In the past decade, a number of proteins located in meiotic chromosomes have been identified, as well as their coding genes, and some regulatory and triggering phenomena have been at least partly explained, especially the widely occurring double strand breaks (DSB) in DNA, which appear as a primary event for meiosis in many organisms. However, the structural, detailed study of the protein assemblies of meiotic chromosomes is far from being exhausted. Besides its grounding foundation for molecular schemes of meiosis, the fine structure of meiotic chromosomes conveys a good deal of esthetic pleasure in the display of elaborate patterns which combine beautifully with each other to make the meiotic process a fine melting pot of new "chromosome generations".

Acknowledgements

This work was supported by Conicet and UBACYT. The helpful cooperation of Dr. M.I. Pigozzi and Lic. M. Rahn is thanked and the technical help of C. Deparci is acknowledged.

References

Ashley TA, Plug W, Xu J, Solari AJ, Reddy G, Golub EI, Ward DC (1995). Dynamic changes in Rad51distribution on chromatin during meiosis in male and female vertebrates. *Chromosoma* 104: 19–28.

Carpenter ATC (1975). Electron microscopy of meiosis in *Drosophila melanogaster* females II. The recombination nodule- A recombination associated structure at pachytene? *Proc Natl Acad Sci USA* 72: 3,186–3,189.

Counce SJ, Meyer GF (1973). Differentiation of the synaptonemal complex and the kinetochore in *Locusta* spermatocytes studied by whole mount electron microscopy. *Chromosoma* 44: 231–253.

del Mazo J, Gil-Alberdi L (1986). Multistranded organization of the lateral elements of the synaptonemal complex in the rat and mouse. *Cytogenet Cell Genet* 41: 219–224.

Dietrich AJJ, van Marle J, Heyting C, Vink AC (1992). Ultrastructural evidence for a triple structure of the lateral element of the synaptonemal complex. *J Struct Biol* 109: 196–200.

Dobson MJ, Pearlman RE, Karaiskakis A, Spyropoulos B, Moens PB (1994). Synaptonemal complex proteins: occurrence, epitope mapping and chromosome disjunction. *J Cell Sci* 107: 2,749–2,760.

Dong H, Shirleen Roeder G (2000). Organization of the yeast Zip1 protein within the central region of the synaptonemal complex. *J Cell Biol* 148: 417–426.

Dresser MJ (2000). Meiotic chromosome behavior in *Saccharomyces cerevisiae* and (mostly) mammals. *Mutat Res* 451: 107–127.

Egel R (2000). Fission yeast on the brink of meiosis. *BioEssays* 22: 854–860.

Eijpe M, Heyting C, Gross B, Jessberger R (2000). Association of mammalian SMC1 and SMC3 proteins with meiotic chromosomes and synaptonemal complexes. *J Cell Sci* 113: 673–682.

Gardner MJ (1999). The genome of the malaria parasite. *Curr Opin Genet Dev* 9: 704–708.

Gillies CB (1989). *Fertility and chromosome pairing: Recent studies in plants and animals.* Boca Raton, FL: CRC Press.

Goldstein P (1987). Multiple synaptonemal complexes (polycomplexes): origin, structure and function. *Cell Biol Int Rep* 11: 759–796.

Kohli J, Bahler J (1994). Homologous recombination in fission yeast: absence of crossover interference and synaptonemal complex. *Experientia* 50: 295–306.

Liu J, Li Y, Brundell E, Björkroth E, Daneholt B, Höög C (1996). Localization of the N-terminus of SCP1 to the central element of the synaptonemal complex and evidence for

direct interactions between the N-termini of SCP1 molecules organized head-to-head. *Exp Cell Res* 226: 11–19.
Meyer GF (1961). The fine structure of spermatocyte nuclei of *Drosophila melanogaster*. *Proc Eur Reg Conf El Microscopy* 2: 951–954.
Moens PJ, Spyropoulos B (1995). Immunocytology of chiasmata and chromosomal disjunction at mouse meiosis. *Chromosoma* 104: 175–182.
Moses MJ (1956). Chromosomal structures in crayfish spermatocytes. *J Biophys Biochem Cytology* 2: 215–218.
Moses MJ (1968). Synaptinemal complex. *Ann Rev Genet* 2: 363–412.
Moses MJ, Poorman PA (1981). Synaptonemal complex analysis of mouse chromosomal rearrangements. II. Synaptic adjustment in a tandem duplication. *Chromosoma* 81: 519–535.
Moses MJ, Slatton GH, Gambling TM, Starmer CF (1977). Synaptonemal complex karyotyping in spermatocytes of the Chinese hamster (*Cricetulus griseus*). III. Quantitative evaluation. *Chromosoma* 60: 345–375.
Muller HJ (1964). The relation of recombination to mutational advance. *Mutat Res* 1: 2–9.
Pigozzi MI (1999). Origin and evolution of the sex chromosomes in birds. *Biocell* 23: 79–95.
Pigozzi MI, Solari AJ (1998). First demonstration of the substructure of recombination nodules. *Biocell* 22: 177–186.
Pigozzi MI, Solari AJ (1999a). Recombination nodule mapping and chiasma distribution in spermatocytes of the pigeon, Columba livia. *Genome* 42: 308–314.
Pigozzi MI, Solari AJ (1999b). The ZW pairs of two paleognath birds from two orders show transitional stages of sex chromosome differentiation. *Crom Res* 7: 541–551.
Rahn MI, Solari AJ (1986). Recombination nodules in the oocytes of the chicken, *Gallus domesticus*. *Cytogenet Cell Genet* 43: 187–193.
Rasmussen SW, Holm PH (1979). Chromosome pairing in autotetraploid *Bombyx mori* females. Mechanism for exclusive bivalent formation. *Carlsberg Res Commun* 44: 101–125.
Roeder GS (1997). Meiotic chromosomes: it takes two to tango. *Genes Dev* 11: 2,600–2,621.
Schalk JAC, Dietrich AJJ, Vink ACG, Offenberg HH, van Aalderen M, Heyting C (1998). Localization of SCP2 and SCP3 protein molecules within synaptonemal complexes of the rat. *Chromosoma* 107: 540–548.
Schmekel K, Wahrman J, Skoglund U, Daneholt B (1993). The central region of the synaptonemal complex in *Blaps cribosa* studied by electron microscope tomography. *Chromosoma* 102: 669–681.
Speed RM (1984). Meiotic configurations in female trisomy 21 fetuses. *Hum Genet* 66: 176–180.
Solari AJ (1970a). The spatial relationship between the X and Y chromosomes during meiotic prophase in mouse spermatocytes. *Chromosoma* 29: 217–236.
Solari AJ (1970b). The behaviour of chromosomal axes during diplotene in mouse spermatocytes. *Chromosoma* 31: 217–230.
Solari AJ (1972). Ultrastructure and composition of the synaptonemal complex in spread and negatively stained spermatocytes of the golden hamster and the albino rat. *Chromosoma* 39: 237–263.
Solari AJ (1973). In: S. Segal, R. Crozier, P. Corfman and P. Condliffe, eds., *The Regulation of Mammalian Reproduction*, C.J. Thomas, Springfield, pp. 91–99.
Solari AJ (1977). Ultrastructure of the synaptic autosomes and the ZW bivalent in chicken oocytes. *Chromosoma* 64: 155–165.
Solari AJ (1979). Autosomal synaptonemal complexes and sex chromosomes without axes in *Triatoma infestans* (Reduviidae; Hemiptera). *Chromosoma* 72: 225–240.
Solari AJ (1980). Synaptonemal complexes and associated structures in microspread human spermatocytes. *Chromosoma* 81: 315–337.

Solari AJ (1981). Chromosomal axes during and after diplotene. In: *International Cell Biology 1980–1981*, HG Schweiger, ed., pp. 178–186, Berlin-Heidelberg: Springer-Verlag.
Solari AJ (1993). *Sex chromosomes and sex determination in vertebrates*. Boca Raton, FL: CRC Press.
Solari AJ (1995). Mitosis and genome partition in trypanosomes. *Biocell* 19: 65–84.
Solari AJ (1998). Structural analysis of meiotic chromosomes and synaptonemal complexes in higher vertebrates. In: Berrios M, ed., *Nuclear Structure and function, Methods in Cell Biology*, vol. 53. San Diego, CA: Academic Press, pp. 235–256.
Solari AJ, Bianchi NJ (1975). The synaptic behavior of the X and Y chromosomes in the marsupial *Monodelphis dimidiata. Chromosoma* 52: 11–25.
Solari AJ, Fechheimer NS (1988). Quadrivalent formation in a tetraploid chicken oocyte. *Genome* 30: 900–902.
Solari AJ, Moses MJ (1973). The structure of the central region in the synaptonemal complex of hamster and cricket spermatocytes. *J Cell Biol* 56: 145–152.
Solari AJ, Tandler CJ (1991). Presence of a centromeric filament during meiosis. *Genome* 34: 888–894.
Solari AJ, Tres LL (1970). The three-dimensional reconstruction of the XY chromosomal pair in human spermatocytes. *J Cell Biol* 45: 43–53.
Solari AJ, Vilar O (1978). Multiple complexes in human spermatocytes. *Chromosoma* 66: 331–340.
Solari AJ, Thorne MH, Sheldon BL, Gillies CB (1991). Synaptonemal complexes of triploid (ZZW) chickens: Z-Z pairing predominates over Z-W pairing. *Genome* 34: 718–726.
Solari AJ, Merani MS, Burgos MH (1993). Dissociation of the synaptonemal complex in the XY body of *Galea musteloides* (Rodentia, Caviidae). *Biocell* 17: 25–37.
Tibayrenc M, Ayala FJ (1999). Evolutionary genetics of Trypanosoma and Leishmania. *Microbes Infect* 1: 465–472.
von Wettstein D, Rasmussen SW, Holm PB (1984). The synaptonemal complex in genetic segregation. *Annu Rev Genet* 18: 331–413.
Walker MY, Hawley RS (2000). Hanging on to your homolog: the roles of pairing, synapsis and recombination in the maintenance of homolog adhesion. *Chromosoma* 109: 3–9.
Westergaard M, Von Wettstein D (1972). The synaptonemal complex. *Annu Rev Genet* 6: 71–110.
Wettstein R, Sotelo RJ (1967). Electron microscope serial reconstruction of the spermatocyte I nuclei at pachytene. *J. Microscopie* 6: 557–576.
Wettstein R, Sotelo RJ (1971). The molecular architecture of the synaptonemal complex. *Adv Cell Mol Biol* 1: 109–152.
Wolf KW, Mesa A (1993). Synaptonemal polycomplexes in spermatids: a characteristic trait of Orthoptera? *Chromosome Res* 1: 181–188.
Zickler D, Kleckner N (1999). Meiotic chromosomes: integrating structure and function. *Annu Rev Genet* 33: 603–754.

2.4. The Nuclear Envelope at the Attachment Sites of Mammalian Meiotic Telomeres

Ricardo Benavente, Manfred Alsheimer and Elisabeth von Glasenapp
Department of Cell and Developmental Biology, Biocenter, University of Würzburg, Am Hubland, D-97074 Würzburg, Germany

Despite of the considerable progress made in recent years in the field of meiosis research, relatively little is still known about some highly relevant aspects of this process. For example, an essential step during meiotic prophase is the synapsis of homologous chromosomes. In order to find each other, homologous chromosomes, which initially occupy different nuclear territories, have to move. However, the mechanisms governing long distance chromosome movements are poorly understood (for review see Loidl, 1990; Zickler and Kleckner, 1998; Sybenga, 1999; Scherthan, 2001).

In the past, several studies have provided indirect evidence that the nuclear envelope (NE) may play an important role in meiotic chromosome dynamics (for review see Wilson, 1925; Scherthan, 2001): (a) during early meiotic prophase, chromosomes become attached to the NE through both telomeres. (b) This attachment is dynamic as it allows telomeres to move and to congregate in a pole of the nucleus to form the so-called bouquet. It is also worth mentioning that during and after bouquet resolution the telomeres remain attached to the NE. (c) The NE shows characteristic features at the attachment sites of meiotic telomeres which have been revealed with the help of the electron microscope (von Wettstein *et al.*, 1984). As shown in Figure 1, the attachment of meiotic chromosomes to the NE involves terminal morphological specializations of the synaptonemal complex (SC) called attachment plates. At the level of the attachment plates, the lateral elements of SCs are characteristically thicker. In these sites, the nuclear membranes are more dense than in other regions and associated with fibrils. Some of these fibrils traverse the perinuclear space and are apparently in contact with both the inner and outer nuclear membranes. Other fibrils are observed which are attached to the cytoplasmic side of the outer nuclear membrane. The molecular composition of all these structures is unknown.

M. Schmid and I. Nanda (eds.), Chromosomes Today, Volume 14, 119–126.

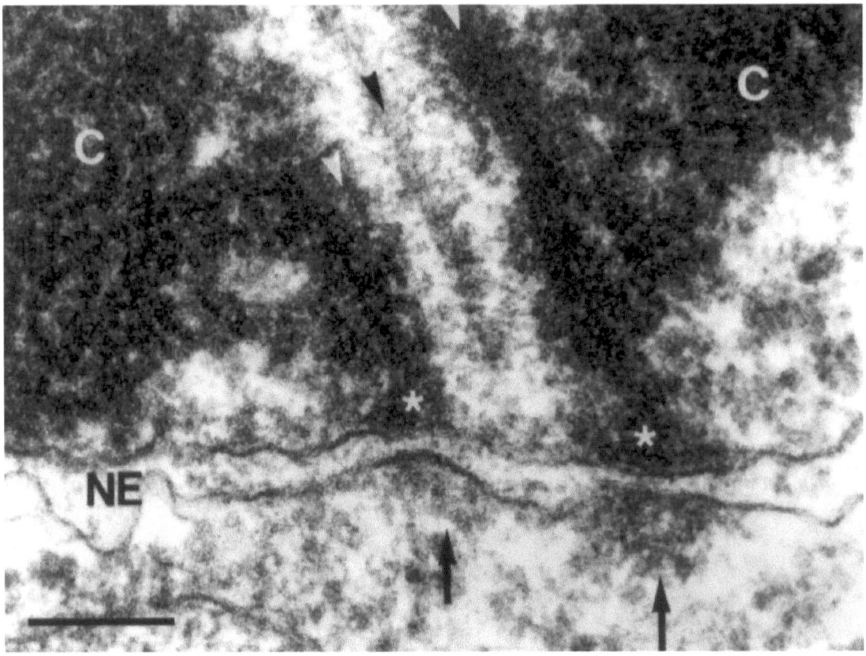

Figure 1. *Electron micrograph of a rat pachytene spermatocyte showing the attachment site of an autosomal bivalent. C, chromatin; NE, nuclear envelope; white arrowheads, lateral elements of the SC; black arrowhead, central element; *, attachment plates; arrows, cytoplasmic fibrils. Bar, 0.2 μm.*

Thus, the peculiarities of the nuclear periphery of meiotic cells and their probable relevance for meiotic process prompted us to investigate the composition of the NE in mammalian meiotic cells in detail. In the following, we will summarize our results obtained on meiotic lamin C2.

The nuclear lamina of mammalian meiotic cells

Evidence has accumulated showing the existence of important differences in the composition and organization of the NE between somatic and mammalian meiotic cells. Most of these observations deal with the nuclear lamina, i.e., the structural component of the NE. Major components of the nuclear lamina are the lamins, a family of structural proteins belonging to the larger family of intermediate filament proteins (McKeon *et al.*, 1986).

Lamins are fibrillar molecules. They show a characteristic α-helical domain that forms coiled coil structures flanked by non-helical N- and C-termini (see Moir *et al.*, 1995; Stuurman *et al.*, 1998; Figure 2). It can be distinguished between A-type and B-type lamins (Krohne and Benavente, 1986). Somatic B-type lamins (B1 and B2) are

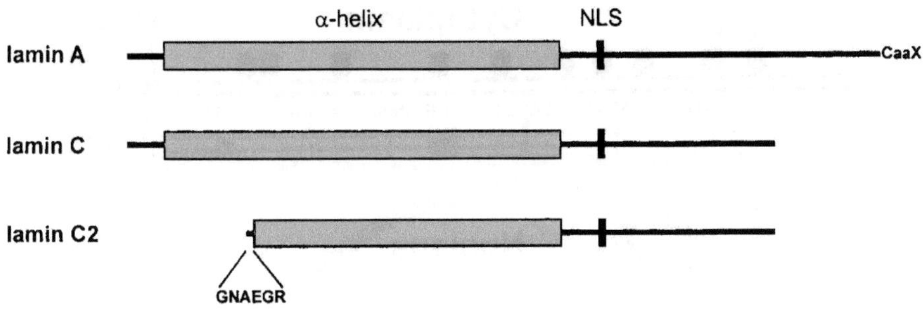

Figure 2. Schematic representation of A-type lamins A, C and C2.

ubiquitous proteins. In contrast, the expression of A-type lamins (A and C) in somatic cells is developmentally regulated.

The expression pattern of lamins is quite different in spermatogenic cells. Lamins A, C and B2 are absent in spermatocytes. Instead, these cells express lamin B1 together with the lamins C2 and B3, which are germ line-specific splicing variants of the lamin A and B2 genes, respectively (Smith and Benavente, 1992; Furukawa and Hotta, 1993; Vester *et al.*, 1993; Furukawa *et al.*, 1994; Alsheimer and Benavente, 1996). Furthermore, the amount of lamins per nucleus appears to be lower than in somatic cells (Vester *et al.*, 1993; Alsheimer *et al.*, 1998).

Mammalian lamins C2 and B3 are remarkably shorter than the somatic isoforms. In both cases, short non-helical sequences substitute the non-helical N-terminus of the molecule and part of the helical domain that are typical for somatic lamins (Furukawa and Hotta, 1993; Furukawa *et al.*, 1994; Alsheimer and Benavente, 1996). In the case of lamin C2 this sequence is six amino acids long (GNAEGR in rat and mouse; Figure 2). It is known from previous investigations on somatic lamins (for review see Stuurman *et al.*, 1998) that the domains that are absent in lamins C2 and B3 are involved in the dimerization and/or formation of complex structures. Therefore, it has been postulated that meiotic lamins C2 and B3 would provide the NE with a more flexible status which could be necessary to allow meiotic chromosome movements (e.g., Smith and Benavente, 1992; Furukawa and Hotta, 1993; Alsheimer and Benavente, 1996). The differences in lamin composition would also explain the lower stability of spermatocyte nuclei that has been observed before (see Stick and Schwarz, 1982).

Discontinuous distribution of meiotic lamin C2 at the nuclear periphery of mammalian spermatocytes

Trying to understand how meiotic chromosomes are associated with the nuclear periphery we focused our investigations on the composition of the NE at the attachment sites of meiotic telomeres (Alsheimer *et al.*, 1999). An interesting result of these investigations was the discovery that lamin C2 is distributed in the form of discontinuous domains at the NE of spermatocytes, contrasting with the situation of lamins of

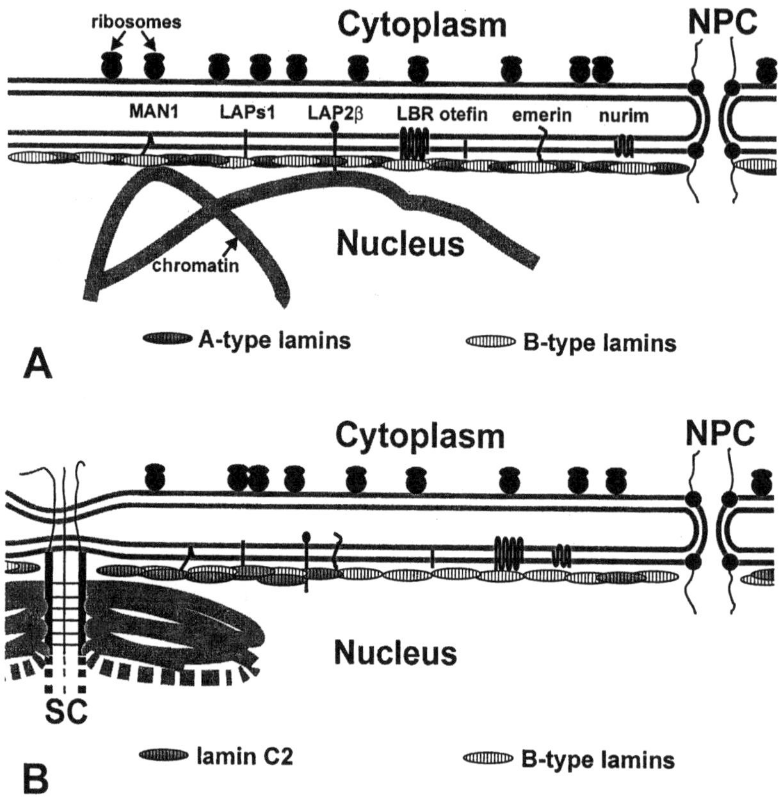

Figure 3. Schematic representation of the nuclear envelope of somatic (A) and meiotic (B) cells.

somatic cells as well as lamin B1 and LAPs2 of spermatocytes, which are distributed in a continuous, ring-like pattern. (At present, the distribution of lamin B3 has not been investigated in detail.) As shown by confocal microscopy and immunolabelling at the electron microscopical level, the attachment sites of meiotic chromosomes at the nuclear periphery are embedded in these lamin C2-enriched domains (see Figure 3). No meiotic telomeres were found in regions of the NE lacking lamin C2. Lamin C2 is the first example of a NE protein of meiotic cells found to be enriched at the attachment sites of chromosomes (Alsheimer *et al.*, 1999).

The function of lamin C2 is unclear. However, it is conceivable that the lamin C2-containing domains represent local reinforcements of the NE that are important for proper meiotic chromosome dynamics (Alsheimer *et al.*, 1999).

Sequence requirements for NE association of meiotic lamin C2

To understand the contribution of lamin C2 to meiotic NE envelope structure and function it is necessary to learn more about the properties of this protein. The targeting and polymerization mechanisms are of particular interest.

The functional significance of somatic nuclear lamin protein domains has been investigated intensively for more than a decade (Moir *et al.*, 1995; Stuurman *et al.*, 1998). Besides the features described above, somatic lamins have a nuclear localization signal (NLS) and the so-called CaaX box (C, cysteine; a, aliphatic; X, any amino acid; Figure 2) in the C-terminal domain.

An essential step for lamin association to the inner nuclear membrane is the isoprenylation of the cysteine residue of the CaaX box (Holtz *et al.*, 1989; Krohne *et al.*, 1989; Kitten and Nigg, 1991; Hennekes and Nigg, 1994; Frimbach-Kraft and Stick, 1995). An exception is lamin C (a short splicing variant of the lamin A gene) which lacks a CaaX box (Figure 2). Its targeting to the nuclear periphery during interphase requires the presence of lamin A (Horton *et al.*, 1992; Pugh *et al.*, 1997). Like lamin C, lamin C2 also lacks a C-terminal CaaX box (Figure 2). Therefore, one interesting aspect was to determine the sequence requirements for NE association of lamin C2, i.e., the only A-type lamin expressed in spermatocytes. For this purpose, we have performed transfection experiments using COS-7 cells. In these experiments we observed lamin C2 to be located at the nuclear periphery where it formed discontinuous plaques (Figure 4; Alsheimer *et al.*, 2000). This result suggested that the formation of discontinuous domains at the nuclear periphery represents most likely a property of lamin C2 (for comparison see Alsheimer *et al.*, 1999). These experiments also indicated that, in contrast to lamin C, lamin C2 contains in itself the information necessary for proper NE association. This point was clarified in transfection experiments, using different lamin C2 constructs. Thus, we were able to demonstrate that the N-terminal hexapetide GNAEGR (Figure 2) is essential for NE association: (a) when this hexapeptide was deleted, lamin C2 was diffusely distributed within nuclei of transfected COS-7 cells; (b) The hexapeptide GNAEGR was sufficient to target somatic lamin C to the NE, provided it was associated at its N-terminus (Alsheimer *et al.*, 2000). Mass spectrometry of lamin C2 peptides as well as the analysis of the fate of mutated lamin C2 molecules provided compelling evidence that the N-terminal glycine is myristoylated (Alsheimer *et al.*, 2000). These results are of high interest because, as demonstrated before, the hydrophobic myristoylglycine confers to the protein affinity for membranes (Towler *et al.*, 1988). To our knowledge lamin C2 is the first example of a member of the intermediate filament protein family that is modified by myristoylation. Taken together, we conclude that NE localization of lamin C2 is achieved by a mechanism that is novel for lamins. It is tempting to speculate that the different mechanisms of association of lamin isoforms with the NE (i.e., myristoylation or isoprenylation) may be of relevance in order to generate local differences in the composition and organization of the nuclear periphery (Alsheimer *et al.*, 2000).

Disappearence of lamin C2 at the end of meiotic prophase

The functional analysis of meiotic proteins is often complicated due to the fact that mammalian spermatocytes are not well suited for manipulations under *in vitro* conditions. A few years ago, Wiltshire *et al.* (1995) developed an *in vitro* model system that allows the investigation of the pachytene/metaphase I transition of mammalian spermatocytes. This *in vitro* model system is based on the observation that the

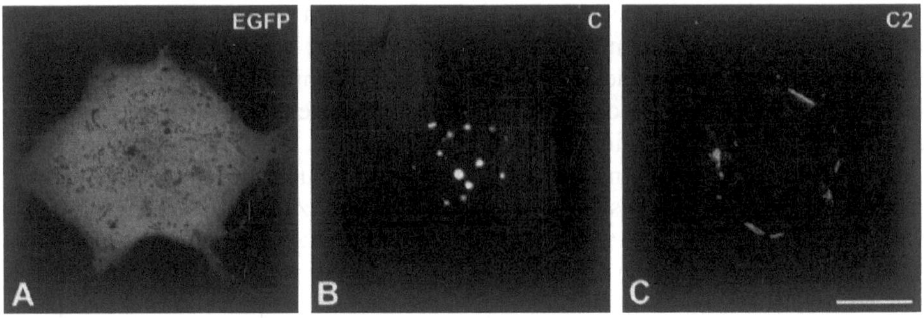

Figure 4. Expression of green fluorescent protein (EGFP; A), somatic lamin C (B) and meiotic lamin C2 (C) in transfected COS-7 cells as investigated with confocal microscopy. Lamins C and C2 were expressed as EGFP fusion proteins. (A) EGFP is diffusely distributed in the cell. (B) In the absence of equivalent amounts of lamin A, lamin C forms aggregates in the nuclear interior (see Alsheimer et al., *2000). (C) Lamin C2 is located at the nuclear periphery where it forms discontinuous plaques. Bar, 10 μm.*

phosphatase inhibitor okadaic acid accelerates the second half of the meiotic prophase. Thus, pachytene spermatocytes reach metaphase I within 6 h, instead of the 3–4 days required *in vivo* (Wiltshire *et al.*, 1995).

By using this *in vitro* model system, we have investigated the fate of NE proteins at the end of the meiotic prophase. For this purpose, pachytene spermatocytes were cultivated in the presence of okadaic acid and harvested at different time points. After cell fractionation and immunoblotting we were able to show that the pachytene/metaphase I transition was accompanied by changes at the level of the protein components of the NE. As expected from previous studies on somatic cells (i.e., Dechat *et al.*, 1998), lamin B1 and the integral NE proteins LAPs2 β and γ became most likely phosphorylated and were recovered in the supernatant fraction (von Glasenapp and Benavente, 2000).

Under the same experimental conditions, lamin C2 behaved in a different way: it became undetectable during the transition from pachytene to metaphase I (von Glasenapp and Benavente, 2000). This result is consistent with our previous immunocytochemical observations showing that lamin C2 is not detectable in postmeiotic cells of mammalian spermatogenesis (Alsheimer and Benavente, 1996). A possible mechanism for the lamin C2 disappearance would be its selective degradation at the end of prophase using a pathway similar to the one described before for other meiosis-specific nuclear proteins (see Tarsounas *et al.*, 1997). These results indicate that, in contrast to other NE proteins of the spermatocyte, lamin C2 is not recycled, and not involved in nuclear reassembly during telophase I.

Future work

At present, very little is known about the mechanisms involved in meiotic chromosome attachment to the NE. The observations summarized here, deal with lamin C2, i.e., the first example of a NE protein found to be enriched at the attachment sites of meiotic telomeres. The next step in our investigations will be the identification of proteins interacting with lamin C2. Furthermore, a matter of particular interest would be the identification and characterization of additional protein components of the nuclear periphery, in particular those of the morphological specializations already described at the attachment sites of SCs (Figure 1).

Acknowledgements

We thank Rosie Rudd for correcting the manuscript. Supported by a grant of the Deutsche Forschungsgemeinschaft (Be 1168/4-3) to R.B.

References

Alsheimer M, Benavente R (1996). Change of karyoskeleton during mammalian spermatogenesis: expression pattern of nuclear lamin C2 and its regulation. *Exp Cell Res* 228: 181–188.

Alsheimer M, Fecher E, Benavente R (1998). Nuclear envelope remodelling during rat spermiogenesis: distribution and expression pattern of LAP2/thymopoietins. *J Cell Sci* 111: 2,227–2,234.

Alsheimer M, von Glasenapp E, Hock R, Benavente R (1999). Architecture of the nuclear periphery of rat pachytene spermatocytes: distribution of nuclear envelope proteins in relation to synaptonemal complex attachment sites. *Mol Biol Cell* 10: 1,235–1,245.

Alsheimer M, von Glasenapp E, Schnölzer M, Heid H, Benavente R (2000). Meiotic lamin C2: the unique amino-terminal hexapeptide GNAEGR is essential for nuclear envelope association. *Proc Natl Acad Sci USA* 97: 13,120–13,125.

Dechat T, Gotzmann J, Stockinger A, Harris CA, Talle MA, Siekierka JJ, Foisner R (1998). Detergent-resistance of LAP2α in interphase nuclei and phosphorylation-dependent association with chromosomes early in nuclear assembly implies functions in nuclear structure dynamics. *EMBO J* 17: 4,887–4,902.

Frimbach-Kraft I, Stick R (1995). Analysis of nuclear lamin isoprenylation in Xenopus oocytes: isoprenylation of lamin B3 precedes its uptake into the nucleus. *J Cell Biol* 129: 17–24.

Furukawa K, Hotta Y (1993). cDNA cloning of a germ cell-specific lamin B3 from mouse spermatocytes and analysis of its ectopic expression in somatic cells. *EMBO J* 12: 97–106.

Furukawa K, Inagaki H, Hotta Y (1994). Identification and cloning of an mRNA coding for a germ cell-specific A-type lamin in mice. *Exp Cell Res* 212: 426–430.

Hennekes H, Nigg EA (1994). The role of isoprenylation in membrane attachment of nuclear lamins. A single point mutation prevents proteolytic cleavage of lamin A precursors and confers membrane binding properties. *J Cell Sci* 107: 1,019–1,029.

Holtz D, Tanaka RA, Hartwig J, McKeon F (1989). The CaaX motif of lamin A functions in conjunction with the nuclear localization signal to target assembly to the nuclear envelope. *Cell* 59: 969–977.

Horton H, McMorrow I, Burke B (1992). Independent expression and assembly properties of heterologous lamins A and C in murine embryonal carcinomas. *Eur J Cell Biol* 57: 172–183.

Kitten G, Nigg EA (1991). The CaaX motif is required for isoprenylation, carboxyl methylation, and nuclear membrane association of lamin B2. *J Cell Biol* 113: 13–23.

Krohne G, Benavente R (1986). The nuclear lamins. A multigene family in evolution and differentiation. *Exp Cell Res* 162: 1–10.

Krohne G, Waizenegger I, Höger, TH (1989). The conserved carboxy-terminal cysteine of nuclear lamins is essential for lamin association with the nuclear envelope. *J Cell Biol* 109: 2,003–2,011.
Loidl J (1990). The initiation of meiotic chromosome pairing: the cytological view. *Genome* 33: 759–778.
McKeon F, Kirschner M, Caput D (1986). Homologies in both primary and secondary structure between nuclear envelope and intermediate filament proteins. *Nature* 319: 463–468.
Moir RD, Spann TP, Goldman RD (1995). The dynamic properties and possible functions of nuclear lamins. *Int Rev Cytol* 162B: 141–182.
Scherthan H (2001). A bouquet makes ends meet. *Nat Rev Mol Cell Biol* 2: 621–627.
Smith A, Benavente R (1992). Identification of a short lamin protein selectively expressed during meiotic stages of rat spermatogenesis. *Differentiation* 52: 55–60.
Stick R, Schwarz H (1982). The disappearance of the nuclear lamina during spermatogenesis: an electron microscopic and immunofluorescence study. *Cell Diff* 11: 235–243.
Stuurman N, Hein S, Aebi U (1998). The nuclear lamins: their structure, assembly, and interactions. *J Struct Biol* 122: 42–66.
Sybenga J (1999). What makes homologous chromosomes find each other in meiosis? A review and an hypothesis. *Chromosoma* 108: 209–219.
Tarsounas M, Pearlman RE, Gasser PJ, Park MS, Moens PB (1997). Protein–protein interactions in the synaptonemal complex. *Mol Biol Cell* 8: 1,405–1,414.
Towler DA, Gordon JI, Adams SP, Glaser L (1988). The biology and enzymology of eukaryotic protein acylation. *Ann Rev Biochem* 57: 69–99.
Vester B, Smith A, Krohne G, Benavente R (1993). Presence of a nuclear lamina in pachytene spermatocytes of the rat. *J Cell Sci* 104: 557–563.
von Glasenapp E, Benavente R (2000). Fate of meiotic lamin C2 in rat spermatocytes cultured in the presence of okadaic acid. *Chromosoma* 109: 117–122.
von Wettstein D, Rasmussen SW, Holm PB (1984). The synaptonemal complex in meiotic segregation. *Annu Rev Genet* 18: 331–413.
Wilson EB (1925). The *Cell in Development and Heredity*. New York: Macmillan.
Wiltshire T, Park C, Caldwell KA, Handel MA (1995). Induced premature G2/M-phase transition in pachytene spermatocytes includes events unique to meiosis. *Dev Biol* 169: 557–567.
Zickler D, Kleckner N (1998). The leptotene–zygotene transition of meiosis. *Annu Rev Genet* 32: 619–697.

2.5. Interphase Cytogenetics in Understanding Chromosome and Telomere Dynamics During Prophase I: Implications for Meiotic Telomere Movements

Harry Scherthan

Max-Planck-Inst. für Molekulare Genetik, Ihnestr. 73, D-14195 Berlin, Germany

Interphase cytogenetics (fluorescent *in situ* hybridization (FISH) to interphase nuclei; Cremer *et al.*, 1986) has matured to a powerful tool capable to delineate whole chromosomes and subregions in nuclei, which usually lack chromosome-specific landmarks accessible to conventional staining methods. The possibility to simultaneously label several targets in interphase nuclei has led to widespread use of this technique in diagnostic settings and investigations on three-dimensional (3D) genome and chromosome organization during the cell cycle. Somatic interphase nuclei undergo architectural changes during development and differentiation. The most striking alterations of chromosome architecture and distribution occur during the extended prophase of the first of the two meiotic divisions. This article will focus on the contribution of interphase cytogenetics to the current view of chromosome organization and telomere dynamics during first meiotic prophase. It will be discussed that mammalian KU70 dsDNA end-binding protein may not play a structural role at the meiotic telomere. I will discuss a putative pathway linking the meiotic telomere to cell cycle progression, cytoskeleton and nuclear envelope that is based on recently disclosed two-hybrid interactions of the meiosis-specific telomere protein Ndj1p of budding yeast. Other aspects of meiotic chromosome biology are dealt with elsewhere (e.g., Loidl, 1990; Kleckner, 1996; Roeder, 1997; Moens *et al.*, 1998; Smith and Nicolas, 1998; Zickler and Kleckner, 1998; Villeneuve and Hillers, 2001; Cowan *et al.*, 2001).

Historical

Over a century ago chromosomes were identified to undergo cycles of condensation and decondensation in the course of the cell cycles that ultimately give rise to tissues,

M. Schmid and I. Nanda (eds.), Chromosomes Today, Volume 14, 127–147.

organs and individuals. Initially, attempts to identify chromosomes in the interphase nucleus and to understand their fate during the cell cycle failed, due to the absence of sensitive staining methods that could discriminate sub-cellular structures and simultaneously prevent fixation-induced artefacts. Improvements of staining methods revealed chromosome structures in cells and nuclei of species with favourable cytology like Salamanders, cartilaginous fish and parasites, which often display huge genomes and few chromosomes per complement (see Wilson, 1925; Paweletz, 2000). The early cytological findings led to the concept that the chromosome number had to be reduced prior to syngamy in order to maintain a species-specific genome size and chromosome number (*Reduktionstheilung* (reduction division); Weismann, 1887) thereby compensating for the genome doubling at meiosis. At the turn of the nineteenth century chromosomes were recognized as the carriers of genetic information (Boveri, 1903; Sutton, 1903), a discovery that further boosted research in chromosome behaviour and structure (for review see Wilson, 1925; Darlington, 1931).

Research into the problem of the reduction of chromosome number during gametogenesis or spore formation showed that a protracted premeiotic S-phase is followed by the contraction of chromosome inside the meiotic prophase nucleus (which is basically equivalent to G2). This process renders them visible as fine intertwined threads (leptonema; von Winniwarter, 1901). Three-dimensional EM reconstruction of meiocytes from species of different kingdoms revealed the formation of thread-like axial elements along replicated chromatids, which attach their ends (telomeres) to the nucleoplasmic side of the nuclear envelope (von Wettstein *et al.*, 1984). Once attached, chromosome ends move in the prophase I nucleus to crowd in a limited area of the nuclear envelope (Bouquet stage; Scherthan, 2001; Yamamoto and Hiraoka, 2001).

During the bouquet stage homologous chromosomes initiate synaptic pairing (zygonema) by assembly of transverse filaments between aligned axial elements (which are then called lateral elements) (von Wettstein *et al.*, 1984; Zickler and Kleckner, 1998). Antisera to proteins of the axial cores of meiotic chromosomes like SCP3/Cor1 of mammals (Dobson *et al.*, 1994; Lammers *et al.*, 1994), Hop1p and Red1 of yeast (Smith and Roeder, 1997), Him3 of *Caenorhabditis elegans* (Zetka *et al.*, 1999) and/or to transverse filament proteins of the synaptonemal complex (SC) like SCP1/Syn1 of mammals (Meuwissen *et al.*, 1992; Dobson *et al.*, 1994) or Zip1 of yeast; (Sym *et al.*, 1993) allow to rapidly monitor axial element formation and synaptic pairing in meiocytes (see, Roeder, 1997; Ashley and Plug, 1998). The latter is detected by immunofluorescence (IF) as thick fluorescent threads which are diagnostic of the SC (Moses, 1956; Fawcett, 1956) that connects all homologue pairs along their length at pachynema (von Wettstein *et al.*, 1984; Heyting, 1996). Comparison of the early images with immunostained meiocytes suggests that the chromosomal structures depicted by the early cytologists most likely represented axial cores and SCs (Figure 1).

Chromosome architecture of meiotic chromosomes

Since the days of Boveri, chromosome research has seen dramatic technical advances in interphase chromosome staining (Lichter *et al.*, 1991; Trask, 1991) that have disclosed a substantial difference between the architecture of mitotic and meiotic

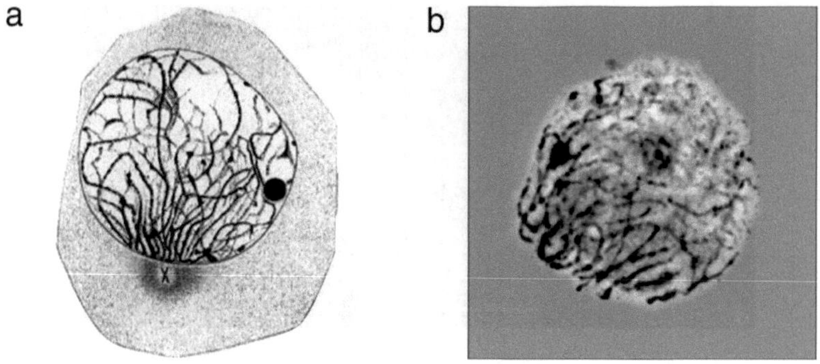

Figure 1. *(a) Zygotene bouquet of* Myxine glutinosa *as revealed by heavy metal staining. Note the centrosome at the bouquet basis (reproduced from Schreiner and Schreiner, 1905). (b) Colour inverted grey scale image of a late leptotene bouquet in a human spermatocyte. Axial cores are marked by immunofluorescence with anti-SCP3 antibodies (black threads). Nuclear DNA (DAPI) is displayed as a light grey stain underlying the nuclear outline.*

interphase chromosomes. In nuclei of somatic cells, chromosomes have been found to occupy discrete and rather compact territories of variable size and surface structure (for recent reviews see e.g., Chevret *et al.*, 2000; Cremer and Cremer, 2001). Upon entry into first meiotic prophase chromosomes develop longitudinal axes, along which the DNA attaches in loops (see, Moens *et al.*, 1998) ultimately leading to formation of extended chromosome territories which may facilitate homologue pairing (e.g., Schwarzacher, 1997; Scherthan *et al.*, 1998; Bass *et al.*, 2000; Stack and Anderson, 2001). Many advances in understanding meiotic chromosome structure have been obtained on nuclei spread on glass slides (e.g., Counce and Meyer, 1973; Heng *et al.*, 1996; Ashley and Plug, 1998). This is because the analysis by 3D EM reconstruction or optical sectioning is demanding and time-consuming. It has been noted that 3D nuclear architecture can be maintained to some (sometimes considerable) extend in spreads or swollen nuclei (e.g., Trelles-Sticken *et al.*, 1999; Croft *et al.*, 1999). However, it should be noted that one has to be cautious when extending these observations to other species. For instance the telomere clustering of the bouquet arrangement of budding yeast is maintained both in undisrupted nuclei and in spreads with the bouquet base being wider in the latter (Trelles-Sticken *et al.*, 1999). However, spreading disintegrates telomere clustering in bouquet nuclei of the mouse, while this 3D architecture is clearly seen in undisrupted nuclei from the same sample (Figure 2).

Sister chromatid cohesion and meiotic chromosome architecture

Once a cell is committed to undergo meiosis it enters a prolonged pre-meiotic phase of DNA replication (Monesi, 1962; Holm, 1977; Cha *et al.*, 2000) during which meiosis-specific functions are laid down, which subsequently contribute to chromosome pairing and recombination (Burgess *et al.*, 1999; Cha *et al.*, 2000; Borde *et al.*, 2000). After the extended premeiotic S-phase, sister chromatids of replicated chromosomes

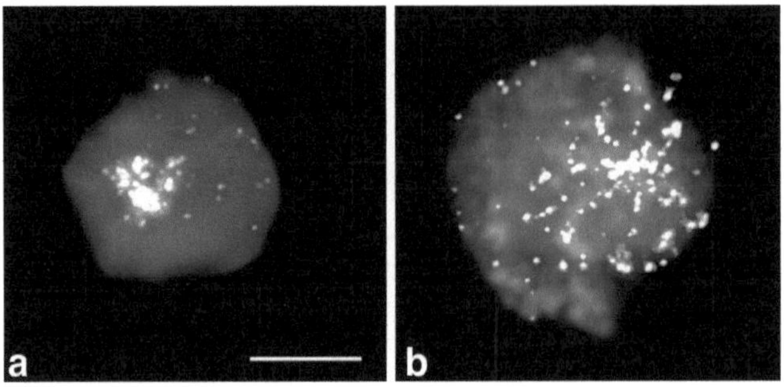

Figure 2. *Bouquet nuclei of the* Atm$^{-/-}$ *mouse (see Pandita* et al.*, 1999). (a) A structurally preserved nucleus of a testes suspension displays most telomere signals (white) clustered at the top of the nucleus. (b) A bouquet nucleus after detergent spreading that exhibits telomere signals dispersed over most of the nuclear area. Fluorescence microscopy, grey scale images; the bar in a represents 10 μm.*

are connected by a multi-subunit protein complex (cohesin) that contains the structural maintenance of chromosome proteins Smc1, Smc3, Scc3 and the meiosis-specific cohesin Rec8 (which largely replaces mitotic Scc1/Mcd1/Rad21) (for review see, van Heemst and Heyting, 2000; Nasmyth, 2001). Cohesion is important for homologue segregation as well as aspects of recombination and double stranded DNA breaks (DSB) repair (Strunnikov and Jessberger, 1999; van Heemst and Heyting, 2000; Lee and Orr-Weaver, 2001).

Orthologues to budding yeast Rec8p (Klein *et al.*, 1999) have been isolated in *Schizosaccharomyces pombe* (Watanabe and Nurse, 1999), *Arabidopsis thaliana* (Bhatt *et al.*, 1999), *C. elegans* (Pasierbek *et al.*, 2001) and man – the latter being strongly expressed in the testes (Parisi *et al.*, 1999). The mammalian meiotic cohesin complex contains Smc1 and 3 (Eijpe *et al.*, 2000) and STAG3 that interacts with the former two (Pezzi *et al.*, 2000). Smc1 has been found to interact with the Scp2 and Scp3 AE proteins (Eijpe *et al.*, 2000). At the light microscope level, cohesins form filaments along meiotic prophase chromosomes, which resemble the axial cores (Eijpe *et al.*, 2000). The meiotic cohesin complex may contribute to axial element morphogenesis, since it occurs in spermatocytes of mice that lack the SCP3 axial element protein (Pelttari *et al.*, 2001) and lateral elements are missing in Rec8 mutants (Klein *et al.*, 1999; Pasierbeck *et al.*, 2001). During meiosis I and II divisions of yeast Rec8-mediated cohesion is lost in two steps (see Klein *et al.*, 1999; Buonomo *et al.*, 2000), while mammalian STAG3 is only present up to MI (Prieto *et al.*, 2001). Recent analysis of SCP3 knockout mice (Yuan *et al.*, 2000) revealed SCP3-independent cohesin axes (Pelttari *et al.*, 2001), which suggests that the SCP3 axial core protein (Lammers *et al.*, 1994) might not be required for sister chromatid cohesion (Pelttari *et al.*, 2001).

At the FISH level, sister chromatid cohesion in mammalian nuclei is reflected by the presence of two closely spaced signal doublets, which appear after replication (Selig *et al.*, 1992; Scherthan *et al.*, 1996). In mitotic yeast cells, strong cohesion leaves signal doublets detectable only in a fraction of nuclei (Guacci *et al.*, 1994), which is also noted after premeiotic S-phase (Scherthan *et al.*, 1992; Weiner and Kleckner, 1994). As studied by chromosome painting, leptotene chromosomes are seen as separated extended territories in the meiocyte (Figure 3) (e.g., Goldman and Hultén 1992; Scherthan *et al.*, 1992, 1998; Cheng and Gartler, 1994; Bass *et al.*, 2000). Chromosome painting to three-dimensionally preserved human spermatocytes I has shown that chromosomes 3 are remote from each other and display separate sister-chromatids around the centromeric region (Figure 3c).

Sister-centromeres were tightly associated creating one signal, as were the terminal parts of the leptotene chromosome 3 arms (Scherthan *et al.*, 1998). Tight centromere cohesion may relate to the properties of heterochromatin (e.g., Warburton and Cooke,

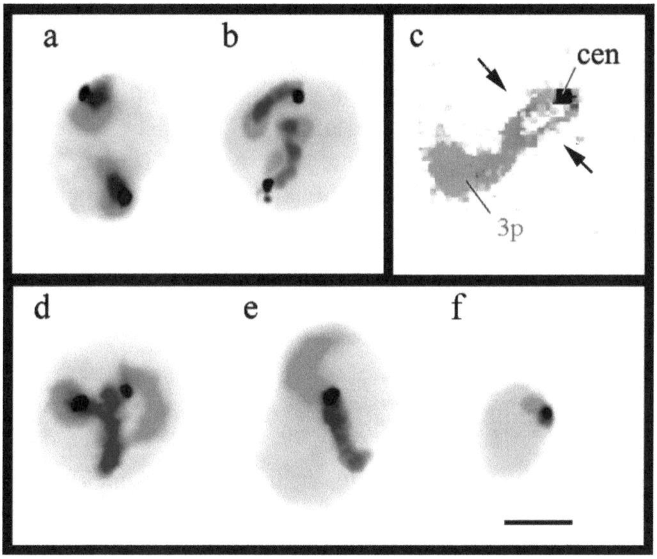

Figure 3. *Changes of chromosome morphology and positioning during first meiotic prophase as revealed by three-colour chromosome painting of human chromosomes 3 in testes tissue section nuclei (inverted grey scale images of colour pictures). (a) Nucleus of a spermatogonium that exhibits separated chromosome 3 territories (pericentromeres: black; 3p: dark grey; 3q: mediocre grey; DAPI: light grey, which gives the nuclear outline; see Scherthan* et al., *1998 for labelling scheme and colour images). (b) Leptotene nucleus displays a pair of elongated but separated leptotene chromosomes with their arms folded back onto each other (projection of the nucleus). (c) Optical section obtained with a confocal laser scanning microscope through a leptotene nucleus which shows separated sister chromatid signal tracks (arrows) between the centromere region (cen) and the terminal part of 3p. (d) Zygotene nucleus that shows bent separate q-arms and centromeres while both 3p arms are paired distally (one signal track). (e) Pachytene nucleus that displays a single signal track which represents the chromosome 3 bivalent that has the chromosome ends at the nuclear periphery (focal plane at the nuclear equator). (f) Haploid sperm nucleus displays a single compacted chromosome 3 territory. Bar represents 5 μm; a, b, d–f, digital fluorescence microscopy.*

1997) and a high concentration of cohesin binding sites (Blat and Kleckner, 1999; Megee *et al.*, 1999; Tanaka *et al.*, 1999). These observations suggest that the median region of leptotene chromosome-arms display a different quality of cohesion than the centromeric and telomeric arm regions. Lacking cohesion at internal sites of human leptotene chromosomes may relate to the absence of axial elements as seen by serial section EM analysis (Boiko, 1983). This suggests that axial element proteins like SCP3 or COR1 are required to reinforce the meiotic cohesin axes (Meuwissen *et al.*, 1992; Dobson *et al.*, 1994) to form relatively rigid leptotene/zygotene chromosome cores. Since leptotene chromosomes carry DSBs (Mahadeviah *et al.*, 2001; and references therein), such rigid axial cores, besides other aspects (see above), may support the movements of telomeres and chromosomes in the crowed prophase I nucleus. Cohesion at yeast meiotic telomeres is mediated by the *Saccharomyces cerevisiae* Ndj1/Tam1p telomere protein (Chua and Roeder, 1997; Conrad *et al.*, 1997).

In somatic cells, chromosome territories are relatively compact and their positioning relatively fixed (Marshall *et al.*, 1997; Edelman *et al.*, 2001), which may represent a structural means to direct repair of DSBs among sister chromatids. Meiotic chromosome territories, on the other hand, are elongated (Figure 3) and have their chromatin loops attached to well developed axial cores (see, Moens *et al.*, 1998), a feature which may facilitate extensive interactions (and recombination) with their neighbouring kins. The latter likely contribute to efficient homology search and alignment during the chromosome/telomere movements that lead to tight clustering during the bouquet stage (Scherthan *et al.*, 1998).

Chromosome pairing and recombination

In most sexually reproducing eukaryotes a high level of meiotic recombination provides for physical connections between homologous chromosomes and co-orientation of sister kinetochores. This ensures that homologues are properly oriented in the MI spindle and can segregate from each other during anaphase I (Hawley, 1988; Lee and Orr-Weaver, 2001). In some specialized cases, mechanisms have evolved to compensate for the absence of the SC or recombination. Where recombination is lacking, like in achiasmate female meiosis of the *Drosophila* and the *Bombyx mori* female, heterochromatin or a specialized SC has been recruited to mediate homologue association, respectively (Dernburg *et al.*, 1996; Rasmussen, 1977; Walker and Hawley, 2000). Recently it has been found that meiosis-specific heterochromatin associations are established *de novo* and this requires the previous resolution of non-homologous pericentromeric heterochromatin associations (Martinez-Perez *et al.*, 2001; Peters *et al.*, 2001). In recombination (DSB)-deficient mutants of *C. elegans* and *Drosophila* SCs still form to connect homologues (Dernburg *et al.*, 1998; McKim *et al.*, 1998). The fission yeast *S. pombe*, on the other hand, lacks SCs (and interference) but is highly recombination proficient (Kohli and Bahler, 1994). Homologue contacts seem to be stabilized by specialized chromosome regions (pairing centres) that promote SC formation in *C. elegans* (Albertson *et al.*, 1997).

In contrast to higher eukaryotes (see above) synapsis in budding yeast is dependent on the initiation of recombination (Chua and Roeder, 1998; Zickler and Kleckner, 1999). Budding yeast, which forms SCs between its 16 homologue pairs exhibits high levels of meiotic recombination, which exceed that of spontaneous events by a factor of up to 10^4 (Keeney, 2001). Recombination initiates by programmed occurrence of DSBs during leptotene (Mahadevaiah *et al.*, 2001; Hunter *et al.*, 2001) which are generated by the transesterase Spo11 (Bergerat *et al.*, 1997; Keeney *et al.*, 1997) that is highly conserved among eukaryotes (for review see Keeney, 2001). The 5′ ends of DSBs are then resected by the RAD50/MRE11/XRS2 complex (XRS2 is replaced by Nbs1 in humans) to produce long 3′ overhangs that are bound by the RecA orthologue RAD51 and its meiosis-specific relative DMC1 (see, Pâques and Haber, 1999; Masson and West, 2001). Single strands invade a chromatid on the homologue (Hunter and Kleckner, 2001) and are further processed and ligated to result in reciprocal crossovers at pachytene (Schwacha and Kleckner, 1997; Pâques and Haber, 1999). Chromosome pairing which initiates *de novo* after premeiotic S-phase represents an immediate prerequisite for this process.

Bouquet formation and homologue pairing

Homologous chromosome pairing has been observed to commence in the bouquet stage (Gelei, 1921; Scherthan *et al.*, 1996; Chikashige *et al.*, 1997; Bass *et al.*, 2000). Budding yeast strains which lack the meiotic telomer protein Ndj1p display a significant delay in homologue pairing (Chua and Roeder, 1997; Trelles-Sticken *et al.*, 2000), with physically smaller chromosomes suffering particularly from the absence of bouquet formation (Chua and Roeder, 1997; E. Trelles-Sticken and Harry Scherthan, unpublished data), presumably since they will have less of an opportunity for chance encounters with their homologues. Thus, the phenotype of this bouquet-defective mutant supports the hypothesis that the bouquet topology serves a catalytic activity that is to promote the physical proximity and alignment of pairing-competent leptotene chromosomes (Loidl, 1990; Yamamoto and Hiraoka, 1998; Scherthan, 2001). Yeast has a "cDNA" genome that is largely devoid of repeats, which should facilitate homology recognition. Hence, it can be anticipated that the catalytic action of meiotic telomere clustering will be more required in species with larger chromosome numbers and more repetitive DNA or in the absence of SCs. In line with this assumption, it has been observed that the bouquet stage is more abundant in cattle ($2n = 60$) than in mouse ($2n = 40$) spermatogenesis – i.e., in the species with more chromosome pairs and otherwise similar genome sizes and similar chromosome morphology (Pfeifer *et al.*, 2001).

Recombination and bouquet formation

Ongoing research into the bouquet stage of meiosis revealed that the Rad51 strand-exchange protein assembles in foci between homologous chromosomes during the bouquet stage at leptotene/zygotene when homologue pairing commences (Franklin *et al.*, 1999; E. Trelles-Sticken and Harry Scherthan, unpublished observations). Meiotic telomere clustering occurs during asynaptic meiosis of fission yeast

(Chikashige *et al.*, 1994), achiasmate (non-recombinant) female meiosis of the dipteran *B. mori* (which is a dipteran like *Drosophila*) (Rasmussen, 1977), as well as in recombination-deficient mutants of the yeast *S. cerevisiae* (Trelles-Sticken *et al.*, 1999) and asynaptic mutants of rye (Mikhailova *et al.*, 2001). This suggests that bouquet formation is highly conserved and independent of recombination *per se*. A prevalence of bouquet spermatocytes in recombination mutants of mouse and yeast has been noted (Trelles-Sticken *et al.*, 1999; Pandita *et al.*, 1999).

It seems likely that progress of recombination may be integrated with telomere behaviour, since disruption of a core component of the DNA damage signalling pathway, ATM (mutated in individuals with the disorder Ataxia telangiectasia; see Kastan and Lim, 2000) leads to a dramatic accumulation of bouquet nuclei in ATM-deficient mouse meiosis (Pandita *et al.*, 1999; Scherthan *et al.*, 2000a). Bouquet stage is short lived in male wild-type mouse meiosis (Scherthan *et al.*, 1996). The recent discovery that the telomere repeat-binding protein TRF1 (Chong *et al.*, 1995) is a substrate for the ATM kinase (Kishi *et al.*, 2001) provides for a direct link between telomeres and ATM. Moreover, ATM has been shown to phosphorylate p95/Nbs1 in response to DNA damage (Lim *et al.*, 2000). Nbs1 is a component of the Rad50/Mre11/Nbs1 DNA repair complex (Rad50/Mre11/Xrs2 in yeast, see Pâques and Haber, 1999). Furthermore, the second mammalian telomere repeat binding protein TRF2 (Bilaud *et al.*, 1997; Broccoli *et al.*, 1997) has been found to directly interact with Mre11, possibly functioning in T-loop formation at the telomere (Zhu *et al.*, 2000). Nbs1, Mre11, TRF1 and -2 localize to meiotic telomeres (Lombard and Guarente, 2000; Scherthan *et al.*, 2000b, Harry Scherthan and J. Petrini, unpublished data). The approx. 20-fold enrichment of bouquet stage nuclei in ATM-deficient meiosis (Scherthan *et al.*, 2000a) thus may be the result of abrogated crosstalk between meiotic telomeres and the recombination pathway which is under checkpoint control (Roeder and Bailis, 2000). In yeast Tel1 and Mec1, which are ATM related DNA-damage checkpoint kinases, have been found to be required for telomere replication and meiosis (Naito *et al.*, 1998; Ritchie *et al.*, 1999) and the pachytene checkpoint (Roeder and Bailis, 2000).

A telomere/recombination and/or synapsis crosstalk could be involved in controlling telomere movements in response to signal(s) from homologue pairing and/or progress of recombination. The p53 pathway downstream of ATM, however, seems not to be involved in signalling to meiotic telomeres, because p53-deficient mice show normal bouquet frequencies (Scherthan *et al.*, 2000a). Interestingly, the checkpoint kinase chk2 of *C. elegans* has recently been found to be required for nuclear reorganization and possibly chromosome polarization during meiotic prophase (MacQueen and Villeneuve, 2001). Chk2 of mammals is expressed in the testes (Bartkova *et al.*, 2001) and its related kinase Chk1 has been localized to meiotic chromosomes (Flaggs *et al.*, 1997). It will thus be interesting to investigate bouquet formation in mammalian meiosis compromised in further downstream targets of ATM.

Telomere/cytoskeleton crosstalk?

Pairing of meiotic telomeres of *S. cerevisiae* requires the Ndj1/Tam1 telomere protein that is specifically expressed during meiotic prophase irrespective whether or not recombination is initiated (Chua and Roeder, 1997; Conrad *et al.*, 1997). Diploid *ndj1Δ* strains induced to undergo meiosis show absence of interference (random distribution of crossovers), homologue non-disjunction and a precocious separation of sister chromatids (Chua and Roeder, 1997), particularly at telomeres (Conrad *et al.*, 1997). Genetic analysis suggests a defect in meiotic telomere redistribution (Rockmill and Roeder, 1998). Telomere movements during leptotene/zygotene stage (Chikashige *et al.*, 1994; E. Trelles-Sticken, U. Nehrbass and H.S. unpublished) lead to the transient accumulation of telomeres in a limited region of the nuclear envelope near the microtubule organizing centre (Figure 4) (MTOC in animals and algae, spindle pole body (SPB) in fungi), which suggests a role of the cytoskeleton in telomere movements (Gelei, 1921; Loidl, 1990).

Recently it has been noted that certain microtubule-disrupting drugs abrogate telomere clustering in rye (Franklin and Cande, 1999). Disruption of SPB components (Shimanuki *et al.*, 1997) or cytoplasmic dynein has been shown to impair bouquet formation and nuclear/chromosome motility in the asynaptic meiosis of the fission yeast *S. pombe* (Yamamoto *et al.*, 1999; Niwa *et al.*, 2000). Most interestingly, the human telomere protein TRF1/Pin2 was recently found to directly interact with MTs of the metaphase spindle and to polymerize MTs *in vitro*, which provides for a link between the cytoskeleton, chromosome segregation and telomeres (Nakamura *et al.*, 2001). Since telomere movements at meiosis occur within the intact nuclear membrane, it will be interesting to determine whether and how a TRF1 and MTs interact in meiotic cells and whether this may play a role in bouquet formation.

In *S. cerevisiae*, deletion of the microtubule-associated motor Kar3p (Meluh and Rose, 1993) has been shown to impair recombination and synapsis (Bascom-Slack and Dawson, 1996). An unpublished analysis of a Ndj1-HA expressing haploid *kar3Δ* yeast strain revealed meiotic telomere clustering in the absence of Kar3p (E. Trelles-Sticken, J. Loidl and Harry Scherthan, unpublished data). This suggests that the

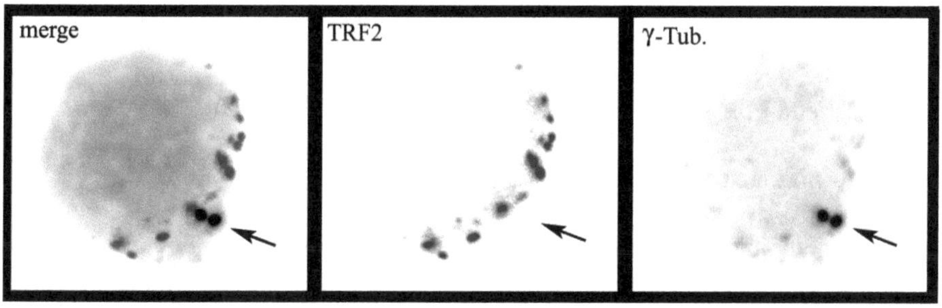

Figure 4. *Human bouquet leptotene spermatocyte from a human testis suspension revealing the spatial relationship of the duplicated centrosome (black dots, arrow, as stained with mouse anti-gamma-Tubulin, γ-Tub.) and TRF2-tagged human leptotene telomeres (TRF2; medium grey in the merge). The nuclear outline is given by DAPI staining in the merge (light grey). Chromosome ends, which are marked by TRF2 signals, group close to the cytoplasmic centrosome at a sector of the nuclear periphery. The centrosome (two black dots in the γ-Tub. channel; arrowed) is situated at the bouquet basis.*

primary meiotic defect in *kar3Δ* strains lies in premeiotic events (Keeney 2001; Scherthan, 2001). This would be consistent with the observation that nocodazole treatment only partially inhibits meiotic telomere clustering (E. Trelles-Sticken, J. Loidl and Harry Scherthan, unpublished data).

Telomere attachment

FISH and IF analysis has shown that the ends of the axial cores of the SC harbour telomeric DNA repeats and associated proteins (Moens and Pearlman, 1990; Klein *et al.*, 1992; Scherthan *et al.*, 2000b; Trelles-Sticken *et al.*, 2000). While the integrity of the very terminal telomere repeats are important for meiosis in *K. lactis* (Maddar *et al.*, 2001), complete loss of telomeric DNA repeats has been found to be incompatible with meiosis in *S. pombe* (Naito *et al.*, 1998). Telomere repeats of *S. pombe* are bound by an orthologue of the human TRF1 protein: Taz1 (Cooper *et al.*, 1997) which is required for telomere clustering in meiotic prophase (Cooper *et al.*, 1998; Nimmo *et al.*, 1998). Another telomere protein, *S. pombe sp*Rap1, binds to telomeres by interaction with Taz1 (Chikashige and Hiraoka, 2001; Kanoh and Ishikawa, 2001). This contrasts with the situation in *S. cerevisae* where scRap1 (Shore and Nasmyth, 1987) binds directly to G-rich repeats via its Myb domain (Longtine *et al.*, 1989). In mammals, telomere repeats are bound by the orthologues of *S. pombe* Taz1, TRF1 and TRF2 (see Li *et al.*, 2000; Scherthan *et al.*, 2000b) with the hRap1 protein interacting with telomere repeats through TRF2 (Li *et al.*, 2000). In *S. pombe*, the orthologous spRap1 protein has been found to link telomeres to the SPB (which resides in the nuclear envelope) via its interaction with Taz1 as the repeat binding factor (Chikashige and Hiraoka 2001). This supports the hypothesis that binding of mammalian meiotic telomeres to the nuclear envelope could similarly involve hRap1/TRF2. Notably, a telomere repeat-binding 70 kDa protein that is related to mammalian TRF2 was isolated from the meiotic nuclear envelope of frog oocytes (Podgornaya *et al.*, 2000).

In *S. cerevisae* meiotic telomeres harbour *sc*Rap1p (Klein *et al.*, 1992) and Ndj1p that seems to be required for perinuclear telomere localization and bouquet formation (Rockmill and Roeder, 1998; Trelles-Sticken *et al.*, 2000). Thus, it is tempting to speculate that Ndj1 has been recruited to mediate a function in *S. cerevisiae* prophase I, which is carried out by Rap1 in meiosis of *S. pombe* and mammals.

Ku may not play a structural role at mammalian meiotic telomeres

Telomere maintenance and perinuclear telomere positioning in *S. cerevisiae* involves Hdf1/Hdf2 (Ku70/Ku80) (see Gasser, 2000) which are abundant nuclear DSB-binding proteins important for non-homologous end joining (NHEJ) pathway of double strand break repair (Critchlow and Jackson, 1998; Pâques and Haber, 1999). Ku70 has been detected at somatic mammalian telomeres (Hsu *et al.*, 1999) and is required for protection of chromosome ends from fusion (Bailey *et al.*, 1999). However, it seems unlikely that tethering of meiotic telomeres to the nuclear envelope involves mammalian Ku proteins, because these are down-regulated during the leptotene/zygotene stages of mammalian meiosis (Goedecke *et al.*, 1999). In agreement with this report,

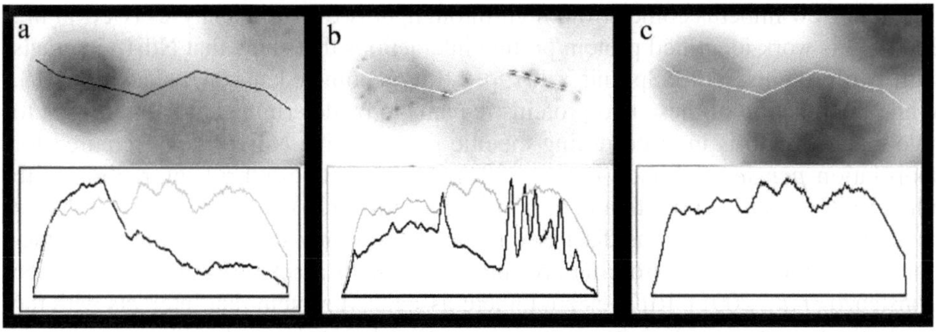

Figure 5. *Analysis of colocalization of fluorescent signals of (a) Ku70 (stained with mouse anti-hKu70 antibodies (BABCO)) and (b) telomeric proteins (rabbit anti-TRF1, FITC) and nuclear DNA (c; DAPI) in human testes nuclei. Fluorescence profile analysis along polygon lines drawn across telomeric signals (dark dots in (b)) and adjacent chromatin (grey in c; DAPI) in colour inverted raw digital images. In the case of colocalization of proteins there are signal peaks at corresponding locations in the red and green channel. Ku is abundant throughout the chromatin of the premeiotic nucleus to the upper left, while the neighbouring spermatocyte shows a dearth of Ku signal as indicated in the intensity profile given below (black line: Ku intensity, grey line DAPI intensity (nuclear DNA)). This agrees with previous work (Goedecke* et al., *1999). (b) Telomere fluorescence creates distinct peaks in the green channel (black line in the intensity graph, grey line: DAPI intensity). Signal peaks which match the TRF1 telomere signals in (b) are absent in the Ku channel, which argues against a structural role of human Ku (e.g., at the attachment plaque) at the meiotic telomere. DAPI staining in (b) shows the presence of chromatin at all areas analysed. Colour inverted grey scale images (for details see Scherthan* et al., *2000b).*

a colocalization study with TRF1 in structurally preserved meiocytes (Harry Scherthan and M. Jerratsch, unpublished) failed to detect Ku70 at TRF1-tagged meiotic leptotene/zygotene telomeres (Figure 5). Other proteins required for the peripheral localization of telomeres in vegetative budding yeast cells are the myosin like proteins Mlp1 and Mlp2 (Galy *et al.*, 2000) that extend from the nuclear pore (Strambio-de-Castillia *et al.*, 1999). This could suggest a role of nuclear pores in meiotic telomere attachment. However, the clustering of nuclear pores (Fawcett and Chemes, 1979) and telomeres during human meiotic prophase I occurs in largely separate sectors of the nuclear envelope (Scherthan *et al.*, 2000b), which argues against a direct role of nuclear pores and associated proteins in attachment of meiotic telomeres. It cannot be excluded, however, that a subset of nuclear pore components or proteins is recruited to serve meiosis-specific functions.

Interconnecting the meiotic telomere

It seems likely that Ndj1p of *S. cerevisiae* is involved in tethering and/or moving meiotic telomeres to and along the nuclear envelope (Rockmill and Roeder, 1998; Trelles-Sticken *et al.*, 2000). However, it is unclear at present how Ndj1p mediates this function. New impetus comes from high-throughput two-hybrid analyses (Uetz *et al.*, 2000). This work identified protein/protein interactions that show that Ndj1p interacts with two proteins, Ydr383p and Yjl019p (http://genome.c.kanazawa-u.ac.jp/Y2H/). While the function of these two proteins remains to be determined, it is interesting to note that Ydr383p is induced during meiotic growth. Like Ndj1p is not required for sporulation *per se* (Rabitsch *et al.*, 2001). It shows two-hybrid interaction with Ylr315p (Uetz *et al.*, 2000) that in turn interacts with Rif2p (Ito *et al.*, 2001). It should be noted that the interactions discussed below delimit only selected string of interactions since most of the proteins are interacting with a set of other proteins (see Constanzo *et al.*, 2001). Rif2p is a telomere-associated protein that interacts with the yeast telomeric repeat DNA binding protein scRap1 (Hardy *et al.*, 1992, see above). Besides this (more structural?) interaction route which links Ndj1p to TG_{1-3} telomere repeats, Ydr383p also interacts with Srp1p, a multifaceted protein that interacts with Nup1p and Nup2p at nuclear pores and that is required for nuclear import/export, nuclear and nucleolar structure and organization of microtubules (Yano *et al.*, 1994; Kussel and Frasch, 1995). This sting of interactions could indicate a link between the meiotic telomere, the nuclear envelope and cytoskeleton.

The second Ndj1-interacting protein, Yjl019p, was found to interact with Est1p in a two-hybrid assay (Ito *et al.*, 2001). Est1 represents a component of the telomerase complex and is required for functional telomerase (e.g., Diede and Gottschling, 1999; Evans and Lundblad, 1999) that synthesizes telomeric DNA repeats (see Shore, 2001). Most interestingly, mammalian spermatogonia (meiotic stem cells) that carry repeat-depleted telomeres (which occur in late-generation telomerase knockout mice (Blasco *et al.*, 1997)) are prevented from entering meiotic prophase (Hemann *et al.*, 2001).

Est1p also interacts with Cdc13p which represents a conserved telomeric single strand binding protein (de Lange, 2001), the absence of which interferes with somatic homologue association (Burgess *et al.*, 1999) due to defects in telomere replication (Garvik *et al.*, 1995). Furthermore, Yjl019p among others interacts with Cdc5p, a multi facetted kinase that is involved in control of DNA replication (Hardy and Pautz, 1996) and G2/M transit (Shirayama *et al.*, 1998). Cdc5 is required for completion of recombination and both meiotic divisions (Simchen *et al.*, 1981; Sharon and Simchen, 1990). Most interestingly, Cdc5 localizes to the spindle pole body and is involved in signalling to the cytoskeleton during cytokinesis (Luca *et al.*, 2001; Song *et al.*, 2000). In mammalian cells the polo-like kinase Cdc5 is involved in controlling G2/M progression in response to DNA damage (Smits *et al.*, 2000). Thus, this indicates a regulatory link between meiotic telomeres, recombination and cytoskeletal reorganization during meiosis.

Furthermore, there seems to be a connection between Ndj1p of the meiotic telomere and the recombination process, since a string of interactions (trough Yil019p and Htz1p) links Ndj1p with Mec3p. Mec3p is a checkpoint kinase of the Rad24 group recently reported to be involved in meiotic check point function (see Roeder and Bailis, 2000). Interestingly, the *C. elegans* homologue of the yeast Rad17 DNA-damage checkpoint protein, MRT2, is required for telomere integrity and germ line immortality (Ahmed and Hodgkin, 2000). Additionally, a link between telomeres and chromatin remodelling seems plausible, since recent research links telomeres with

Mec3p and Set1p which interact (Corda *et al.*, 1999). SET domain proteins of the trithorax family are involved in DNA damage response, maintenance of telomeres and chromatin remodelling (Jenuwein *et al.*, 1998). Interestingly, the disruption of the murine Suv39h1 and -2 SET domain histone methyltransferases impairs meiotic chromosome pairing and leads to spermatogenic failure (Peters *et al.*, 2001). In yeast Set1p is required for silencing at telomeres as is the pachytene checkpoint gene Dot1 (San-Segundo and Roeder, 2000). Dot1 interacts with Mec3p and are required for the pachytene checkpoint that monitors meiotic recombination (Roeder and Bailis, 2000). The protein interaction network emanating from Ndj1p provide a first view how a meiotic telomere could integrate and modulate information from cellular processes as diverse as the replication status of telomeric DNA, repeat length, prophase I progression, nuclear pores and cytoskeleton dynamics. This speculative network around the meiotic telomere now clearly demands experimental investigation to work out the important players and their connections in the crowded bouquet act.

Acknowledgements

I am grateful to E. Trelles-Sticken and B. Liebe for comments on this MS, and to E.T-S. and M. Jerratsch for communication of data prior to publication, and C. Heyting and T. de Lange for antibodies. I apologize to those whose work could only be cited through reviews. This work was supported by the DFG (SCHE 350/8.3).

References

Ahmed S, Hodgkin J (2000). MRT-2 checkpoint protein is required for germline immortality and telomere replication in *C. elegans. Nature* 403: 159–164.

Albertson DG, Rose AM, Villeneuve AM (1997). In: Riddle DL, Blumenthal T, Meyer BJ and Priess JR eds. "*C. elegans* II". Plainview, NY: Cold Spring Harbour Laboratory Press, pp. 47–78.

Albini SM, Jones GH (1988). Synaptonemal complex spreading in *Allium cepa* and *A. fistulosum*. II. Pachytene observations. The SC karyotype and correspondence of late recombination nodules and chiasma. *Chromosoma* 95: 324–338.

Anderson LK, Offenberg HH, Verkuilen WM, Heyting C (1997). RecA-like proteins are components of early meioitc nodules in Lilly. *Proc Natl Acad Sci USA* 94: 6868–6873.

Ashley T, Plug A (1998). Caught in the act: deducing meiotic function from protein immunolocalization. *Curr Top Dev Biol* 37: 201–239.

Bailey SM, Meyne J, Chen DJ, *et al.* (1999). DNA double-strand break repair proteins are required to cap the ends of mammalian chromosomes. *Proc Natl Acad Sci USA* 96: 14,899–14,904.

Bartkova J, Falck J, Rajpert-De Meyts E, *et al.* (2001). Chk2 tumour suppressor protein in human spermatogenesis and testicular germ-cell tumours. *Oncogene* 20: 5,897–5,902.

Bascom-Slack CA, Dawson DS (1997). The yeast motor protein, Kar3p, is essential for meiosis I. *J Cell Biol* 139: 459–467.

Bass HW, Riera-Lizarazu O, Ananiev EV, *et al.* (2000). Evidence for the coincident initiation of homolog pairing and synapsis during the telomere-clustering (bouquet) stage of meiotic prophase. *J Cell Sci* 113: 1,033–1,042.

Bergerat A, de Massy B, Gadelle D, *et al.* (1997). An atypical topoisomerase II from Archaea with implications for meiotic recombination. *Nature* 386: 414–417.

Bhatt, AM, Lister C, Page T, Fransz P, Findlay K, Jones GH, Dickinson HG, Dean C (1999). The *DIF1* gene of *Arabidopsis* is required for meiotic chromosome segregation and belongs to the *REC8/RAD21* cohesin gene family. *Plant J* 19: 463–472.
Bilaud T, Brun C, Ancelin K, *et al.* (1997). Telomeric localization of TRF2, a novel human telobox protein. *Nat Genet* 17: 236–239.
Bishop DK (1994). RecA homologs Dmc1 and Rad51 interact to form multiple nuclear complexes prior to meiotic chromosome synapsis. *Cell* 79: 1,081–1,092.
Blasco MA, Lee HW, Hande MP, *et al.* (1997). Telomere shortening and tumor formation by mouse cells lacking telomerase RNA. *Cell* 91: 25–34.
Blat Y, Kleckner N (1999). Cohesins bind to preferential sites along yeast chromosome III, with differential regulation along arms versus the centric region. *Cell* 98: 249–259.
Borde V, Goldman AS, Lichten M (2000). Direct coupling between meiotic DNA replication and recombination initiation. *Science* 290: 806–809.
Boveri T (1903). "Über die Konjugation der chromatischen Kernsubstanz". In: ed. Korschelt E, Verhandlungen der dt. Zoolog. Ges. 13 Jahresvers. zu Würzburg, Leipzig: W. Engelmann, pp.10–33.
Broccoli D, Smogorzewska A, Chong L, de Lange T (1997). Human telomeres contain two distinct Myb-related proteins, TRF1 and TRF2. *Nat Genet* 17: 231–235.
Buonomo SBC, Clyne RK, Fuchs J, Loidl J, Uhlmann F, Nasmyth K (2000). Disjunction of homologous chromosomes in meiosis I depends on proteolytic cleavage of the meiotic cohesin Rec8p by separin. *Cell* 103: 387–398.
Burgess SM, Kleckner N, Weiner BM (1999). Somatic pairing of homologs in budding yeast: existence and modulation. *Genes Dev* 13: 1,627–1,641.
Carpenter, A (1975). Electron microscopy of meiosis in *Drosophila melanogaster* females. *Chromosoma* 51: 157–182.
Carpenter AT (1994). Chiasma function. *Cell* 77: 957–962.
Cha RS, Weiner BM, Keeney S, Dekker J, Kleckner N (2000). Progression of meiotic DNA replication is modulated by interchromosomal interaction proteins, negatively by Spo11p and positively by Rec8p. *Genes Dev* 14: 493–503.
Cheng EY, Gartler SM (1994). A fluorescent *in situ* hybridization analysis of X chromosome pairing in human female meiosis. *Hum Genet* 94: 389–394.
Chevret E, Volpi EV, Sheer D (2000). Form and function in the human interphase chromosome. *Cytogenet Cell Genet* 90: 13–21.
Chikashige Y, Hiraoka Y (2001). Telomere binding of the Rap1 protein is required for meiosis in fission yeast. *Curr Biol* 11: 1,618–1,623.
Chikashige Y, Ding DQ, Funabiki H, *et al.* (1994). Telomere-led premeiotic chromosome movement in fission yeast. *Science* 264: 270–273.
Chikashige Y, Ding DQ, Imai Y, *et al.* (1997). Meiotic nuclear reorganization: switching the position of centromeres and telomeres in the fission yeast *Schizosaccharomyces pombe. EMBO J* 16: 193–202.
Chong L, van Steensel B, Broccoli D, *et al.* (1995). A human telomeric protein. *Science* 270: 1,663–1,667.
Chu S, Herskowitz I (1998). Gametogenesis in yeast is regulated by a transcriptional cascade dependent on Ndt80. *Mol. Cell* 1: 685–696.
Chua, PR, Roeder GS (1997). Tam1, a telomere-associated meiotic protein, functions in chromosome synapsis and crossover interference. *Genes Dev* 11: 1,786–1,800.
Chua PR, Roeder GS (1998). Zip2, a meiosis-specific protein required for the initiation of chromosome synapsis. *Cell* 93: 349–359.
Conrad MN, Dominguez AM, Dresser ME (1997). Ndj1p, a meiotic telomere protein required for normal chromosome synapsis and segregation in yeast. *Science* 276: 1,252–1,255.

Costanzo MC, Crawford ME, Hirschman, *et al.* (2001). YPD™, PombePD™, and WormPD™: model organism volumes of the BioKnowledge™ library, an integrated resource for protein information. *Nucleic Acids Res* 29: 75–79.
Cooper JP, Nimmo ER, Allshire RC, Cech TR (1997). Regulation of telomere length and function by a Myb-domain protein in fission yeast. *Nature* 385: 744–747.
Cooper JP, Watanabe Y, Nurse P (1998). Fission yeast Taz1 protein is required for meiotic telomere clustering and recombination. *Nature* 392: 828–831.
Corda Y, Schramke V, Longhese MP, *et al.* (1999). Interaction between Set1p and checkpoint protein Mec3p in DNA repair and telomere functions. *Nat Genet* 21: 204–208.
Counce SJ, Meyer GF (1973). Differentiation of the synaptonemal complex and the kinetochore in Locusta spermatocytes studied by whole mount electron microscopy. *Chromosoma* 44: 231–253.
Cowan CR, Carlton PM, Cande WZ (2001). The polar arrangement of telomeres in interphase and meiosis. rabl organization and the bouquet. *Plant Physiol* 125: 532–538.
Cremer T, Cremer C (2001). Chromosome territories, nuclear architecture and gene regulation in mammalian cells. *Nat Rev Genet* 2: 292–301.
Cremer T, Landegent J, Brückner A, *et al.* (1986). Detection of chromosome aberrations in the human interphase nucleus by visulization of specific target DNAs with radioactive and non radioactive *in situ* hybridization techniques: diagnosis of trisomy 18 with probe L1.84. *Hum Genet* 74: 346–352.
Critchlow SE, Jackson SP (1998). DNA end-joining: from yeast to man. *Trends Biochem Sci* 23: 394–398.
Croft JA, Bridger JM, Boyle S, *et al.* (1999). Differences in the localization and morphology of chromosomes in the human nucleus. *J Cell Biol* 145: 1,119–1,131.
Darlington CW (1931). Meiosis. *Biol Rev* 6: 221–264.
de Lange T (2001). Telomere capping – one strand fits all. *Science* 292: 1,075–1,076.
Dernburg AF, McDonald K, Moulder G, *et al.* (1998). Meiotic recombination in *C. elegans* initiates by a conserved mechanism and is dispensable for homologous chromosome synapsis. *Cell* 94: 387–398.
Dernburg AF, Sedat JW, Hawley RS (1996). Direct evidence of a role for heterochromatin in meiotic chromosome segregation. *Cell* 86: 135–146.
Diede SJ, Gottschling DE (1999). Telomerase-mediated telomere addition *in vivo* requires DNA primase and DNA polymerases alpha and delta. *Cell* 99: 723–733.
Dilworth DJ, Suprapto A, Padovan JC, Chait BT, Wozniak RW, Rout MP, Aitchison JD (2001). Nup2p dynamically associates with the distal regions of the yeast nuclear pore complex. *J Cell Biol* 153:1,465–1,478.
Dobson MJ, Pearlman RE, Karaiskakis A, Spyropoulos B, Moens PB (1994). Synaptonemal complex proteins: occurrence, epitope mapping and chromosome disjunction. *J Cell Sci* 107(Pt 10): 2,749–2,760.
Edelmann P, Bornfleth H, Zink D, Cremer T, Cremer C (2001). Morphology and dynamics of chromosome territories in living cells. *Biochim Biophys Acta* 1551: M29–M39.
Eijpe M, Heyting C, Gross B, Jessberger R (2000). Association of mammalian SMC1 and SMC3 proteins with meiotic chromosomes and synaptonemal complexes. *J Cell Sci* 113: 673–682.
Evans SK, Lundblad V (1999). Est1 and Cdc13 as comediators of telomerase access. *Science* 286: 117–120.
Fawcett DW (1956). The fine structure of chromosomes in the meiotic prophase of vertebrate spermatocytes. *J Biophys Biochem Cytol* 2: 403.
Fawcett DW, Chemes HE (1979). Changes in distribution of nuclear pores during differentiation of the male germ cells. *Tissue and Cell* 11: 147–162.

Flaggs G, Plug AW, Dunks KM, *et al.* (1997). Atm-dependent interactions of a mammalian chk1 homolog with meiotic chromosomes. *Curr Biol* 7: 977–986.

Franklin AE, Cande WZ (1999). Nuclear organization and chromosome segregation. *Plant Cell* 11: 523–534.

Franklin AE, McElver J, Sunjevaric I, Rothstein R, Bowen B, Cande WZ (1999). Three-dimensional microscopy of the Rad51 recombination protein during meiotic prophase. *Plant Cell* 11: 809–824.

Galy V, Olivo-Marin JC, Scherthan H, *et al.* (2000). Nuclear pore complexes in the organization of silent telomeric chromatin. *Nature* 403: 108–112.

Garvik B, Carson M, Hartwell L (1995). Single-stranded DNA arising at telomeres in cdc13 mutants may constitute a specific signal for the RAD9 checkpoint. *Mol Cell Biol* 15: 6,128–6,138.

Gasser SM (2000). A sense of the end. *Science* 288: 1,377–1,379.

Gelei J (1921). Weitere Studien über die Oogenese des *Dendrocoelum lacteum*. III. Die Konjugation der Chromosomen in der Literatur und meine Befunde. *Archiv f Zellforsch* 16: 300–365.

Goedecke W, Eijpe M, Offenberg HH, *et al.* (1999). Mre11 and Ku70 interact in somatic cells, but are differentially expressed in early meiosis. *Nat Genet* 23: 194–198.

Goldman AS, Hulten MA (1992). Chromosome *in situ* suppression hybridisation in human male meiosis. *J Med Genet* 29: 98–102.

Gotta M, Laroche T, Formenton A, *et al.* (1996). The clustering of telomeres and colocalization with Rap1, Sir3, and Sir4 proteins in wild-type *Saccharomyces cerevisiae. J Cell Biol* 134: 1,349–1,363.

Guacci V, Hogan E, Koshland D (1994). Chromosome condensation and sister chromatid pairing in budding yeast. *J Cell Biol* 125: 517–530.

Hardy CF, Pautz A (1996). A novel role for Cdc5p in DNA replication. *Mol Cell Biol* 16: 6,775–6,782.

Hardy CF, Sussel L, Shore D (1992). A RAP1-interacting protein involved in transcriptional silencing and telomere length regulation. *Genes Dev* 6: 801–814.

Hawley RS (1988). "Exchange and chromosomal segregation in eukaryotes". In: Kucherlapati R and Smith GR, eds. *Genetic Recombination.* Washington DC: American Society of Microbiology, pp. 497–527.

Hemann MT, Rudolph KL, Strong MA, *et al.* (2001). Telomere dysfunction triggers developmentally regulated germ cell apoptosis. *Mol Biol Cell* 12: 2,023–2,030.

Heng HH, Chamberlain JW, Shi XM, *et al.* (1996). Regulation of meiotic chromatin loop size by chromosomal position. *Proc Natl Acad Sci USA* 93: 2,795–2,800.

Heyting C (1996). Synaptonemal complexes: structure and function. *Curr Opin Cell Biol* 8: 389–396.

Holm PB (1977). Three-dimensional reconstruction of chromosome pairing during the zygotene stage of meiosis in *Lillium longiflorum* (Thunb.) *Carlsb Res Comm* 42: 103–151.

Hsu H-L, Gilley D, Blackburn EH, Chen DJ (1999). Ku is associated with the telomere in mammals. *Proc Natl Acad Sci USA* 96: 12,454–12,458.

Hunter N, Kleckner N (2001). The single-end invasion: an asymmetric intermediate at the double-strand break to double-holliday junction transition of meiotic recombination. *Cell* 106: 59–70.

Hunter N, Valentin BG, Lichten M, Kleckner N (2001). Gamma-H2AX illuminates meiosis. *Nat Genet* 27: 236–238.

Kanoh J, Ishikawa F (2001). spRap1 and spRif1, recruited to telomeres by Taz1, are essential for telomere function in fission yeast. *Curr Biol* 11: 1,624–1,630.

Jenuwein T, Laible G, Dorn R, Reuter G (1998). SET domain proteins modulate chromatin domains in eu- and heterochromatin. *Cell Mol Life Sci* 54: 80–93.

Jin Q-W, Trelles-Sticken E, Scherthan H, Loidl J (1998). Yeast nuclei display prominent centromere clustering that is reduced in nondividing cells and in meiotic prophase. *J Cell Biol* 141: 21–29.

Kastan MB, Lim DS (2000). The many substrates and functions of ATM. *Nat Rev Mol Cell Biol* 1: 179–186.

Keeney S (2001). Mechanism and control of meiotic recombination initiation. *Curr Top Dev Biol* 52: 1–53.

Keeney S, Giroux CN, Kleckner N (1997). Meiosis-specific DNA double-strand breaks are catalyzed by Spo11, a member of a widely conserved protein family. *Cell* 88: 375–384.

Kishi S, Zhou XZ, Ziv Y, *et al.* (2001). Telomeric protein Pin2/TRF1 as an important ATM target in response to double strand DNA breaks. *J Biol Chem* 276: 29,282–29,291.

Kleckner N (1996). Meiosis: how could it work? *Proc Natl Acad Sci USA* 93: 8,167–8,174.

Klein F, Mahr P, Galova M, *et al.* (1999). A central role for cohesins in sister chromatid cohesion, formation of axial elements, and recombination during yeast meiosis. *Cell* 98: 91–103.

Klein F, *et al.* (1992). Localization of Rap1 and topoisomerase II in nuclei and chromosomes of yeast. *J Cell Biol* 117: 935–948.

Kohli J, Bahler J (1994). Homologous recombination in fission yeast: absence of crossover interference and synaptonemal complex. *Experientia* 50: 295–306.

Kussel P, Frasch M (1995). Yeast Srp1, a nuclear protein related to *Drosophila* and mouse pendulin, is required for normal migration, division, and integrity of nuclei during mitosis. *Mol Gen Genet* 248: 351–363.

Lammers JH, Offenberg HH, van Aalderen M, *et al.* (1994). The gene encoding a major component of the lateral elements of synaptonemal complexes of the rat is related to X-linked lymphocyte-regulated genes. *Mol Cell Biol* 14: 1,137–1,146.

Lee JY, Orr-Weaver TL (2001). The molecular basis of sister-chromatid cohesion. *Annu Rev Cell Dev Biol* 17: 753–777.

Li B, Oestreich S, de Lange T (2000). Identification of human Rap1: implications for telomere evolution. *Cell* 101: 471–483.

Lichter P, Boyle AL, Cremer T, Ward DC (1991). Analysis of genes and chromosomes by non-isotopic *in situ* hybridization. *Genet Anal Tech Appl* 8: 24–35.

Lim DS, Kim ST, Xu B, *et al.* (2000). ATM phosphorylates p95/Nbs1 in an S-phase checkpoint pathway. *Nature* 404: 613–617.

Loidl J (1990). The initiation of meiotic chromosome pairing: the cytological view. *Genome* 33: 759–778.

Lombard DB, Guarente L (2000). Nijmegen breakage syndrome disease protein and MRE11 at PML nuclear bodies and meiotic telomeres. *Cancer Res* 60: 2,331–2,334.

Longtine MS, Wilson NM, Petracek ME, Berman J (1989). A yeast telomere binding activity binds to two related telomere sequence motifs and is indistinguishable from RAP1. *Curr Genet* 16: 225–239.

Luca FC, Mody M, Kurischko C, *et al.* (2001). *Saccharomyces cerevisiae* Mob1p is required for cytokinesis and mitotic exit. *Mol Cell Biol* 21: 6,972–6,983.

Maddar H, Ratzkovsky N, Krauskopf A (2001). Role for telomere cap structure in meiosis. *Mol Biol Cell* 12: 3,191–3,203.

Mahadevaiah SK, *et al.* (2001). Recombinational DNA double-strand breaks in mice precede synapsis. *Nat Genet* 27: 271–276.

Marshall WF, Straight A, Marko JF, *et al.* (1997). Interphase chromosomes undergo constrained diffusional motion in living cells. *Curr Biol* 7: 930–939.

Martinez-Perez E, Shaw P, Moore G (2001). The Ph1 locus is needed to ensure specific somatic and meiotic centromere association *Nature* 411: 204–207.

Martinez-Perez E, Shaw P, Reader S, Aragon-Alcaide L, Miller T, Moore G (1999). Homologous chromosome pairing in wheat. *J Cell Sci* 112: 1,761–1,769.

Masson JY, West SC (2001). The Rad51 and Dmc1 recombinases: a non-identical twin relationship. *Trends Biochem Sci* 26: 131–136.

McEachern MJ, Krauskopf A, Blackburn EH (2000). Telomeres and their control. *Annu Rev Genet* 34: 331–358.

McKim KS, Green-Marroquin BL, Sekelsky JJ, *et al.* (1998). Meiotic synapsis in the absence of recombination. *Science* 279: 876–878.

MacQueen AJ, Villeneuve AM (2001). Nuclear reorganization and homologous chromosome pairing during meiotic prophase require *C. elegans* chk-2. *Genes Dev* 15: 1,674–1,687.

McLaren A (1983). Primordial germ cells in mice. *Bibliotheca Anat* 24: 59–66.

Megee PC, Mistrot C, Guacci V, Koshland D (1999). The centromeric sister chromatid cohesion site directs Mcd1p binding to adjacent sequences. *Mol Cell* 4: 445–450.

Meluh PB, Rose MD (1990). KAR3, a kinesin-related gene required for yeast nuclear fusion. *Cell* 60: 1,029–1,041.

Meuwissen RL, Offenberg HH, Dietrich AJ, Riesewijk A, van Iersel M, Heyting C (1992). A coiled-coil related protein specific for synapsed regions of meiotic prophase chromosomes. *EMBO J* 11: 5,091–5,100.

Mikhailova EI, Sosnikhina SP, Kirillova GA, *et al.* (2001). Nuclear dispositions of subtelomeric and pericentromeric chromosomal domains during meiosis in asynaptic mutants of rye (*Secale cereale* L.). *J Cell Sci* 114:1,875–1,882.

Moens PB, Pearlman RE (1990). Telomere and centromere DNA are associated with the cores of meiotic prophase chromosomes. *Chromosoma* 100: 8–14.

Moens PB, Pearlman RE, Heng HH, Traut W (1998). Chromosome cores and chromatin at meiotic prophase. *Curr Top Dev Biol* 37: 241–262.

Monesi V (1962). Autoradiographic study of DNA synthesis and the cell cycle in spermatogonia and spermatocytes of the mouse testis using tritiated thymidine. *J Cell Biol* 14: 1–18.

Moore JES (1895). On the structural changes in the reproductive cells in the spermatogenesis of Elasmobranchs. *Quart J of Micros Sci* 38: 275–313.

Moses MJ (1956). Chromosomal structures in crayfish spermatocytes. *J Biophys Biochem Cytol* 2: 215–218.

Naito T, Matsuura A, Ishikawa F (1998). Circular chromosome formation in a fission yeast mutant defective in two ATM homologues. *Nat Genet* 20: 203–206.

Nakamura M, Zhou XZ, Kishi S, Kosugi I, Tsutsui Y, Lu KP (2001). A specific interaction between the telomeric protein Pin2/TRF1 and the mitotic spindle. *Curr Biol* 11: 1,512–1,516.

Nasmyth K (2001). Disseminating the genome: joining, resolving, and separating sister chromatids during mitosis and meiosis. *Annu Rev Genet* 35: 673–745.

Nimmo ER, Pidoux AL, Perry PE, Allshire RC (1998). Defective meiosis in telomere-silencing mutants of *Schizosaccharomyces pombe. Nature* 392: 825–828.

Niwa O, Shimanuki M, Miki F (2000). Telomere-led bouquet formation facilitates homologous chromosome pairing and restricts ectopic interaction in fission yeast meiosis. *EMBO J* 19: 3,831–3,840.

Pâques F, Haber JE (1999). Multiple pathways of recombination induced by double-strand breaks in *Saccharomyces cerevisiae. Microbiol Mol Biol Rev* 63: 349–404.

Pandita TK, Westphal CH, Anger M, *et al.* (1999). Atm inactivation results in aberrant telomere clustering during meiotic prophase. *Mol Cell Biol* 19: 5,096–5,105.

Parisi S, McKay MJ, Molnar M, *et al.* (1999). Rec8p, a meiotic recombination and sister chromatid cohesion phosphoprotein of the Rad21p family conserved from fission yeast to humans. *Mol Cell Biol* 19: 3,515–3,528.

Pasierbek P, Jantsch M, Melcher M, *et al.* (2001). A *Caenorhabditis elegans* cohesion protein with functions in meiotic chromosome pairing and disjunction. *Genes Dev* 15: 1,349–1,360.

Paweletz N (2001). Walther Flemming: a pioneer of mitosis research. *Nat Rev Mol Cell Biol* 2: 72–75.

Pelttari J, Hoja MR, Yuan L, *et al.* (2001). A meiotic chromosomal core consisting of cohesin complex proteins recruits DNA recombination proteins and promotes synapsis in the absence of an axial element in mammalian meiotic cells. *Mol Cell Biol* 21: 5,667–5,677.

Peters AHM, O'Carroll D, Scherthan H, *et al.* (2001). Loss of the suv39h histone methyltransferases impairs mammalian heterochromatin and genome stability. *Cell* 107: 323–337.

Pezzi N, Prieto I, Kremer L, *et al.* (2000). STAG3, a novel gene encoding a protein involved in meiotic chromosome pairing and location of STAG3-related genes flanking the Williams–Beuren syndrome deletion. *FASEB J* 14: 581–592.

Pfeifer C, Thomsen PD, Scherthan H (2001). Centromere and telomere redistribution precedes homologue pairing and terminal synapsis initiation during prophase I of cattle spermatogenesis. *Cytogenet Cell Genet* 93: 304–314.

Plug AW, Peters AH, Keegan KS, Hoekstra MF, de Boer P, Ashley T (1998). Changes in protein composition of meiotic nodules during mammalian meiosis. *J Cell Sci* 111: 413–423.

Podgornaya OI, *et al.* (2000). Nuclear envelope associated protein that binds telomeric DNAs. *Mol Reprod Dev* 57: 16–25.

Prieto I, Suja JA, Pezzi N, *et al.* (2001). Mammalian STAG3 is a cohesin specific to sister chromatid arms in meiosis I. *Nat Cell Biol* 3: 761–766.

Primig M, Williams RM, Winzeler EA, *et al.* (2000). The core meiotic transcriptome in budding yeasts. *Nat Genet* 26: 415–423.

Rabitsch KP, Toth A, Galova M, *et al.* (2001). A screen for genes required for meiosis and spore formation based on whole-genome expression. *Curr Biol* 11: 1,001–1,009.

Rasmussen SW (1977). Meiosis in *Bombyx mori* females. *Philos Trans R Soc Lond B Biol Sci* 277: 343–350.

Ritchie KB, Mallory JC, Petes TD (1999). Interactions of TLC1 (which encodes the RNA subunit of telomerase), TEL1, and MEC1 in regulating telomere length in the yeast *Saccharomyces cerevisiae. Mol Cell Biol* 19: 6,065–6,075.

Rockmill B, Roeder GS (1998). Telomere-mediated chromosome pairing during meiosis in budding yeast. *Genes Dev* 12: 2574–2586.

Roeder S (1997). Meiosis: it takes two to tango. *Genes Dev* 11: 2,600–2,621.

Roeder GS, Bailis JM (2000). The pachytene checkpoint. *Trends Genet* 16: 395–403.

San-Segundo PA, Roeder GS (2000). Role for the silencing protein Dot1 in meiotic checkpoint control. *Mol Biol Cell* 11: 3,601–3,615.

Scherthan H (2001). A bouquet makes ends meet. *Nat Rev Mol Cell Biol* 2: 621–627.

Scherthan H, Jerratsch M, Dhar S, *et al.* (2000a) Meiotic telomere distribution and Sertoli cell nuclear architecture are altered in Atm- and Atm-p53-deficient mice. *Mol Cell Biol* 20: 7,773–7,783.

Scherthan H, Jerratsch M, Li B, *et al.* (2000b). Mammalian meiotic telomeres: protein composition and redistribution in relation to nuclear pores. *Mol Biol Cell* 11: 4,189–4,203.

Scherthan H, Eils R, Trelles-Sticken E, *et al.* (1998). Aspects of three-dimensional chromosome reorganization during the onset of human male meiotic prophase. *J Cell Sci* 111: 2,337–2,351.

Scherthan H, *et al.* (1996). Centromere and telomere movements during early meiotic prophase of mouse and man are associated with the onset of chromosome pairing. *J Cell Biol* 134: 1,109–1,125.

Scherthan H, Loidl J, Schuster T, Schweizer D (1992). Meiotic chromosome condensation and pairing in *Saccharomyces cerevisiae* studied by chromosome painting. *Chromosoma* 101: 590–595.

Schreiner A, Schreiner KE (1905). Über die Entwickelung der männlichen Geschlechtszellen von *Myxine glutinosa* (L). II. Die Centriolen und ihre Vermehrungsweise. *Arch d Biol* 21: 315–353.

Schwacha A, Kleckner N (1997). Interhomolog bias during meiotic recombination: meiotic functions promote a highly differentiated interhomolog-only pathway. *Cell* 90: 1,123–1,135.

Schwarzacher T (1997). Three stages of meiotic homologous chromosome pairing in wheat: cognition, alignment and synapsis. *Sexual Plant Reprod* 10: 324–331.

Selig S, Okumura K, Ward DC, Cedar H (1992). Delineation of DNA replication time zones by fluorescence *in situ* hybridization. *EMBO J* 11: 1,217–1,225.

Sharon G, Simchen G (1990). Centromeric regions control autonomous segregation tendencies in single-division meiosis of *Saccharomyces cerevisiae. Genetics* 125: 487–494.

Shimanuki M, *et al.* (1997). A novel fission yeast gene, *kms1+*, is required for the formation of meiotic prophase-specific nuclear architecture. *Mol Gen Genet* 254: 238–249.

Shore D (2001). Telomeric chromatin: replicating and wrapping up chromosome ends. *Curr Opin Genet Dev* 11: 189–198.

Shore D, Nasmyth K (1987). Purification and cloning of a DNA binding protein from yeast that binds to both silencer and activator elements. *Cell* 51: 721–732.

Simchen G, Kassir Y, Horesh-Cabilly O, Friedmann A (1981). Elevated recombination and pairing structures during meiotic arrest in yeast of the nuclear division mutant cdc5. *Mol Gen Genet* 184: 46–51.

Smith KN, Nicolas A (1998). Recombination at work for meiosis. *Curr Opin Genet Dev* 8: 200–211.

Smith AV, Roeder GS (1997). The yeast Red1 protein localizes to the cores of meiotic chromosomes. *J Cell Biol* 136: 957–967.

Smits VA, Klompmaker R, Arnaud L, Rijksen G, Nigg EA, Medema RH (2000). Polo-like kinase-1 is a target of the DNA damage checkpoint. *Nat Cell Biol* 2: 672–676

Song S, Grenfell TZ, Garfield S, Erikson RL, Lee KS (2000). Essential function of the polo box of Cdc5 in subcellular localization and induction of cytokinetic structures. *Mol Cell Biol* 20: 286–298.

Stack SM, Anderson LK (2001). A model for chromosome structure during the mitotic and meiotic cell cycles. *Chromosome Res* 9: 175–198.

Strambio-de-Castillia C, Blobel G, Rout MP (1999). Proteins connecting the nuclear pore complex with the nuclear interior. *J Cell Biol* 144: 839–855.

Strunnikov AV, Jessberger R (1999). Structural maintenance of chromosomes (SMC) proteins: conserved molecular properties for multiple biological functions. *Eur J Biochem* 263: 6–13.

Sullivan DS, Biggins S, Rose MD (1998). The yeast centrin, cdc31p, and the interacting protein kinase, Kic1p, are required for cell integrity. *J Cell Biol* 143: 751–765.

Sutton WS (1903). The chromosomes in heredity. *Biol Bull* 4: 231–251.

Sym M, Engebrecht J, Roeder GS (1993). ZIP1 is a synaptonemal complex protein reqired for meiotic chromosome synapsis. *Cell* 72: 365–378.

Tanaka T, Cosma MP, Wirth K, Nasmyth K (1999). Identification of cohesin association sites at centromeres and along chromosome arms. *Cell* 98: 847–858.

Tarsounas M, Morita T, Pearlman RE, Moens PB (1999). RAD51 and DMC1 form mixed complexes associated with mouse meiotic chromosome cores and synaptonemal complexes. *J Cell Biol* 147: 207–220.

Terasawa M, Shinohara A, Hotta Y, Ogawa H, Ogawa T (1995). Localization of RecA-like recombination proteins on chromosomes of the lily at various meiotic stages. *Genes Dev* 9: 925–934.
Trask BJ (1991). Fluorescence *in situ* hybridization: applications in cytogenetics and gene mapping. *Trends Genet* 7: 149–154.
Trelles-Sticken E, Dresser ME, Scherthan H (2000). Meiotic telomere protein Ndj1p is required for meiosis specific telomere distribution and bouquet formation in budding yeast. *J Cell Biol* 151: 95–106.
Trelles-Sticken E, Loidl J, Scherthan H (1999). Bouquet formation in budding yeast: initiation of recombination is not required for meiotic telomere clustering. *J Cell Sci* 112: 651–658.
Uetz P, Giot L, Cagney G, *et al.* (2000). A comprehensive analysis of protein–protein interactions in *Saccharomyces cerevisiae. Nature* 403: 623–627.
van Heemst D, Heyting C (2000). Sister chromatid cohesion and recombination in meiosis. *Chromosoma* 109: 10–26.
Villeneuve AM, Hillers KJ (2001). Whence meiosis? *Cell* 106: 647–650.
von Wettstein D, Rasmussen SW, Holm PB (1984). The synaptonemal complex in genetic segregation. *Annu Rev Genetics* 18: 331–413.
von Winniwarter H (1900). Rechereches sur l'ovogenèse et l'organogenèse de l'ovaire des Mammifères (lapin et homme). *Arch Biol* 17: 33–199.
Walker MY, Hawley RS (2000). Hanging on to your homolog: the roles of pairing, synapsis and recombination in the maintenance of homolog adhesion. *Chromosoma* 109: 3–9.
Warburton PE, Cooke HJ (1997). Hamster chromosomes containing amplified human alpha-satellite DNA show delayed sister chromatid separation in the absence of *de novo* kinetochore formation. *Chromosoma* 106: 149–159.
Watanabe Y, Nurse P (1999). Cohesin Rec8 is required for reductional chromosome segregation at meiosis. *Nature* 400: 461–464.
Weiner BM, Kleckner N (1994). Chromosome pairing via multiple interstitial interactions before and during meiosis in yeast. *Cell* 77: 977–991.
Weismann A (1887). *Über die Zahl der Richtungskörper und über ihre Bedeutung für die Vererbung*. Jena: Gustav Fischer.
Wilson EB (1925). *The Cell in Heredity and Development*, 3rd edn. New York: MacMillan.
Yamamoto A, Hiraoka Y (2001). How do meiotic chromosomes meet their homologous partners?: lessons from fission yeast. *Bioessays* 23: 526–533.
Yamamoto A, West RR, McIntosh JR, Hiraoka Y (1999). A cytoplasmic dynein heavy chain is required for oscillatory nuclear movement of meiotic prophase and efficient meiotic recombination in fission yeast. *J Cell Biol* 145: 1,233–1,249.
Yano R, Oakes ML, Tabb MM, Nomura M (1994). Yeast Srp1p has homology to armadillo/plakoglobin/beta-catenin and participates in apparently multiple nuclear functions including the maintenance of the nucleolar structure. *Proc Natl Acad Sci USA* 91: 6,880–6,884.
Yuan L, Liu JG, Zhao J, *et al.* (2000). The murine SCP3 gene is required for synaptonemal complex assembly, chromosome synapsis, and male fertility. *Mol Cell* 5: 73–83
Zetka MC, Kawasaki I, Strome S, Muller F (1999). Synapsis and chiasma formation in *Caenorhabditis elegans* require HIM-3, a meiotic chromosome core component that functions in chromosome segregation. *Genes Dev* 13: 2,258–2,270.
Zhu XD, Kuster B, Mann M, Petrini JH, Lange T (2000). Cell-cycle-regulated association of RAD50/MRE11/NBS1 with TRF2 and human telomeres. *Nat Genet* 25: 347–352.
Zickler D, Kleckner N (1998). The leptotene-zygotene transition of meiosis. *Annu Rev Genet* 32: 619–697.
Zickler D, Kleckner N (1999). Meiotic chromosomes: integrating structure and function. *Annu Rev Genet* 33: 603–754.

3. Cancer Cytogenetics

3.1. Molecular Cytogenetics in the Study of Cancer

Jorn Koch
Cancercytogenetics Laboratory, Tage Hansens Gade 2, DK-8000 Aarhus C., Denmark

Introduction

Cancer is a genetic disease in which cells accumulate enough damage in their genome to both deregulate their growth and to allow unlimited division. Until recent years our ability to study and describe these genetic defects relied largely on the analysis of banded chromosomes. Chromosome banding is generally an efficient low-cost screening technique, and its use has identified the importance of losses and gains of chromosomes and of specific translocations in the natural history of cancer. However, it has a number of limitations, which have made the development of complementary techniques necessary. First, it requires that the malignant cells can be brought to divide outside the human body, i.e., in culture. This requirement is often not possible with neoplasms other than the leukaemias and lymphomas, and consequently the body of chromosome banding data on cancer cells is mainly based on studies of cells from these diseases, although they constitute only a small fraction of all human cancers. Second, the resolution of banding analysis is for obvious reasons roughly one chromosome band, and since a typical band contains millions of bases and hundreds of genes, that is a rather low resolution. Third, the technique only identifies two qualities – dark and light bands – and so two dark bands or two light bands may be mistaken for each other. Revision of chromosome banding data using molecular cytogenetics has shown that such mistakes are in fact rather common.

Molecular cytogenetics relies on the specific base pairing of DNA, which enables hybridization probes to recognize and bind their complementary sequence in the chromosomes. By labelling the probes to make them visible under the microscope, the location and number of hybridization targets is revealed by the specific binding of the probes. The information produced by molecular cytogenetics is much more detailed than that produced by chromosome banding, and consequently a variety of

M. Schmid and I. Nanda (eds.), Chromosomes Today, Volume 14, 151–155.

techniques providing a wide range of different data have been developed. Thus, some molecular genetic techniques can be used to screen the genome for aberrations at low resolution, while others probe small targets; some techniques provide qualitative data (i.e., the presence and location of the target), whereas others produce quantitative data (i.e., how much target is there). In the following I will give an overview of the molecular cytogenetic techniques available today and briefly discuss what the individual techniques can and cannot be used for.

Reaction principles in molecular cytogenetics

Two basic reaction principles exist – either a labelled probe is hybridized to the cell preparation (*in situ* hybridization; ISH) or an unlabelled probe is hybridized and serves as a primer for DNA polymerase mediated DNA synthesis at the site of hybridization (PRimed IN Situ labelling (PRINS)). ISH was first described more than 40 years ago by Gall and Pardue to detect ribosomal genes (Gall and Pardue, 1969), but did not find large scale application until the more user friendly non-radioactive methodology was developed in the 1980s, notably in the form of the fluorescent *in situ* hybridization technique (FISH) (Pinkel *et al.*, 1986). Since then, the technique has developed almost explosively, giving rise to a range of technical variants each adapted for particular purposes, and each having both strengths and weaknesses as detailed later.

PRINS was first described a few years after FISH, in order to provide efficient detection of small tandem repeat target sequences while at the same time discriminating small sequence variations (Koch *et al.*, 1989). The PRINS technique provides the highest sensitivity for the detection of such targets, it has the ability to discriminate single base variations, and the reaction can be made quantitative (Therkelsen *et al.*, 1995; Krejci and Koch, 1998). However, the actual use of the technique has been sporadic, probably largely due to its inefficiency for detection of single copy target sequences. While this limitation can reportedly be overcome by using a cocktail of probes and enhancing signals with tyramide (Kadandale *et al.*, 2000a,b), PRINS will probably not see its full use until the potential for analysis of single base variations in single copy genes is realized, e.g., through the use of the circular padlock probes which makes endless DNA synthesis, and thereby detection of single probes on single targets, possible.

Individual FISH techniques

Locus specific FISH was the original FISH technique. It employs cloned DNA fragments hybridizing to the site in the chromosome harbouring that target DNA fragment. The approach primarily produces specific qualitative data – is the fragment present in the genome, and is it at the usual site? Quantitative data has the form of signal counts – how many signals does the probe give in a particular cell, and thus, how many copies of the target sequence does the cell have? Locus specific FISH is applicable to interphase nuclei (interphase cytogenetics), where it reports how many hybridization targets a given nucleus contains, and how they are placed relative to each other. This particular application is probably the most rapidly expanding and most widely used type of FISH right now.

The possibility of identifying the location of individual target sequence is commonly used for the study of specific reciprocal translocations, such as the bcr/abl translocation found in almost all cases of chronic myeloid leukaemia (CML) and about 20% of cases of acute lymphocytic leukaemia (ALL). In this approach, two probes labelled in different colours are used, each recognizing a different translocation partner. In normal cells, the two coloured signals are randomly distributed and so will rarely be co-localized. However, in cells with the translocation the two signals are juxta-positioned as evidence of the translocation product. It is possible to quickly screen a large number of nuclei to determine the signal pattern and thereby the fraction of cells with the translocation. Determining the fraction of cells with the translocation is very useful for measuring the response to treatment. Counting signals on interphase nuclei is, e.g., used to identify the fraction of cells with an important numerical aberration, e.g., monosomy 5 or monosomy 7.

The interphase cytogenetic approach is extendable to paraffin embedded tissue sections. This application has received much interest recently as individual tissue samples can be organized into a so-called tissue array where hundreds of tumour samples are simultaneously hybridized with a particular probe (Kononen *et al.*, 1998). The tissue array approach complements another recent invention, the expression chip. In one technique multiple genes in one tumour sample are screened while in the other multiple samples are screened for one gene. Genetic aberrations revealed by chip technology can thus be tested for biological significance with the tissue arrays.

The major limitation of the locus specific FISH is that it provides very specific answers to very specific questions. Testing for the bcr/abl translocation thus tells nothing about any other genomic rearrangements in the target cells. Consequently, the technique requires that one knows what to look for, and it is thus often necessary to first do a screening with chromosome banding or a screening-type FISH technique.

Comparative genomic hybridization (CGH) was the next FISH techniques to be developed (Kallioniemi *et al.*, 1992). It employs labelled whole genomes, which are hybridized together in a competitive reaction. The typical setup is that DNA from cancer cells is labelled with one colour and DNA from normal cells is labelled with another after which the two types of DNA are mixed and hybridized to normal metaphase chromosomes. The colour at any specific chromosomal locus will then reflect the relative amount of DNA from that locus found in the two genomes; if a chromosomal segment is amplified in the tumour, the colour of the tumour DNA will dominate, and if a segment is deleted the colour of the normal DNA will dominate. The result is somewhat quantitative, in that the level of colour dominance reflects the level of amplification or deletion. CGH thus screens the tumour genome for chromosomal gains and loses.

The strengths of the technique are: (1) that it does not require dividing tumour cells, and thus is applicable to the study of any type of cancer; (2) that because the technique screens the entire genome, there is no need for prior knowledge about what genetic defects might be found. A weakness of the technique is that it employs DNA from pooled tumour cells, the data therefore representing an average over these cells. A population of tumour cell where half of the cells lack a chromosome 5 and the other half have an extra chromosome 5 will thus appear to have no chromosome 5 aberrations.

This limitation extends to the dilution effect of normal cells within the tumour, and as a rule of thumb the tumour sample needs to contain at least 50% tumour cells to be informative. It is possible to work around these limitations through sorting of tumour cells and PCR mediated amplification of DNA from small numbers of cells, but such detours obviously make the CGH approach less attractive. Balanced aberrations such as polyploidies and balanced translocations are also not recognized by CGH, as there is no gain or loss of individual chromosomes or chromosome segments.

In conclusion, CGH is a useful technique for screening tumour genomes in cells which cannot produce metaphase cells in the laboratory, and which therefore cannot be screened by chromosome banding. However, CGH findings must always be tested with another methodology. Furthermore, CGH is rather laborious and time-consuming and is therefore more suitable as a research technique than as a diagnostic technique.

Combinatorial labelling in FISH exists in two forms, either the *multicolour FISH (M-FISH)* or the *Spectral KaryotYping (SKY)* technique. (Speicher *et al.*, 1996; Schröck *et al.*, 1996.) The two methods primarily differ in how signals are recorded and processed by the computerized capture and analysis equipment, whereas everything else is similar between the techniques. Consequently M-FISH and SKY will here be described as if they were one technique with M-FISH serving as the example. In M-FISH a probe represents a fragment of the genome, typically a chromosome. This probe is then labelled with either a colour or a combination of colours. Five colours are used; so five chromosomes are identified by having one of these colours, whereas unique combinations of two or more colours identify the remaining 19 chromosomes. The probes for the individual chromosomes are then pooled and hybridized together to metaphase spreads of cancer cells, identifying all the cancer chromosomes through the colour combination they obtain.

The strength of this technique is that it can screen the whole cancer genome for any chromosomal translocation in one reaction, and the technique has through this ability probably produced more revised karyotypes than any other FISH technique. However the technique is both laborious and expensive, and is has important limitations. First, it requires cancer cells in mitosis, and is thus applicable to largely the same fraction of cancers as chromosome banding. Second, the resolution is at roughly the same level as that possible with chromosome banding and CGH, so it provides mainly low-resolution screening data. Finally, since the technique detects translocations between non-homologous chromosomes, translocations between homologous chromosomes (e.g., forming isochromosomes) or rearrangements within a chromosome (inversions, deletions, etc.) may not be recognized. Thus as with chromosome banding and CGH, M-FISH is a screening technique with blind spots and results should be corroborated by testing with other methodologies.

Practical application of molecular cytogenetics in the study of cancer

As should now be apparent, the various molecular cytogenetic techniques are – in spite of all their individual strengths – generally much stronger when combined than when used separately. CGH and M-FISH can thus be used to screen the genome for regions that should be studied in more detail with locus specific approaches such as either

FISH or PRINS. Additionally, combining CGH and M-FISH removes several of the blind spots inherent to the two techniques individually. Balanced translocations overlooked by CGH are thus readily detectable by M-FISH, whereas chromosomal deletions overlooked by M-FISH will show up in CGH. For many research purposes it is therefore preferable to employ the full range of molecular cytogenetics techniques to optimize the data yield.

For diagnostic purposes locus specific FISH is often sufficient. First, it is faster and less expensive than CGH and M-FISH, and time and money are of an essence in handling cancer. Second, the information currently requested for the diagnosis of cancers, and for evaluating disease progress and response to treatment, is best provided by locus specific FISH. Possibly, at some point in the future, PRINS may add to this as a fast low-cost alternative capable of revealing single base mutations in important genes.

References

Gall J, Pardue ML (1969). Formation and detection of RNA–DNA hybrid molecules in cytological preparations. *Proc Natl acad Sci USA* 63: 378–383.

Kadandale JS, Tunca Y, Tharapel AT (2000a). Chromosomal localization of single copy genes SRY and SOX3 by primed *in situ* labeling (PRINS). *Microb Comparat Genomes* 5: 71–74.

Kadandale JS, Wachtel SS, Tunca Y, Wilroy RS Jr, Martens PR, Tharapel AT (2000b). Localization of SRY by primed *in situ* labeling in XX and XY sex reversal. *Amer J Med Genet* 95: 71–74.

Kallioniemi A, Kallioniemi O-P, Sudar D, Rutovitz R, Gray JW, Waldman F, Pinkel D (1992). Comparative genomic hybridization for molecular cytogenetics analysis of solid tumors. *Science* 258: 818–821.

Koch J, Kolvraa S, Gregersen N, Bolund L (1989). Oligonucleotide-priming methods for the chromosome-specific labeling of alpha satellite *in situ*. *Chromosoma* 98: 259–265.

Kononen J, Bubendorf L, Kallioniemi A, Bärlund M, Schraml P, Leighton S, Torhorst J, Mihatsch MJ, Sauter G, Kallioniemi O-P (1998). Tissue microarrays for high-throughput molecular profiling of tumor specimens. *Nat Med* 4: 844–847.

Krejci K, Koch J (1998). Improved detection and comparative sizing of human chromosomal telomeres *in situ*. *Chromosoma* 107: 198–203.

Pinkel D, Straume T, Gray JW (1986). Cytogenetic analysis using quantitative, high-sensitivity, fluorescence hybridization. *Proc Natl acad Sci USA* 83: 2,934–2,938.

Schröck E, du Manoir S, Veldman T, Schoell B, Wienberg J, Ferguson-Smith MA, Ning Y, Ledbetter DH, Bar-Am I, Soenksen D, Garini Y, Ried T (1996). Multicolor spectral karyotyping of human chromosomes. *Science* 494–497.

Speicher MJ, Gwyn-Ballard S, Ward DC (1996). Karyotyping human chromosomes by combinatorial multi-fluor FISH. *Nat Genet* 368–375.

Therkelsen AJ, Nielsen A, Koch J, Hindkjær J, Kølvraa S (1995). Staining of human telomeres with primed *in situ* labeling (PRINS). *Cytogenet Cell Genet* 68: 115–118.

3.2. Chromosomal Translocations in Leukaemia: Emerging Networks

Stefan K. Bohlander
Department of Medicine III, Grosshadern, University of Munich and GSF, Clinical Cooperative Group "Leukemia", Munich, Germany

The molecular dissection of the genetic causes of leukaemia has been helped by the relative ease with which leukaemic cells can be subjected to cytogenetic analysis. Almost 40 years ago it was noted that a peculiar small chromosome was a constant feature of chronic myeloid leukaemia (CML) (Nowell and Hungerford, 1960). This so-called Philadelphia chromosome was initially thought to be a partially deleted G-group chromosome. However, it became apparent in the early seventies when the first chromosome banding techniques were developed that the Philadelphia chromosome was a derivative chromosome 22 and the result of a reciprocal translocation between chromosomes 9 and 22 (Rowley, 1973). It still was not known whether this recurring translocation was the result of the malignant transformation process or whether it was indeed the cause of this process. When the breakpoints of the t(9;22)(q34;q11) were analysed at the molecular level in the early 1980s it was shown that they were located in the introns of two genes: the Abelson tryosine kinase gene (*ABL*) on 9q34 and a gene called break point cluster region (*BCR*) on 22q11 (Heisterkamp *et al.*, 1983). The breaking and rejoining of the two chromosomes created two chimeric genes: the BCR/ABL chimeric gene on the Philadelphia chromosome (the der(22)) and the ABL/BCR gene on the derivative chromosome 9. These chimeric genes give rise to chimeric or fusion mRNAs which in turn can be translated into fusion proteins (Heisterkamp *et al.*, 1985; Shtivelman *et al.*, 1985). The open reading frame of the 3′ fusion partner is used correctly so that the resulting protein are true chimeras containing at their N-terminus the amino acid sequence of the 5′ gene and in their C-terminal region the amino acid sequence of the 3′ gene. The generation of transgenic mice that expressed the BCR/ABL fusion protein proved that this protein was indeed capable of causing leukaemia (Heisterkamp *et al.*, 1990). Recently,

M. Schmid and I. Nanda (eds.), Chromosomes Today, Volume 14, 157–168

experiments using mice with inducibly BCR/ABL transgenes showed that BCR/ABL is not only required for the initiation of the malignant transformation but also for the maintenance of the transformed state (Huettner *et al.*, 2000).

Tumour cytogeneticists have described several hundred recurring chromosomal translocations in leukaemias (Mitelman, 1994). Starting with the t(9;22), more and more of these translocation have been analysed at the molecular level. There are basically two types of molecular mechanisms used by translocations to alter gene function: (1) the formation of fusion genes as described above and (2) the inappropriate expression of oncogenes the juxtaposition of strong enhancers/promoters (e.g. overexpression of *MYC* in Burkitt's lymphoma by the t(8;14) translocation (Taub *et al.*, 1982)).

In the following I will concentrate on the first type of mechanism, i.e. the formation of fusion genes. For every translocation analysed at the molecular level two genes are identified. In theory then, the number of newly identified fusion genes at translocation breakpoints should be twice the number of translocations analysed.

Even before the molecular analysis of translocations became possible it was clear that this simple relationship between the number of translocations and the number of genes affected might not be true. There were some chromosomal bands (e.g. 11q23 or 12p13) that were involved in translocations with several different partner bands (Rowley, 1990). So in theory, one gene might form fusion genes with a whole number of other genes. Alternatively, there might be slight differences in the breakpoint locations in for instances 11q23 and a different gene at 11q23 would be affected every time a different partner band was observed. However, when different 11q23 rearrangements were analysed only one gene, the myeloid lymphoid leukaemia (MLL) gene, was found to form fusion genes with a larger number of partner genes (Rubnitz *et al.*, 1996).

This observation meant that the total number of genes involved in leukaemic fusions would be less than twice the number of translocations. However, results from the last five years indicate that the number of genes involved in leukaemic fusions might be even less than the number of different translocations. How can this be possible and what implications does this observation have for our understanding of leukaemogenesis?

Chromosomal band 12p13 had been known to be involved in many different balanced leukaemic translocations (Mitelman, 1994). Fluorescence *in situ* hybridization (FISH) studies indicated that there was one locus and possibly one gene that was involved in the majority of these balanced translocations (Kobayashi *et al.*, 1994). In 1994, Gary Gilliland and co-workers cloned the fusion gene that resulted from a t(5;12)(q31;p13) translocation found in the leukaemic cells of a patient with chronic myelomonocytic leukaemia. They identified the platelet-derived growth factor receptor β (*PDGFRB*) from 5q31 and an hitherto unknown ets family transcription factor they called translocation ets leukaemia (*TEL* gene; renamed *ETV6*: ets translocation variant gene 6) on 12p13 (Golub *et al.*, 1994). Hoping to identify many novel fusion partner genes of *ETV6*, several research groups started to analyse other balanced translocations involving chromosomal band 12p13.

To everyone's great surprise, several genes that had already been known to form fusion genes in other leukaemic translocations were identified.

One of the first other fusion partners of *ETV6* to be identified was the transcription factor *RUNX1* (*CBFA2, AML1*) in the leukaemic blasts of a child with acute lymphoblastic leukaemia (ALL) of B-cell origin (Golub *et al.*, 1995). This was quite surprising, given that conventional karyotyping did not detect a t(12;21)(p13;q22) in this case. This translocation gave rise to a fusion mRNA encoding an N-terminal ETV6 C-terminal RUNX1 fusion protein and to a reciprocal fusion mRNA. Subsequent molecular analysis of cases with childhood B-cell ALL revealed that the ETV6/RUNX1 fusion was present in up to 30% of the cases making this translocation the most common single genetic alteration in childhood leukaemia (Shurtleff *et al.*, 1995). These analyses also showed that the critical fusion product is the ETV6/RUNX1 fusion and not the reciprocal fusion because the RUNX1/ETV6 mRNA was not detectable in all cases. Futhermore, FISH studies showed that the non-rearranged *ETV6* allele is deleted in up to 60% of the cases that have an ETV6/RUNX1 fusion (Romana *et al.*, 1996). *RUNX1* had been identified before as the fusion partner of the *CBFA2T1* (*ETO*) gene in the common t(8;21)(q24;q22) in acute myeloid leukaemia (AML) and as the partner of *MDS1-EVI1* gene in the t(3;21)(q26;q22) translocation in myelodysplastic syndrome. In these cases the critical fusion product was RUNX1/CBFA2T1 or the RUNX1/MDS1-EVI1 fusion (Miyoshi *et al.*, 1991; Nucifora and Rowley, 1995).

In rare cases of myeloproliferative disorders Peeters and co-workers were also able to identify a ETV6/MDS1-EVI1 fusion as the result of a t(3;12)(q26;p13) translocation (Peeters *et al.*, 1997b).

In some cases of myeloid leukaemia and myloproliferative disorder a t(12;22)(p13;q11) translocation is found which results in the fusion of *ETV6* with the *MN1* gene (Buijs *et al.*, 1995). *MN1* is a gene that had previously been identified as being rearranged in a (4;22) translocation in meningeoma (Lekanne Deprez *et al.*, 1995). The (4;22) translocation apparently does not lead to a fusion of *MN1* to another gene.

Another fusion partner of *ETV6* was found to be the well-known Abelson tyrosine kinase gene which results in an ETV6/ABL fusion protein. This fusion is less common than the ETV6/RUNX1 fusion and is found in some cases of AML (Golub *et al.*, 1996; Papadopoulos *et al.*, 1995). Again *ABL* – as discussed above – was already known as a fusion partner of *BCR* in CML and ALL.

In addition to *PDGFRB* and *ABL* three other tyrosine kinases have been found as fusion partners of *ETV6*: (1) the t(9;12)(p24;p13) found in childhood T-cell ALL results in the fusion of the Janus activated kinase *JAK2* with *ETV6* (Lacronique *et al.*, 1997; Peeters *et al.*, 1997a); (2) the t(12;15)(p13;q25) found in congential fibrosarcoma, congenital mesoblastic nephroma and in AML results in the fusion of the neurotrophin-3 receptor gene (*NTRK3*) with *ETV6* (Eguchi *et al.*, 1999; Knezevich *et al.*, 1998; Rubin *et al.*, 1998); (3) recently two cases of AML and one case of T-cell ALL have been described in which a fusion of *ETV6* with the *ABL*-related tyrosine kinase gene (*ARG* or *ABL2*) was found (Bohlander *et al.*, 2000a; Cazzaniga *et al.*, 1999; Iijima *et al.*, 2000) Interestingly, *PDGFRB* has also been shown to form a fusion with the Huntingtin interacting protein 1 (*HIP1*) in cases of chronic myelomonocytic leukaemia (Ross *et al.*, 1998) and with the *H4* gene in cases of atypical CML and t(5;10)(q33;q22) (Kulkarni *et al.*, 2000; Schwaller *et al.*, 2001). We have recently been able to identify a BCR/JAK2 fusion in a case of CML with a t(9;22)(p24;q11) (Griesinger *et al.*, 2000).

The laboratory of Peter Marynen reported four cases of very immature AML (M0) in which a t(4;12)(q11-q12;p13) resulted in the fusion of *ETV6* to a novel gene called *BTL* (Brx-like translocated in leukaemia) (Cools *et al.*, 1999).

The caudal-related homoeobox gene *CDX2* has been reported to form ETV6/CDX2 fusion mRNAs as the result of a t(12;13)(p13;q25) translocation found in AMLs. However, it is not clear whether the ETV6/CDX2 fusion plays an important role in these leukaemias or whether the upregulation of *CDX2* is the leukaemogenic effect of these translocations (Chase *et al.*, 1999).

In addition to the 10 fusion partner genes of *ETV6* described above there are some rare translocations that disrupt the *ETV6* locus without giving rise to a productive fusion, i.e. either no fusion mRNAs can be identified or the fusion mRNAs are out of frame. These are the t(6;12)(q23;p13) and the t(5;12)(q31;p13) translocations, leading to a juxtaposition of *ETV6* to the *STL* locus and the *ACS2* locus, respectively (Suto *et al.*, 1997; Yagasaki *et al.*, 1999).

It is also quite interesting to note that the heterodimeric transcription factor RUNX1-CBFB is not only involved in the t(8;21), the t(3;21), and the t(12;21) resulting in the formation of RUNX1/CBFA2T1, RUNX1/MDS1-EVI1 and ETV6/RUNX1, respectively, but also in the inv(16) and t(16;16) which result in the fusion of *CBFB* with *MYHII* in AML-M4 eo type leukaemias (Liu *et al.*, 1993).

Figure 1 illustrates this complex network of leukaemic fusion genes that has emerged around *ETV6* and *RUNX1*. Both *ETV6* and *RUNX1* are located at central nodes in this network. That these two genes must play important roles in leukaemogenesis is also supported by the fact that the non-rearranged allele of *ETV6* is frequently deleted in ALL blasts that contain an ETV6/RUNX1 fusion (Romana *et al.*, 1996) and the observation that haploinsufficiency of *RUNX1* is the cause of a rare form of familial platelet disorder with a predisposition to develop AML (Song *et al.*, 1999).

To arrive at a better understanding of the physiological role of ETV6 – and with the hope to identify other potential members of this fusion gene network – we searched for ETV6 interacting proteins using the yeast two-hybrid approach.

One of the ETV6 interacting proteins identified in this search was the histone acetyltransferase (HAT) TIP60 (Bohlander *et al.*, 1998), a protein that had first been described as an interacting partner and coactivator of the HIV1 Tat transactivator (Kamine *et al.*, 1996). It is not surprising that ETV6, which is a transcription factor, should interact with a HAT, a group of proteins that have been widely described as coactivators. What is interesting about TIP60 is the fact that it is a HAT containing an about 200 amino acid long protein domain – the so-called MYST domain – which is highly conserved in evolution down to *Drosophila, Caenorhabditis elegans* and *Saccharomyces cerevisiae*. The MYST domain contains a conserved C_2HC zinc finger and an acetyl transferase site (Hilfiker *et al.*, 1997; Reifsnyder *et al.*, 1996). There are two other human proteins known that contain a MYST domain: MOZ and MORF (Borrow *et al.*, 1996; Champagne *et al.*, 1999).

MOZ (ZNF220) is the myelomonocytic zinc finger protein originally identified as the fusion partner of cyclic AMP-response element binding protein (*CBP*) in the t(8;16) translocation found in AML patients (Borrow *et al.*, 1996). *MORF* has recently been also identified as the fusion partner of *CBP* in some rare cases of AML harbouring

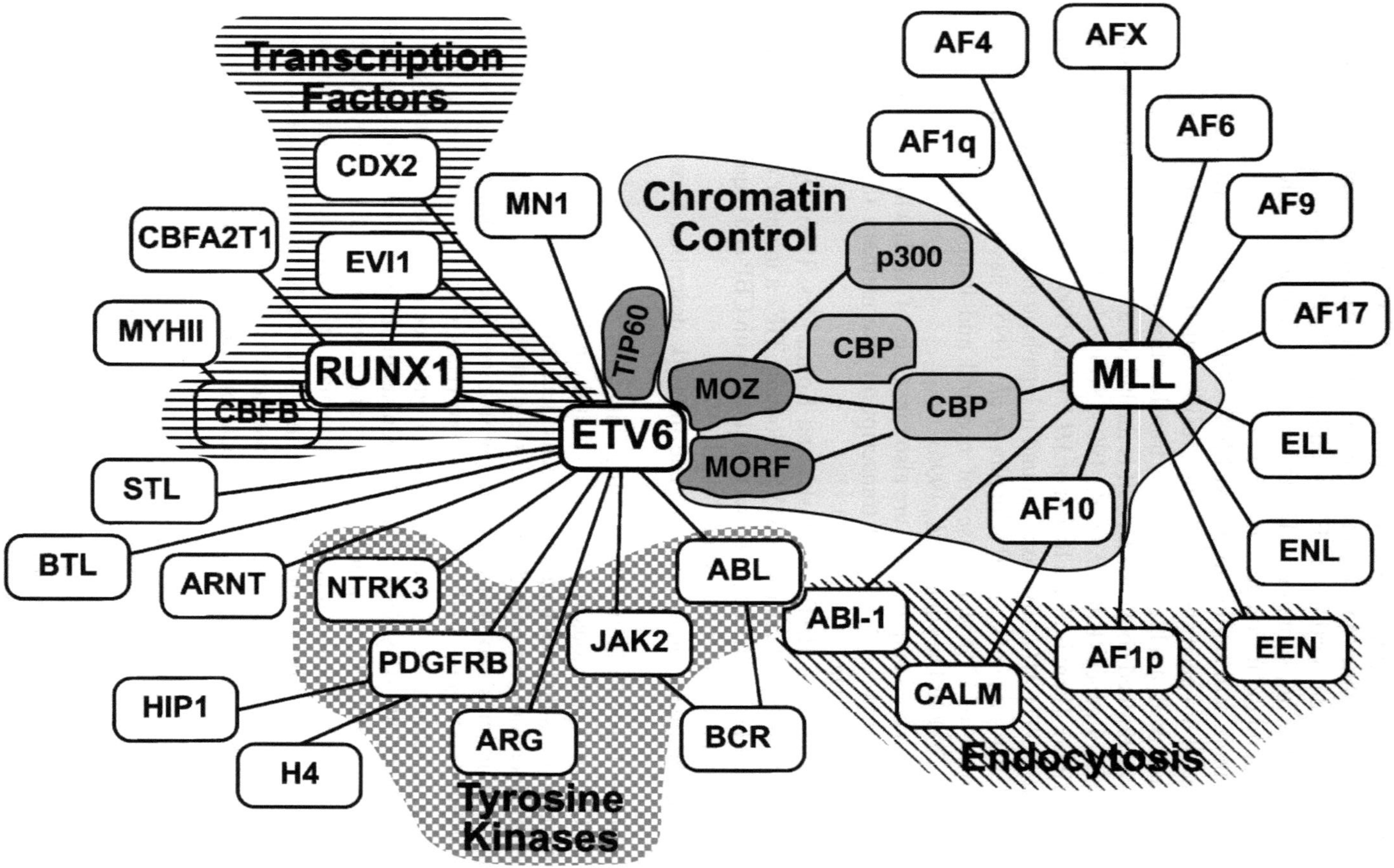

Figure 1. *Diagrammatic representation of fusion gene network. Genes involved in leukaemic translocations are shown in boxes. Boxes connected by a line denote fusion genes. MYST domain HATs are in dark grey boxes, other HATs in light grey boxes. Not all fusions of the MLL gene are shown.*

a t(10;16)(q22;p13) (Panagopoulos *et al.*, 2001). The ETV6 interaction domain of TIP60 can be mapped to the MYST domain of TIP60 and we could show that there is also interaction between portions of ETV6 and the MYST domains of MOZ and MORF (Bohlander *et al.*, submitted). Using these MYST domain interactions as a bridge one can extend the fusion gene network.

Not only are there leukaemias in which MOZ/CBP or MORF/CBP fusions are found but *CBP* itself has been shown to form a fusion with the human trithorax homologue *MLL* in cases with a t(11;16)(q23;p13) translocation (Sobulo *et al.*, 1997; Taki *et al.*, 1997). As mentioned above *MLL* itself is the hub of a fusion gene network containing more than 14 genes (Rubnitz *et al.*, 1996; So *et al.*, 1997; Taki *et al.*, 1998; Thirman *et al.*, 1993). There is also a fusion of *MLL* with *P300* which is closely related to *CBP* (Ida *et al.*, 1997). *MLL* fusions are found in lymphoid and myeloid leukaemias. One of the *MLL* partner genes, *AF10* (Chaplin *et al.*, 1995), has also been identified as a fusion partner of *CALM* in cases of myeloid and lymphoid malignancies (Bohlander *et al.*, 2000b; Dreyling *et al.*, 1996).

To make matters more complex there are also rare cases of AML with an inv(8) that leads to a fusions between *MOZ* and transcription intermediary factor 2 (*TIF2*) (Carapeti *et al.*, 1998; Liang *et al.*, 1998).

CBP and P300 are both HATs (Bannister and Kouzarides, 1996; Ogryzko *et al.*, 1996). However, they do not contain a MYST domain. TIF2 a member of the steroid receptor coactivator family has been shown to interact with CBP and requires CBP for coactivation (Leo and Chen, 2000; Voegel *et al.*, 1998).

A schematic summary of the fusion gene network described here is given in Figure 1.

What can we learn from this "emerging" network of fusion genes found in leukaemia? One of the obvious conclusions is that there does not appear to be a very large number of genes that can be altered to cause leukaemia. The network in Figure 1 shows 38 translocations but only 38 genes instead of the expected 76 genes (i.e. if every translocation joins two genes that are not involved in any other translocation). There are about 30,000 genes in the human genome (Hattori *et al.*, 2000; International Human Genome Sequencing Consortium, 2001) but it seems that only a very small subset of less than 0.4% of genes is altered in leukaemic translocations. This implies that the number of signal transduction pathways and developmental pathways which can be altered to produce leukaemia is also rather small. It is probably smaller than the number of genes affected by translocations because some pathways will be affected at different levels, i.e. at different proteins. For instance the heterodimeric transcription factor RUNX1/CBFB is affected by translocations targeting *RUNX1* and *CBFB* separately.

Another aspect of the fusion gene network is the observation that there are some genes that are involved in more than one translocation. These genes occupy the central nodes of the network. These are most prominently *RUNX1, ETV6* and *MLL* and to a lesser extent *MDS1-EVI1, BCR, ABL, PDGFRB, JAK2, MOZ, CBP* and *AF10*. It has been shown for several of these genes and can be assumed for the others that they play central roles in normal haematopoietic differentiation and lineage commitment (Hess *et al.*, 1997; Okuda *et al.*, 1996; Wang *et al.*, 1997, 1998, 1996). Functionally these important players can be roughly grouped into genes encoding transcription factors (*CBFA2*, *ETV6*), tyrosine kinases (*ABL*, *PDGFRB*, *JAK2*) and coactivators or chromatin control proteins (*MOZ*, *CBP*, *MLL*, *AF10*) (Linder *et al.*, 2001). Another functional group in this network are those proteins (CALM, AF1p, EEN, ABI-1) that have recently been shown to play a role in endocytotic processes. CALM is the non-neuronal counterpart of the AP180 clathrin assembly protein and has been shown to bind to clathrin and phosphoinositides (Dreyling *et al.*, 1996; Tebar *et al.*, 1999; Ford *et al.*, 2001). EEN (SH3p8, endophilin II) and ABI-1 bind to dynamin and synaptojanin through their SH3 domains (Floyd and De Camilli, 1998; So *et al.*, 2000). AF-1p (EPS15) is the target of EGF receptor phosphorylation and binds to AP-2 and synaptojanin (Floyd and De Camilli, 1998). Thus the study of leukaemic translocations has offered us invaluable insights into and access to the crucial pathways that control normal and malignant processes in haematopoietic cells.

About half of AMLs do not have detectable balanced chromosomal translocations (Mrózek *et al.*, 1997). While in some of these cases cryptic translocations (as was the case with the t(12;21) in ALL) might be present, this will certainly not be the case for all of them. However, it seems highly likely that we will find mutations in either the genes involved in translocations (i.e. the ones discussed in this review) or in genes that have a function in a limited number of critical pathways. Example for such mutations are the partial tandem duplications of *MLL* and *FLT3* (Caligiuri *et al.*, 1996; Nakao *et al.*, 1996). Another exciting example for the validity of this assumption is the recent dicovery of mutations in the *CEBPA* gene in cases of AML-M2 without the typical t(8;21) translocation (Pabst *et al.*, 2001b). *CEBPA* has been shown to be critical for the differentiation of the granulocytic lineage (Zhang *et al.*, 1997). It could also be shown that the expression of *CEBPA* is downregulated in AML-M2 leukaemias that have a t(8;21) and express the RUNX1/CBFA2T1 fusion protein. The downregulation of *CEBPA* expression is mediated by the RUNX1/CBFA2T1 fusion protein (Pabst *et al.*, 2001a). These findings are also in agreement with the hypothesis that a differentiation block in normal myeloid development can cause AML (Tenen *et al.*, 1997).

This emerging picture offers hope that the mutational spectrum in leukaemias is limited and that the understanding of the crucial differentiation and proliferation pathways will lead to very specific and effective treatments for leukaemia.

Note: this is an update of the review that was published in Cytogenetics Cell Genetics (Bohlander, 2000)

References

Bannister AJ, Kouzarides T (1996). The CBP co-activator is a histone acetyltransferase. *Nature* 384: 641–643.

Bohlander SK (2000). Fusion genes in leukemia: an emerging network. *Cytogenet Cell Genet* 91: 52–56.

Bohlander SK, Janke A, Podleschny M, Griesinger F (2000a). The tyrosine kinase ARG (c-ABL-l) is fused to ETV6 in a mature T-Acute Lymphoblastic Leukemia (ALL) cell line with a t(1;10;12)(q25;p13;p13) and multilineage differentiation potential. *Med Genet* 12: 125 (abstract).

Bohlander SK, Muschinsky V, Schrader K, *et al.* (2000b). Molecular analysis of the CALM/AF10 fusion: identical rearrangements in acute myeloid leukemia, acute lymphoblastic leukemia and malignant lymphoma patients. *Leukemia* 14: 93–99.

Bohlander SK, Putnik J, Bartels S (1998). ETV6 interacts with a histone acetyltransferase. *Med Genet* 10: 103 (abstract).

Borrow J, Stanton VJ, Andresen JM, *et al.* (1996). The translocation t(8;16)(p11;p13) of acute myeloid leukaemia fuses a putative acetyltransferase to the CREB-binding protein. *Nat Genet* 14: 33–41.

Buijs A, Sherr S, van Baal S, *et al.* (1995). Translocation (12;22) (p13;q11) in myeloproliferative disorders results in fusion of the ETS-like TEL gene on 12p13 to the MN1 gene on 22q11. *Oncogene* 10: 1,511–1,519.

Caligiuri MA, Strout MP, Schichman SA, *et al.* (1996). Partial tandem duplication of ALL1 as a recurrent molecular defect in acute myeloid leukemia with trisomy 11. *Cancer Res* 56: 1,418–1,425.

Carapeti M, Aguiar RC, Goldman JM, Cross NC (1998). A novel fusion between MOZ and the nuclear receptor coactivator TIF2 in acute myeloid leukemia. *Blood* 91: 3,127–3,133.

Cazzaniga G, Tosi S, Aloisi A, *et al.* (1999). The tyrosine kinase Abl-related gene ARG is fused to ETV6 in an AML-M4Eo patient with a t(1;12)(q25;p13): molecular cloning of both reciprocal transcripts. *Blood* 94: 4,370–4,373.

Champagne N, Bertos NR, Pelletier N, *et al.* (1999). Identification of a human histone acetyltransferase related to monocytic leukemia zinc finger protein. *J Bio Chem* 274: 28,528–28,536.

Chaplin T, Ayton P, Bernard OA, *et al.* (1995). A novel class of zinc finger/leucine zipper genes identified from the molecular cloning of the t(10;11) translocation in acute leukemia. *Blood* 85: 1,435–1,441.

Chase A, Reiter A, Burci L, *et al.* (1999). Fusion of ETV6 to the caudal-related homeobox gene CDX2 in acute myeloid leukemia with the t(12;13)(p13;q12). *Blood* 93: 1,025–1,031.

Cools J, Bilhou NC, Wlodarska I, *et al.* (1999). Fusion of a novel gene, BTL, to ETV6 in acute myeloid leukemias with a t(4;12)(q11-q12;p13). *Blood* 94: 1,820–1,824.

Dreyling MH, Martinez-Climent JA, Zheng M, Mao J, Rowley JD, Bohlander SK (1996). The t(10;11)(p13;q14) in the U937 cell line results in the fusion of the *AF10* gene and *CALM*, encoding a new member of the AP-3-clathrin assembly protein family. *Proc Natl Acad Sci USA* 93: 4,804–4,809.

Eguchi M, Eguchi IM, Tojo A, *et al.* (1999). Fusion of ETV6 to neurotrophin-3 receptor TRKC in acute myeloid leukemia with t(12;15)(p13;q25). *Blood* 93: 1,355–1,363.

Floyd S, De Camilli P (1998). Endocytosis proteins and cancer: a potential link? *Trends Cell Biol* 8: 299–301.

Ford MG, Pearse BM, Higgins MK, *et al.* (2001). Simultaneous binding of PtdIns(4,5)P2 and clathrin by AP180 in the nucleation of clathrin lattices on membranes. *Science* 291: 1,051–1,055.

Golub TR, Barker GF, Bohlander SK, *et al.* (1995). Fusion of the *TEL* gene on 12p13 to the *AML1* gene on 21q22 in acute lymphoblastic leukemia. *Proc Natl Acad Sci USA* 92: 4,917–4,921.

Golub TR, Barker GF, Lovett M, Gilliland DG (1994). Fusion of PDGF receptor β to a novel *ets*-like gene, *tel*, in chronic myelomonocytic leukemia with t(5;12) chromosomal translocation. *Cell* 77: 307–316.

Golub TR, Goga A, Barker GF, *et al.* (1996). Oligomerization of the ABL tyrosine kinase by the Ets protein TEL in human leukemia. *Mol Cell Biol* 16: 4,107–4,116.

Griesinger F, Podleschny M, Steffens R, Bohlander SK, Woermann B, Haase D (2000). A novel BCR-JAK2 fusion gene is the result of a translocation (9;22)(p24;q11) in a case of CML. *Blood* 96: 353a (abstract).

Hattori M, Fujiyama A, Taylor TD, *et al.* (2000). The DNA sequence of human chromosome 21. *Nature* 405: 311–319.

Heisterkamp N, Jenster G, ten Hoeve J, Zovich D, Pattengale PK, Groffen J (1990). Acute leukaemia in *bcr/abl* trangenic mice. *Nature* 344: 251–253.

Heisterkamp N, Stam K, Groffen J, de Klein A, Grosveld G (1985). Structural organization of the bcr gene and its role in the Ph1 translocation. *Nature* 315: 758–761.

Heisterkamp N, Stephenson J, Groffen J, *et al.* (1983). Localization of the c-abl oncogene adjacent to a translocation breakpoint in chronic myelocytic leukemia. *Nature* 306: 239–242.

Hess JL, Yu BD, Li B, Hanson R, Korsmeyer SJ (1997). Defects in Yolk Sac Hematopoiesis in *Mll*-Null Embryos. *Blood* 90: 1,799–1,806.

Hilfiker A, Hilfiker KD, Pannuti A, Lucchesi JC (1997). mof, a putative acetyl transferase gene related to the Tip60 and MOZ human genes and to the SAS genes of yeast, is required for dosage compensation in *Drosophila. EMBO J* 16: 2,054–2,060.

Huettner CS, Zhang P, Van Etten RA, Tenen DG (2000). Reversibility of acute B-cell leukaemia induced by BCR-ABL1. *Nat Genet* 24: 57–60.

Ida K, Kitabayashi I, Taki T, *et al.* (1997). Adenoviral E1A-associated protein p300 is involved in acute myeloid leukemia with t(11;22)(q23;q13). *Blood* 90: 4,699–4,704.

Iijima Y, Ito T, Oikawa T, *et al.* (2000). A new *ETV6/TEL* partner gene, *ARG* (*ABL* related gene or *ABL2*), identified in an AML-M3 cell line with a t(1;12)(q25;p13) translocation. *Blood* 95: 2,126–2,131.

International Human Genome Sequencing Consortium (2001). Initial sequencing and analysis of the human genome. *Nature* 409: 860–921.

Kamine J, Elangovan B, Subramanian T, Coleman D, Chinnadurai G (1996). Identification of a cellular protein that specifically interacts with the essential cysteine region of the HIV-1 Tat transactivator. *Virology* 216: 357–366.

Knezevich SR, McFadden DE, Tao W, Lim JF, Sorensen PH (1998). A novel ETV6-NTRK3 gene fusion in congenital fibrosarcoma. *Nat Genet* 18: 184–187.

Kobayashi H, Montgomery KT, Bohlander SK, *et al.* (1994). Fluorescence *in situ* hybridization mapping of translocations and deletions involving the short arm of human chromosome 12 in malignant hematologic diseases. *Blood* 84: 3,473–3,482.

Kulkarni S, Heath C, Parker S, *et al.* (2000). Fusion of H4/D10S170 to the platelet-derived growth factor receptor beta in BCR-ABL-negative myeloproliferative disorders with a t(5;10)(q33;q21). *Cancer Res* 60: 3,592–3,598.

Lacronique V, Boureux A, Valle VD, *et al.* (1997). A TEL-JAK2 fusion protein with constitutive kinase activity in human leukemia. *Science* 278: 1,309–1,312.

Lekanne Deprez RH, Riegman PH, Groen NA, *et al.* (1995). Cloning and characterization of MN1, a gene from chromosome 22q11, which is disrupted by a balanced translocation in a meningioma. *Oncogene* 10: 1,521–1,528.

Leo C, Chen JD (2000). The SRC family of nuclear receptor coactivators. *Gene* 245: 1–11.

Liang J, Prouty L, Williams BJ, Dayton MA, Blanchard KL (1998). Acute mixed lineage leukemia with an inv(8)(p11q13) resulting in fusion of the genes for MOZ and TIF2. *Blood* 92: 2,118–2,122.

Linder B, Gerlach N, Jäckle H (2001). The *Drosophila* homolog of the human AF10 is an HP1-interacting suppressor of position effect variegation. *EMBO rep* 2: 221–216.

Liu P, Tarle SA, Hajra A, *et al.* (1993). Fusion between transcription factor CBFβ/PEBP2β and a myosin heavy chain in acute myeloid leukemia. *Science* 261: 1,041.

Mitelman F (1994). *Catalog of Chromosome Aberrations in Cancer.* New York: Alan R. Liss.

Miyoshi H, Shimizu K, Kozu T, Maseki N, Kaneko Y, Ohki M (1991). t(8;21) breakpoints on chromosome 21 in acute myeloid leukemia are clustered within a limited region of a single gene, AML1. *Proc Natl Acad Sci USA* 88: 10,431–10,434.

Mrózek K, Heinonen K, de la Chalpelle A, Bloomfield CD (1997). Clinical significance of cytogenetics in acute myeloid leukemia. *Semin Oncol* 24: 17–31.

Nakao M, Yokota S, Iwai T, *et al.* (1996). Internal tandem duplication of the flt3 gene found in acute myeloid leukemia. *Leukemia* 10: 1,911–1,918.

Nowell PC, Hungerford DA (1960). A minute chromosome in human granulocytic leukemia. *Science* 132: 1,497.

Nucifora G, Rowley JD (1995). AML1 and the 8;21 and 3;21 translocations in acute and chronic myeloid leukemia. *Blood* 86: 1–14.

Ogryzko VV, Schiltz RL, Russanova V, Howard BH, Nakatani Y (1996). The transcriptional coactivators p300 and CBP are histone acetyltransferases. *Cell* 87: 953–939.

Okuda T, van Deursen J, Hiebert SW, Grosveld G, Downing JR (1996). AML1, the target of multiple chromosomal translocations in human leukemia, is essential for normal fetal liver hematopoiesis. *Cell* 84: 321–330.

Pabst T, Mueller BU, Harakawa N, *et al.* (2001a). AML1-ETO downregulates the granulocytic differentiation factor C/EBPα in t(8;21) myeloid leukemia. *Nat Med* 7: 444–451.

Pabst T, Mueller BU, Zhang P, *et al.* (2001b). Dominant-negative mutations of *CEBPA*, encoding CCAAT/enhancer binding protein-α (C/EBPα), in acute myeloid leukemia. *Nat Genet* 27: 263–270.

Panagopoulos I, Fioretos T, Isaksson M, *et al.* (2001). Fusion of the MORF and CBP genes in acute myeloid leukemia with the t(10;16)(q22;p13). *Hum Mol Genet* 10: 395–404.

Papadopoulos P, Ridge SA, Boucher CA, Stocking C, Wiedemann LM (1995). The novel activation of *ABL* by fusion to an *ets*-related gene, *TEL*. *Cancer Res* 55: 34–38.

Peeters P, Raynaud SD, Cools J, *et al.* (1997a). Fusion of TEL, the ETS-variant gene 6 (ETV6), to the receptor-associated kinase JAK2 as a result of t(9;12) in a lymphoid and t(9;15;12) in a myeloid leukemia. *Blood* 90: 2,535–2,540.

Peeters P, Wlodarska I, Baens M, *et al.* (1997b). Fusion of ETV6 to MDS1/EVI1 as a result of t(3;12)(q26;p13) in myeloproliferative disorders. *Cancer Res* 57: 564–569.

Reifsnyder C, Lowell J, Clarke A, Pillus L (1996). Yeast SAS silencing genes and human genes associated with AML and HIV-1 Tat interactions are homologous with acetyltransferases. *Nat Genet* 14: 42–49.

Romana SP, Le Coniat M, Poirel H, Marynen P, Bernard OA, Berger R (1996). Deletion of the short arm of chromosome 12 is a secondary event in acute lymphoblastic leukemia with t(12;21). *Leukemia* 10: 167–170.

Ross TS, Bernard OA, Berger R, Gilliland DG (1998). Fusion of Huntingtin interacting protein 1 to platelet-derived growth factor beta receptor (PDGFbetaR) in chronic myelomonocytic leukemia with t(5;7)(q33;q11.2). *Blood* 91: 4,419–4,426.

Rowley J (1973). A new consistent chromosomal abnormality in chronic myelogenous leukemia identified by quinacrine fluorescence and Giemsa staining. *Nature* 243: 290–293.

Rowley JD (1990). Recurring chromosome abnormalities in leukemia and lymphoma. *Semin Hematol* 27: 122–136.

Rubin BP, Chen CJ, Morgan TW, *et al.* (1998). Congenital mesoblastic nephroma t(12;15) is associated with ETV6-NTRK3 gene fusion: cytogenetic and molecular relationship to congenital (infantile) fibrosarcoma. *Am J Pathol* 153: 1,451–1,458.

Rubnitz JE, Behm FG, Downing JR (1996). 11q23 rearrangements in acute leukemia. *Leukemia* 10: 74–82.

Schwaller J, Anastasiadou E, Cain D, *et al.* (2001). H4(D10S170), a gene frequently rearranged in papillary thyroid carcinoma, is fused to the platelet-derived growth factor receptor beta gene in atypical chronic myeloid leukemia with t(5;10)(q33;q22). *Blood* 97: 3,910–3,918.

Shtivelman E, Lifshitz B, Gale R, Canaani E (1985). Fused transcript of Abl and Bcr genes in chronic myelogenous leukemia. *Nature* 315: 550–554.

Shurtleff SA, Buijs A, Behm FG, *et al.* (1995). Tel/AML1 fusion resulting from a cryptic t(12;21) is the most common genetic lesion in pediatric ALL and defines a subgroup of patients with an excellent prognosis. *Leukemia* 9: 1,985–1,989.

So CW, Caldas C, Liu MM, *et al.* (1997). EEN encodes for a member of a new family of proteins containing an Src homology 3 domain and is the third gene located on chromosome 19p13 that fuses to MLL in human leukemia. *Proc Natl Acad Sci USA* 94: 2,563–2,568.

So CW, So CK, Cheung N, Chew SL, Sham MH, Chan LC (2000). The interaction between EEN and Abi-1, two MLL fusion partners, and synaptojanin and dynamin: implications for leukaemogenesis. *Leukemia* 14: 594–601.

Sobulo OM, Borrow J, Tomek R, *et al.* (1997). MLL is fused to CBP, a histone acetyltransferase, in therapy-related acute myeloid leukemia with a t(11;16)(q23;p13.3). *Proc Natl Acad Sci USA* 94: 8,732–8,737.

Song WJ, Sullivan MG, Legare RD, *et al.* (1999). Haploinsufficiency of CBFA2 causes familial thrombocytopenia with propensity to develop acute myelogenous leukaemia. *Nat Genet* 23: 166–175.

Suto Y, Sato Y, Smith SD, Rowley JD, Bohlander SK (1997). A t(6;12)(q23;p13) results in the fusion of ETV6 to a novel gene, STL, in a B-cell ALL cell line. *Genes Chrom Cancer* 18: 254–268.

Taki T, Ohnishi H, Shinohara K, *et al.* (1998). *ABI-1*, a human homologue to mouse ABL/Interactor 1, fuses the *MLL* gene in acute myeloid leukemia with t(10;11)(p11.2;q23). *Blood* 92: 1,125–1,130.

Taki T, Sako M, Tsuchida M, Hayashi Y (1997). The t(11;16)(q23;p13) translocation in myelodysplastic syndrome fuses the MLL gene to the CBP gene. *Blood* 89: 3,945–3,950.

Taub R, Kirsch I, Morton C, *et al.* (1982). Translocation of the c-myc gene into the immunoglobulin heavy chain locus in human Burkitt lymphoma and murine plasmacytoma cells. *Proc Natl Acad Sci USA* 79: 7,837–7,841.

Tebar F, Bohlander SK, Sorkin A (1999). Clathrin assembly lymphoid myeloid leukemia (CALM) protein: localization in endocytic-coated pits, interactions with clathrin, and the impact of overexpression on clathrin-mediated traffic. *Mol Biol Cell* 10: 2,687–2,702.

Tenen DG, Hromas R, Licht JD, Zhang DE (1997). Transcription factors, normal myeloid development, and leukemia. *Blood* 90: 489–519.

Thirman MJ, Gill HJ, Burnett RC, *et al.* (1993). Rearrangement of the MLL gene in acute lymphoblastic and acute myeloid leukemias with 11q23 chromosomal translocations. *New Engl J Med* 329: 909–914.

Voegel JJ, Heine MJ, Tini M, Vivat V, Chambon P, Gronemeyer H (1998). The coactivator TIF2 contains three nuclear receptor-binding motifs and mediates transactivation through CBP binding-dependent and -independent pathways. *EMBO J* 17: 507–519.

Wang LC, Kuo F, Fujiwara Y, Gilliland DG, Golub TR, Orkin SH (1997). Yolk sac angiogenic defect and intra-embryonic apoptosis in mice lacking the Ets-related factor TEL. *EMBO J* 16: 4,374–4,383.

Wang Q, Stacy T, Binder M, Marín-Padilla M, Sharpe AH, Speck NA (1996). Disruption of the *Cbfa2* gene causes necrosis and hemorrhaging in the central nervous system and blocks definitive hematopoiesis. *Proc Natl Acad Sci USA* 93: 3,444–3,449.

Wang LC, Swat W, Fujiwara Y, *et al.* (1998). The TEL/ETV6 gene is required specifically for hematopoiesis in the bone marrow. *Genes Dev* 12: 2,392–2,402.
Yagasaki F, Jinnai I, Yoshida S, *et al.* (1999). Fusion of TEL/ETV6 to a novel ACS2 in myelodysplastic syndrome and acute myelogenous leukemia with t(5;12)(q31;p13). *Genes Chrom Cancer* 26: 192–202.
Zhang DE, Zhang P, Wang ND, Hetherington CJ, Darlington GJ, Tenen DG (1997). Absence of granulocyte colony-stimulating factor signaling and neutrophil development in CCAAT enhancer binding protein α-deficient mice. *Proc Natl Acad Sci USA* 94: 569–574.

4. Chromosome Structures

4.1. The EVO-DEVO of Pericentromeric DNA in the Mouse*

Silvia Garagna[1], Ernesto Capanna[2], Maurizio Zuccotti[3] and Carlo Alberto Redi[1]

[1]Dipartimento di Biologia Animale, Laboratorio di Biologia dello Sviluppo e Centro di Eccellenza in Biologia Applicata, Universita' di Pavia, Piazza Botta, 9 – 27100 Pavia, Italy; [2]Dipartimento di Biologia Animale e dell'Uomo, Universita' "La Sapienza" Roma, Via Alfonso Borelli, 50 – 00161 Roma, Italy; [3]Dipartimento di Medicina Sperimentale, Sezione di Istologia ed Embriologia, Universita' di Parma, Via Volturno, 39 – 43100 Parma, Italy

EVO–DEVO

The path from genome to function is becoming shorter and shorter thanks to the flood of sequence data, advanced bioinformatics techniques and new approaches to study genome-wide expression profiling and protein molecules. Today, astonishing as it may seem, the challenge of a molecular understanding of very complex developmental and evolutionary processes seems open.

It is recent news that a number of genomes, including those of *Arabidopsis thaliana, Saccharomyces cerevisiae, Caenorhabditis elegans, Drosophila melanogaster* and *Homo sapiens*, have been completely sequenced and it will not be long before the genome sequencing of many other species, including other mammals (e.g., the mouse), will be finished. The *genomic era* will produce profound changes in both biology and medicine, since it will be possible to apply a global approach to the study of gene and protein functions for an understanding of supramolecular processes. During the coming years, "functional genomics" (the understanding of biological functions

*This chapter is dedicated to the memory of Prof. Alfred Gropp, great cytologist and unforgettable friend, on the occasion of the 31st anniversary of his description of the *Mus poschiavinus* karyotype. Many of us still have a vivid remembrance of his enthusiasm for biological research and nature. The milestone contributions he made, till his premature death in October 1983 soon after the Luebeck Chromosome Conference, continue to nourish the scientific community.

M. Schmid and I. Nanda (eds.), Chromosomes Today, Volume 14, 171–185.

thanks to genome composition/organization data) will try to address and answer crucial questions to clarify genome functioning in relation to developmental (e.g., cell differentiation) and evolutionary (e.g., speciation) processes. As shown by the human genome project, the vast majority of the DNA content of a nucleus has no known function, although it is likely that the DNA sequences capable of regulating the expression of structural genes reside in this portion of the genome. To fully understand genome functioning, the linear genome map must be integrated with studies on the epigenetic mechanisms of gene regulation and expression, such as DNA methylation, histone acetylation, chromatin organization and nuclear architecture. To perform with success this challenge, biologists need to enter paradigm shifts of their conceptual and experimental tools to break away from the gene-centric view of life: genes alone do not tell us the whole story, and deciphering biology solely from DNA sequences may be impossible. However, there is still a lack of knowledge of the genomic bases that link evolutionary (EVO) and developmental (DEVO) life processes; moreover, conflicting EVO–DEVO views assign different constraints (Darwinism versus self-organization mechanisms) at different levels of life organization. The eukaryotic genome acts within developmental time scales, while evolutionary processes occur within geological time scales: thus, we have a hazy view of the link between genome functioning and evolution, since it is rather difficult to understand the molecular basis of the specific phenotypic traits that assign different fitness values to individuals.

Several tracks can be followed to attain a better view of the links among genome composition/organization, developmental processes and evolution, in search of a unifying view of life processes at any hierarchical level, as in 1973 Francois Jacob in *The logic of life* was hoping "to interpret the properties of the organism by the structure of its constituent molecules".

We have chosen a *murine track* for an easier understanding of these relationships, because a massive quantity of data, from biochemical to behavioural, has been collected on the mouse in the last century.

The studies described here, concerning several aspects of the pericentromeric satellite DNA (satDNA) biology of the house mouse in relation to karyotype structure, development and microevolutionary processes, consider how the genome composition/organization can influence supragenomic processes of life organization. We try to show that there is a cascade of effects that links genome composition to microevolutionary processes: satDNA features can affect the chromosomal translocation rates, which in turn can alter germ cell cytodifferentiation and thus can trigger the creation of reproductive barriers and speciation events.

Mouse pericentromeric DNA

Most species of the genus *Mus* show a $2n = 40$ (NF = 40) all-telocentric karyotype. The pericentromeric region of the standard telocentric chromosomes (i.e. that of *Mus musculus domesticus*) comprises a large block of major satDNA, about 6 Mb, flanked by a smaller block of minor satDNA, about 600 kb, adjacent to the telomeric repeats, 10–60 kb, present at the chromosomal ends (Hemann and Greider, 2000; Zijlmans *et al.*, 1997). The *domesticus* major satDNA, which is particularly rich in AT nucleotides, makes up nearly 10% of the entire genome and is present in all chromosomes except the Y. Intragenus comparative analyses of pericentromeric DNAs show that, in *domesticus* alone, one of the two satDNA families, the major, is highly amplified and is characterized by the presence of an EcoRI restriction site at the level of the half-monomer repetition unit of 230–240 bp (Garagna *et al.*, 1993; Redi *et al.*, 1990a). The minor satDNA, likely derived from the major satDNA (Wong and Rattner, 1988), has a monomeric unit of tandem repetition of 120 bp and is abundant only in the *spretus* genome (6% of the genome). Fluorometric analysis of the organization of the major and minor satDNA sequences suggests a very low interspersion pattern of satDNA with non-satDNA in the *domesticus* pericentromeric region (Garagna *et al.*, 1993; Redi *et al.*, 1990a): thus, in this species, satDNA sequences are clustered in tandem repeats that are longer than those in other *Mus* genomes. This long-range organization of the cluster repeats can be visualized by means of FIBER-FISH analysis of the major and minor satDNA sequences: there is no intermingling of the telomeric and minor tandem repeats nor of the major and minor satDNAs (Figure 1a,b). In addition, the overall view of the pictures suggests that the telomeric, the minor and the major satDNAs are not intermingled with the non-satDNA, since the distribution of their fluorescent signals appears uninterrupted. The split between the *domesticus* and *musculus* groups is thought to have occurred when mice entered Europe with man during Neolithic times; thus, it is likely that the amplification of the major satDNA sequences, with acquisition of the EcoRI restriction site, occurred no more than 3,000–9,000 years ago. These peculiar features of the *domesticus* pericentromeric satDNA composition and organization are correlated with the proneness of its telocentric chromosomes to give rise to whole arm Robertsonian (Rb) translocations. As is well known (King, 1993), Rb translocations, among the most common chromosome rearrangements in mammals, may contribute to chromosomal differences between species and can characterize chromosomal races within species, such as in the *domesticus* house mouse. In the wild, more than 40 distinct local populations of mice, with more than 120 Rb chromosomes differing in arm combinations (Britton-Davidian *et al.*, 2000; Garagna *et al.*, 1997), have been described since the first discovery by Alfred Gropp in the Poschiavo Valley (Gropp *et al.*, 1970).

We studied the organization of pericentromeric satDNA, both major and minor, of telocentrics and Rb chromosomes derived from wild mice of two chromosomal races in northern Italy, the Poschiavo (POS) and Cremona (CRE) races (Gropp *et al.*, 1982a). In the pericentromeric Rb regions, there are no telomeric sequences, while only 20–60 kb of minor satDNA, organizing a functionally active kinetochore, are retained, sandwiched between two blocks of about 6 Mb each of major satDNA (Garagna *et al.*, 1995). The present FIBER-FISH analysis confirms this structural organization of the pericentromeric region of Rb chromosomes (Figure 1c). We re-examined the geometry of Rb translocations by studying the polarity of major satDNA using the chromosome-orientation fluorescence *in situ* hybridization (CO-FISH) technique. Previous results, obtained by BrdUrd quenching of Hoechst 33258 fluorescence in Rb chromosomes present in the mouse L-cell line (Holmquist and Comings, 1975) and in the POS wild-derived mice (Lin and Davidson, 1974), suggested that the major satDNA maintains the same polarity across the centromeres for all Rb chromosomes. We used a method that combines CO-FISH with primed *in situ* labelling (PRINS) and FISH to detect the very

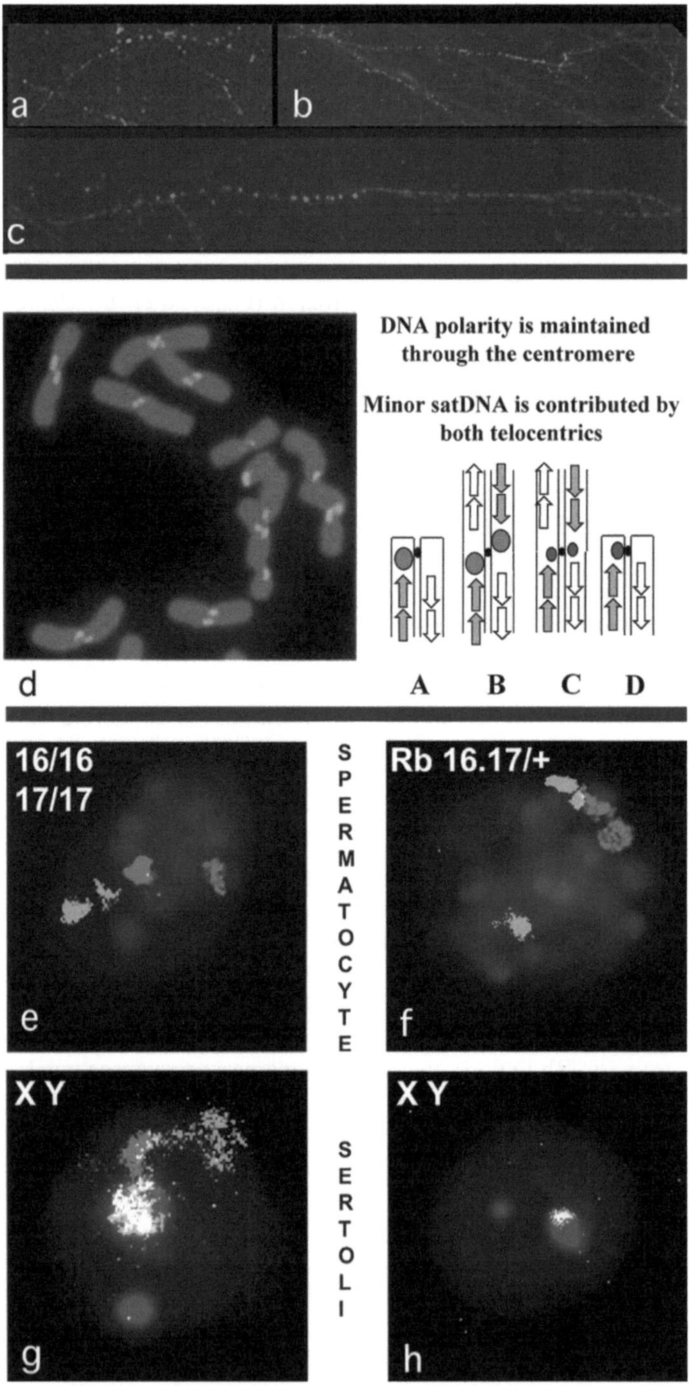
a
b
c
d
DNA polarity is maintained through the centromere
Minor satDNA is contributed by both telocentrics
A
B
C
D
16/16
17/17
e
SPERMATOCYTE
Rb 16.17/+
f
X Y
g
SERTOLI
X Y
h

Figure 1.

small amounts of minor satDNA still present in the centromeric region of the Rb chromosomes. This method (CO-FISH-PRINS-FISH) allows localization of the breakpoints in telocentrics involved in Rb translocations. The ability to detect the two minor satDNA blocks, contributed by both telocentrics, shows that the CO-FISH-PRINS-FISH method has a sensitivity of 10–30 kb, 2–5 times higher than the previously reported limit of 50 kb for the CO-FISH technique alone (Meyne *et al.*, 1994). In Rb chromosomes, the two telocentrics contribute symmetrically with 10–30 kb of minor satDNA each to the newly formed centromeric region (Garagna *et al.*, 2001). Our findings, directly visualizing the two families of satDNA, demonstrate that in mouse chromosomes both major and minor satDNA sequences have exactly the same polarity relative to centromeres and telomeres. In addition, both the major and minor satDNA tandem repeats are oriented head-to-tail in telocentric and Rb chromosomes; their polarity is always the same relative to the centromere, with both blocks of major and minor satDNA showing the same orientation and pointing to the centromere (Figure 1d, (A, B, C, D)). When Rb translocations occur, the overall DNA polarity is maintained through the centromere in Rb chromosomes, i.e. from one telocentric through the joining point of the other telocentric, as shown by the contralateral symmetry of the hybridization signals (Figure 1d). We suggest that all tandem repetitive satDNAs in a species are locked into such symmetry as a universal result of chromosomal evolution. In the house mouse, the very high frequency of Rb chromosomes, and the random involvement of the telocentrics in the translocation process, could be caused by "inherent genomic traits" such as the clustering of heterochromatic regions (Cerda *et al.*, 1999) and the homology of satDNA sequences shared by the telocentric chromosomes (Redi *et al.*, 1990b). Recombination between minor satDNA on non-homologous chromosomes could be mediated by the nicking activity of CENP-B protein which is bound to the minor satDNA itself. In this context, the minor satDNA–CENP-B protein complex appears to be the precise molecular substrate for

---

Figure 1. *(a) FIBER-FISH with telomeric (red) and minor satDNA (green) probes on DNA fibres from an all-telocentric (2n = 40) chromosome mouse. A non-interspersion pattern of the red and green signals is evident; (b) after hybridization with minor (red) and major (green) satDNA probes, a non-interspersion pattern is clearly detectable; (c) In Rb metacentric chromosomes, the minor satDNA (red) is sandwiched between two major satDNA blocks of the original telocentric chromosomes; (d) CO-FISH with major satDNA probe: two contralateral signals are visible in the pericentromeric regions of Rb metacentric chromosomes, while in the telocentric chromosomes signals are evident on a single chromatid only. The sketched CO-FISH hybridization with both the major (A, red dot) and the minor (D, blue dot) satDNAs indicates that there is a head-to-tail orientation of the tandem repeats along the DNA strand. The contralateral disposition relative to the centromere of the hybridization signals (B, red dot, major satDNA; C, blue dot, minor satDNA) in the Rb chromosomes shows that the DNA polarity is maintained throughout the centromere. The black dot represents the centromere; (e) chromosome 16 (green) and 17 (red) localization in spermatocyte nuclei of telocentric homozygous and (f) heterozygous mice (note the changes in the nuclear territories of both chromosome pairs and the total asynapsis of one chromosome 16); (h) in the Sertoli cells of all-telocentric homozygous mice, chromosomes X (green) and Y (red) show decondensed signals, while (g) the signals appear highly condensed in Rb heterozygotes.*

Rb translocations in house mice. In support of this view, all the breakpoints were found within minor satDNA (Garagna *et al.*, 2001).

... its developmental ...

In wild mammals, Rb chromosomes may contribute to reproductive isolation, and thus to speciation, due to detrimental effects on developmental processes (notably germ cell cytodifferentiation) in both male and female carriers. Heterozygotes for Rb chromosomes tend to be infertile or to produce offspring with birth defects. Clear evidence (Gropp *et al.*, 1982b; Redi and Capanna, 1988; Wallace *et al.*, 1992) has shown that both the multiplying mitotic gonia (Redi *et al.*, 1983) and the cytodifferentiation meiotic steps are sensitive to Rb heterozygosities. The reduced fertility of Rb heterozygous carriers of huge multivalent rings is well known, but the same is true for Rb heterozygous carriers of chains and of several multiple independent trivalents (Redi and Capanna, 1988); it is likely that one cause of such impairment is difficulty with meiotic pairing and chromosome segregation. In the last two decades, many researchers (ourselves included) have provided evidence of the varying extent of germ cell losses in many different models, e.g. the mouse, lemur, rock wallaby, grasshopper, dik-dik, *Bos* hybrid, mole rat and shrew. The problem in understanding the causes leading to disruption of the gametogenic process is that each case equally supports the view of failure of saturation of pairing sites (Miklos, 1974) as opposed to interference with X-chromosome inactivation brought about by the Rb structural rearrangement (Lifschytz and Lindsley, 1972). A new conceptual approach is needed to uncouple these conflicting views. As stated above, to understand genome function fully, it is necessary to consider the functioning of the linear genome sequence when integrated into a spatial map of chromosomes in the nucleus. Within the cell nucleus, the chromatin has morphological and functional compartmentalizations (Strouboulis and Wolffe, 1996) related to ribonucleoprotein and deoxyribonucleoprotein activities and to the nuclear organization of gene expression (Misteli and Spector, 1998). Recently, the application of cytochemical techniques to specifically reveal DNA and the protein components of chromatin (reviewed in van der Ploeg, 2000) has provided a clearer view of the changes in nuclear architecture during cell differentiation (Manuelidis, 1997; Bickmore and Bridger, 1999) and gene expression (Misteli and Spector, 1998). Indeed, the nucleus and its architecture are now seen by molecular morphologists as the structure underlying genome function (Lamond and Earnshaw, 1998). Bickmore and coworkers (Bridger *et al.*, 2000) showed that the spatial organization of the genome within the nucleus is dynamic (with a 2–4 h gap at the beginning of G1 when the spatial organization is established) and is related to the functional state of the cell (quiescent/senescent). When the cell exits the cell cycle, the gene-poor human chromosome 18 moves from the nuclear periphery to a more internal site within the nucleus; it returns to the edge after re-entry into the cell cycle (Bridger *et al.*, 2000). In other words, chromosomal territories are not static, pre-assembled structures that carry out particular nuclear processes; rather, their spatial organization has a determinant role in the regular functioning of the various cell machineries (Hamkalo and Elgin, 1991). Strongly supporting this conclusion is the clear causal association between changes in the spatial arrangement of some chromosomal territories and the triggering of phenomena like the induction of trisomy 21 (Nagele *et al.*, 1998) and the pathogenesis of human leukemias due to t(9,22) translocation (Kozubek *et al.*, 1999).

Thus, we investigated the spatial relationship of some chromosome arms in Sertoli and male germ cells of Rb heterozygotes and compared it with those of all-telocentric and Rb homozygotes. The spatial relationships of the chromosome pairs studied (5, 11, 13, 15, 16 and 17) indicate that they occupy a specific territory of the cell nucleus at each of the cell-cycle stages analysed (G_0 for the non-cycling Sertoli cell, G_1 for the long-cycling spermatogonia, M for the pachytene cells) and in the haploid spermatids. The Rb heterozygosities introduce large-scale changes in chromosomal territories (Figure 1e,f). In fact, the Rb fusion itself, while determining the loss of telomeric and most of the minor satellite DNA sequences, sequesters chromosomes from their original subregions; thus the spatial relationships of the chromosome arms in Rb heterozygotes differ from those in the homozygote patterns, with one homologue of each chromosome pair "sequestered" from its original position, as occurs in human chromosomes 18 and 19 when involved in reciprocal translocation (Croft *et al.*, 1999). Our findings suggest that the changes caused in the chromosomal nuclear territories by the presence of Rb heterozygosities has the potential to impair spermatogenesis. The damage to spermatogenesis may occur at any step of male germ cell cytodifferentiation. The Sertoli cells of Rb homozygous fertile animals always show diffuse X and Y FISH signals (Figure 1g), suggesting a decondensed chromatin organization correlated with active gene expression. It is well known that condensation of the X and Y chromosomes in Sertoli cells of man and mouse is dynamic, e.g. the X and Y are condensed in pre-pubertal fertile individuals and infertile men with the Sertoli-cell-only syndrome while they are expanded in adult fertile individuals (Guttenbach *et al.*, 1996 in the house mouse; Kofman-Alfaro *et al.*, 1994, in man). Our finding that the X and Y signals are condensed in most (84%) of the Sertoli nuclei of Rb subfertile mice (Figure 1h) is similar to findings in humans with the Sertoli-cell-only syndrome (Kofman-Alfaro *et al.*, 1994). We suggest that large-scale changes in chromosomal territories, which change the physical interactions of specific genome portions, act as an epigenetic factor in controlling gene expression in somatic and germ cells. In other words, we think we have found a new additional factor (changes of chromosomal territories in Rb structural heterozygotes) that can trigger distorted cytodifferentiation processes leading to germ cell death. By changing the physical interactions of specific genome portions, Rb heterozygosities might disturb gene expression and several cellular machineries, thus being detrimental to germ cell cytodifferentiation: the heterogeneous effects of Rb heterozygosities can be explained by this view if we consider that not all of the 3-D changes will necessarily result in distorted gene expression. Interestingly, primary spermatocytes of structural heterozygotes show deregulated enzymatic activities (Redi *et al.*, 1983) and qualitative differences in protein patterns (Forejt, 1982), and also have a longer prophase I (Redi *et al.*, 1985) so as to complete synapsis and to resolve tangles or interlocks between non-homologues (Rose and Holm, 1993).

If this hypothesis is proved, the mouse Rb model of chromosomal variability could be used for more general studies on the epigenetic regulation of gene activity. In addition, it could be used as a model for studies of impaired spermatogenesis in humans and farm animals.

...and evolutionary...

Notwithstanding the Haldane rule (when structural heterozygosities occur, chromosome-derived subfertility/sterility is limited to the heterogametic sex), oogenesis is also affected by Rb heterozygosities. Impaired developmental processes in germ cells have also been detected in female carriers of Rb heterozygosities (Mittwoch *et al.*, 1984; Garagna *et al.*, 1990), with the production of very few growing oocytes; thus female Rb heterozygotes have a reduced reproductive lifespan. These data indicate the need to reconsider the concepts of genetic epistemology used to formulate the Haldane rule. It might be more correct to compare the male and female cytological steps of meiosis rather than the final condition of their reproductive capacity (fertility), which is based on the total output of germ cells able to develop in conditions of Rb heterozygosity. If this view is assumed (adult males carry out the full gametogenetic process while adult females carry out folliculogenesis; males need millions of cells to be fertile, females just a few), it follows that the validity of the Haldane rule must be limited to the fertility status and it cannot be used to infer the ultimate condition of the gametogenetic process. The fitness of Rb heterozygotes is not only reduced by primary losses of germ cells but also by secondary impairment of fertility. The production of aneuploid gametes (Redi *et al.*, 1984; Redi and Capanna, 1988) due to segregational meiotic disorders causes both pre-zygotic and postzygotic losses because embryonic development is altered (Gropp *et al.*, 1982b; for a review see Miller and Therman, 2001).

It is important to understand the relevance of all these effects on developmental process (gonad formation and function) since, theoretically, they can be crucial for evolutionary processes like speciation.

Complex heterozygotes certainly occur along the contact zones between karyotypically different populations and they likely play an active role in keeping the populations isolated. This role is played not only by male heterozygotes but also by females, mainly through the shortened reproductive lifespan due to reduced oocyte numbers; in addition, some of the few oocytes produced are carriers of aneuploidies and this leads to postzygotic losses due to zygotic chromosomal unbalance. Thus, the fertility loss of structural heterozygotes has the potential to establish reproductive barriers, leading to conditions that can trigger chromosomal microevolutionary processes. However, the contribution of Rb heterozygotes to chromosomal speciation is still hotly debated (reviewed in King, 1993), even though there is no doubt about their active role in creating genetic barriers between populations, thus contributing to genetic isolation and the appearance of cryptic species (and their evolution to full species). Nevertheless, the so-called chromosomal speciation cannot be fully understood simply on the basis of pre-mating and/or postmating isolating barriers due to chromosome-derived subfertility of the structural hybrids. Hybrid subfertility can surely reinforce the isolation of populations in which the chromosomal novelty is already fixed in homozygosis (and the crosses produce hybrid carriers of complex structural heterozygosities; Redi and Capanna, 1988) but cannot explain the fixation of the chromosomal novelty. This is currently the conceptual constraint of chromosomal speciation: the low probability of fixation of the chromosomal novelty under negative heterosis of the structural hybrids. In other words, this is the "thanks to – in spite of" paradox: on the one hand, the hybrid fertility loss should help to establish the postmating barrier but, on the other hand, the hybrid negative heterosis should lead to rapid disappearance of the chromosomal novelty within a panmictic population. This problem was clearly stated by White (1978) and has been widely explored since the end of the 1970s (Lande, 1985; Rieseberg, 2001) thanks to mathematical modelling. One of the main contributions to it has been the understanding that genetic drive can lead to homozygosis; however, to be effective, the deme size must be small, to justify the rapidity of the process (Garagna *et al.*, 1997). Resolving the "thanks to – in spite of" paradox by traditional genetic approaches seems quite difficult. The use of functional genomics, as discussed above, could help in sketching a favourable scenario for Rb chromosome formation and fixation since it shows that: (a) the rate of Rb chromosome formation is related to the genome composition/organization, which can be very high and occurring at the primordial stages of gonad differentiation; (b) not all the Rb chromosomes alter the nuclear architecture (disturbing synapsis or scheduled gene expressions) and thus have a chance of being fixed because do not affect gametogenesis; (c) an Rb chromosome in heterozygosity has the potentiality to catalyse the formation of its homologue (Redi *et al.*, 1990b); (d) the formation of new Rb variants is then "driven", *sensu* Dover (Dover *et al.*, 1982), thanks to whole arm exchanges between Rb chromosomes (Capanna and Redi, 1995); (e) these phenomena are continuously running in the genome of small-sized demes leading to the establishment of several Rb population.

...Activities: the DEVO–EVO volcano

In the present paper, we suggest how molecular and cellular processes can drive supraindividual processes such as speciation. The rationale of this approach is based on the conceptual bases of functional genomics which tries to explain the phenotypic features of cell differentiation able to affect developmental processes. Within this context, individuals acquire different fitness values, creating the conditions for speciation. The study of genome composition/organization in vertebrates, and particularly in mammals, constitutes an emerging field of integrated analysis, bringing together cytogenetics, cytology, developmental biology and ecology. Some nice examples of this type of study have been provided recently; indeed they have revived neglected research areas previously considered impracticable. For example, it is well known that genome sizes are positively correlated with nuclear and cellular volumes (Hertwig, 1903). Thus, very small genomes are well adapted to the high metabolic requirements of flight (Hughes and Hughes, 1995). In birds, running species have much larger genomes than flyers (Tiersch and Wachtel, 1991). Among vertebrates, birds and bats have the highest metabolic rates and also the smallest genome sizes (Capanna and Manfredi Romanini, 1971; Tiersch and Wachtel, 1991): the reduction of genome size is constantly achieved in eukaryotes by quantitative variation in the copy number of repetitive elements, i.e. those elements that build up heterochromatin; cytologically detectable or not, it is the heterochromatin (i.e. repetitive DNA, like satDNA) variation that accounts for the tremendous variation in C-values (Gregory, 2001). A nice comparison here is that between the human genome and the genome of the pufferfish *Fugu rubripes*, which has the smallest genome ever recorded among vertebrates, i.e. 400 Mb or 0.41 pg DNA: "Rather than being contracted, the *Fugu* genome is unexpanded and never acquired large quantities of non-coding DNA sequences" (Brenner *et al.*, 1993). Non-coding DNA sequences can be seen as the nuclear buffer to fit eco-ethological constraints via nucleotypic effects (Redi *et al.*, 2001). The components of genomes that affect the phenotype independently of their coding capacity (Bennet and Leitch, 1997), and that rely on Het variations, are becoming increasingly clear. Vinogradov's paper (1995) on the nucleotypic effect in homeotherms, with its finding that in mammals the body-mass-corrected basal metabolic rate is related to genome size, clearly shows that a sizeable part of the genome (above 25% in humans) affects evolutionary adjustment of the metabolic rate. Strong refutation of the "junk" DNA theories (Redi *et al.*, 2001) comes from the differential scaling of cryptomonal genomes (flagellate single-cell organisms) that is achieved thanks to non-coding (Het) DNA (Beaton and Cavalier-Smith, 1999). Monaghan and Metcalfe (2000) provided another nice example, showing a clear correlation between genome size and relative longevity in birds; this suggests the potential influence of genome size on life-history parameters which can be independent of the information encoded by genes.

The aim of this type of study is to dissect the mammalian genome to identify the sequences responsible for nucleotypic effects. Clarifying the "essential" vertebrate genome, hypothesized by Brenner *et al.* (1993), should be a big help in understanding key problems in biology: structure–function relationships could become more easily studied, and ultimately revealed, thanks to the integration of cell biology and ecology. The challenge to explain symmorphosis ("the quantitative match of design and function parameters", Weibel *et al.*, 1991) exists and it seems that we now have, in non-coding DNA biology (e.g. satDNA, heterochromatin and intergenic region DNAs), the means to meet it (Redi *et al.*, 2001). The genomes of tetraodontoid fish and of bats seem to teach us that high redundancy of this type of DNA is not essential, but that a minimum is indispensable for regular gene expression during development. Increased genome sizes are negatively correlated with the complexity of an organism because they tolerate repetitive sequences. Non-coding DNA has been added during evolution and this acquisition is positively correlated with increasing complexity of functional anatomy. A much broader view of this phenomenon reveals that non-coding repetitive DNA represents a large proportion of the eukaryotic genome, comprising several classes of sequences ranked on the basis of their copy numbers (highly or moderately repetitive and simple sequence). The fraction of non-coding DNA increases with the complexity of an organism, from 15% of the genome in bacteria, to ≅ 30%, 70% and 95% in *S. cerevisiae*, *C. elegans* and *H. sapiens*, respectively (Shabalina *et al.*, 2001). Although the function of non-coding DNA remains poorly understood, this simple comparison suggests that this DNA is functionally important and likely can help us to understand the complexity of an organism. Indirect support of this conclusion is the low variation of the estimated number of coding genes exhibited by organisms as different as yeast, ≅ 6,000 (Goffeau, 2000), *Caenorhabditis*, 19,000 (*C. elegans* sequencing consortium, 1998), *Drosophila*, 14,000 (Adams *et al.*, 2000) and humans, 35,000 (Lander *et al.*, 2001). Generally, the gene density in a genome seems to decrease consistently with the increased complexity of an organism.

Since the complexity of an organism is related neither to its genome size (the well known C-value paradox) nor to its gene number (the new GN paradox; Claverie, 2001), we might look for a relationship with the genome's complexity of composition/organization. Our intuitive perception of organismal complexity is based on simplistic measures of biological complexity, thus forgetting that living beings are the end results of non-linear and epigenetic phenomena (Maleszka *et al.*, 1998). Our analysis links, in a cascade, the origin and evolution of pericentromeric DNA to the genome structuring and architectural properties and then to the consequent ability to influence developmental and evolutionary processes. This highlights a possible unitary view for an understanding of life organization and evolution at any hierarchical level. The origin and amplification of satDNA sequences is not constrained by selection (Kimura, 1968; Dover *et al.*, 1982; Charlesworth *et al.*, 1994); they merely follow their own selfish replication patterns. However, when dispersing through the genome via karyotype restructuring, satDNA could be exposed to selection because the resulting chromosomal rearrangements can affect the mitotic–meiotic machinery involved in cell cytodifferentiation. From the point of view of genetic epistemology, this two-step pattern of the origin, evolution and maintenance of satDNA in the eukaryote genome has the advantage of reconciling the conflicting functionalist and non-functionalist views on satDNA since it assigns different constraints at different levels of life organization. Self-organization (for simple sequence motifs that can develop higher order periodicities based on intrinsic metabolic or geometric properties, Kauffman, 1993; Hancock, 1996) and Darwinism (i.e. evolution by selection, by which it is difficult to explain the molecular bases of development and the generation of novelties, rather than what is likely to survive once it has appeared) are not subject to the same constraints but are two interdependent processes: the former is able to originate novelties for the latter which, by selecting those novelties that best fit the actual ecological requirements for survival, ensures the transmission of the genome features prone to produce the novelties.

The view we have presented on the capacity of non-coding DNA to affect phenotypic traits can be generalized thinking that the vast majority of mammal's genome is composed of this class of DNA and making use of metaphors.

The present-day activities of the Etna volcano provide a good metaphor: the constant magmatic activities (the intrinsic metabolic activities of DNA) produce billions and billions of jets (moulding and shuffling of the genome); sometimes, one of these jets becomes an eruption (a novelty) that reaches the cone of the volcano and has the chance to become a lava-flow (a species) rather than to fall down again into the volcano. The fate of the lava-flow (death or life of the species) depends on the different ecological scenarios present at that time in the arena in which it will be selected for (development), entering a temporal route (evolution) that will change its geological characteristics (functional anatomical changes, speciation) depending on the interactions with the earth's features (DEVO–EVO interactions); cooled lava-flows could represent the fossil record.

This metaphor suggests that in the near future, thanks to the hundreds of ongoing full-genome sequencing projects, humankind will have the opportunity to deeply alter these processes (i.e. genome moulding and origin of novelties) for benefical medical and agricultural applications: biologists will do this by making use of the genomes' gene list, with their proteins, as if they are the elements of the "periodic table of biology" (Lander, 1996; Lander *et al.*, 2001), as now-a-day chemists synthetize new compounds employing the few elements we know.

Acknowledgements

This study was supported by the Telethon Fondazione Onlus, Grant N. A 132 to SG and MZ; the Italian MURST Progetti di rilevante interesse Nazionale COFIN 2000 provided support to SG and MZ. The continuous support by the Fondazione di Piacenza e Vigevano and the Olympus Foundation Sciences for Life to SG, MZ and CAR is much appreciated.

References

Adams MD, Celniker SE, Holt RA, Evans CA, Gocayne JD, Amanatides PG, Scherer SE, Li PW, Hoskins RA, Galle RF, *et al.* (2000). The genome sequence of *Drosophila melanogaster. Science* 287: 2,185–2,195.

Beaton MJ, Cavalier-Smith T (1999). Eukaryotic non-coding DNA is functional: evidence from the differential scaling of cryptomonal genomes. *Proc R Soc Lond B* 266: 2,053–2,059.

Bennet MD, Leitc IJ (1997). Nuclear DNA amounts in angiosperms – 583 new estimates. *Ann Bot* 80: 169–196.

Bickmore WA, Bridger JM (1999). A sense of time and place. *Chromos Res* 7: 425–429.

Brenner S, Elgar G, Sandford R, Macrae A, Venkatesh B, Aparicio S (1993). Characterization of the pufferfish (Fugu) genome as a compact model of vertebrate genome. *Nature* 366: 265–268.

Bridger JM, Boyle S, Kill IR, Bickmore WA (2000). Re-modelling of nuclear architecture in quiescent and senescent human fibroblasts. *Curr Biol* 10: 149–152.

Britton-Davidian J, Catalan J, da Graca Ramalhinho M, Ganem G, Auffray JC, Capela R, Biscoito M, Searle JB, da Luz Mathias M (2000). Rapid chromosomal evolution in island mice. *Nature* 403: 158–159.

Capanna E, Manfredi Romanini MG (1971). Nuclear DNA content and morphology of the karyotype in certain Paleartic microchiroptera. *Caryologia* 24: 471–482.

Capanna E, Redi CA (1995). Whole-arm reciprocal translocation (WART) between Robertsonian chromosomes: finding of a Robertsonian heterozygous mouse with karyotype derived through WARTs. *Chromos Res* 3: 135–137.

C. elegans sequencing consortium (1998). Genome sequence of the nematode *C. elegans*: a platform for investigating biology. *Science* 282: 2,012–2,018.

Cerda MC, Berrios S, Fernandez-Donoso R, Garagna S, Redi CA (1999). Organisation of complex nuclear domains in somatic mouse cells. *Biol Cell* 91: 55–65.

Charlesworth B, Sniegowski P, Stephan W (1994). The evolutionary dynamics of repetitive DNA in eukaryotes. *Nature* 371: 215–220.

Claverie JM (2001). What if there are only 30,000 human genes? *Science* 291: 1,255–1,257.

Croft JA, Bridger JM, Boyle S, Perry P, Teague P, Bickmore WA (1999). Differences in the localization and morphology of chromosomes in the human nucleus. *J Cell Biol* 145: 1,119–1,131.

Dover GA, Brown S, Coen E, Dallas J, Strachan T, Trick M (1982). The dynamics of genome evolution and species differentiation. In: Dover GA and Flavell RB, eds. *Genome Evolution.* New York: Academic Press, pp. 343–372.

Forejt J (1982). X–Y involvement in male sterility caused by autosome translocations – A hypothesis. In: Crosignani PG and Rubin BL, eds. *Genetic Control of Gamete Production and Function.* New York: Academic Press, pp. 135–151.

Garagna S, Redi CA, Zuccotti M, Britton-Davidian J, Winking H (1990). Kinetics of oogenesis in mice heterozygous for Robertsonian translocations. *Differentiation* 42: 167–171.

Garagna S, Redi CA, Capanna E, Andayani N, Alfano RM, Doi P, Viale G (1993). Genome distribution, chromosomal allocation, and organization of the major and minor satellite DNAs in 11 species and subspecies of the genus *Mus*. *Cytogenet Cell Genet* 64: 247–255.

Garagna S, Broccoli D, Redi CA, Searle JB, Cooke HJ, Capanna E (1995). Robertsonian metacentrics of the house mouse lose telomeric sequences but retain some minor satellite DNA in the pericentromeric area. *Chromosoma* 103: 685–692.

Garagna S, Zuccotti M, Redi CA, Capanna E (1997). Trapping speciation. *Nature* 390: 241–242.

Garagna S, Marziliano N, Zuccotti M, Searle JB, Capanna E, Redi CA (2001). Pericentromeric organization at the fusion point of mouse Robertsonian translocation chromosomes. *Proc Natl Acad Sci USA* 98: 171–175.

Goffeau A (2000). Four years of post-genomic life with 6,000 yeast genes. *FEBS Lett* 480: 37–41.

Gregory TR (2001). Animal Genome Size Database. http: //www.genomesize.com

Gropp A, Tettenborn U, Lehmann von E (1970). Chromosomenvariation vom Robertsonschen Typus bei der Tabakmaus, *M. poschiavinus* und ihren Hybriden mit der Laboratoriumsmaus. *Cytogenetics* 9: 9–23.

Gropp A, Winking H, Redi CA, Capanna E, Britton Davidian J, Noack G (1982a). Robertsonian karyotype variation in wild house mice from Rhaeto-Lombardia. *Cytogenet Cell Genet* 34: 67–77.

Gropp A, Winking H, Redi CA (1982b). Consequences of Robertsonian heterozygosity: segregational impairment of fertility versus male-limited sterility. In: Crosignani PG, Rubin BL, Fraccaro M, eds. *Genetic Control of Gamete Production and Function*. London–New York: Academic Press, pp. 115–134.

Guttenbach M, Martinez Esposito MJ, Engel W, Schmid M (1996). Interphase chromosome arrangement in Sertoli cells of adult mice. *Biol Reprod* 54: 980–986.

Hamkalo BA, Elgin SCR (1991). Functional organization of the nucleus. A laboratory guide. *Methods in Cell Biology*. New York: Academic Press, p. 35.

Hancock JM (1996). Simple sequences and the expanding genome. *BioEssays* 18: 421–425.

Hemann MT, Greider CW (2000). Wild-derived inbred mouse strains have short telomeres. *Nucl Acids Res* 28: 4,474–4,478.

Hertwig R (1903). Ueber Korrelation von Zell- und Kerngrösse und ihre Bedeutung für die geschlechtliche Differenzierung und die Teilung der Zelle. *Biologisches Centralblatt* 23: 108–119.

Holmquist GP, Comings DE (1975). Sister chromatid exchange and chromosome organization based on a bromodeoxyuridine Giemsa-C-banding technique (TC-banding). *Chromosoma* 52: 245–259.

Hughes AL, Hughes MK (1995). Small genomes for better flyers. *Nature* 377: 391.

Kauffman SA (1993). *The Origins of Order: Self-Organization and Selection in Evolution*. Oxford: Oxford University Press.

Kimura M (1968). Evolutionary rate at the molecular level. *Nature* 217: 624–626.

King M (1993). *Species Evolution: The Role of Chromosome Change*. Cambridge: Cambridge University Press.

Kofman-Alfaro S, Speed RM, Boyle S, Chandley AC (1994). Condensation behaviour of the human X chromosome in male germ cells and Sertoli cells examined by fluorescence *in situ* hybridization. *Chromos Res* 2: 439–444.

Kozubek S, Lukasova E, Mareckova A, Skalnikova M, Kozubek M, Bartova E, Kroha V, Krahulcova E, Slotova J (1999). The topological organization of chromosomes 9 and 22 in cell nuclei has a determinative role in the induction of t(9,22) translocations and in the pathogenesis of t(9,22) leukemias. *Chromosoma* 108: 426–435.

Lamond AI, Earnshaw WC (1998). Structure and function in the nucleus. *Science* 280: 547–553.
Lande R (1985). The fixation of chromosomal rearrangements in a subdivided population with local extiction and colonization. *Heredity* 54: 323–332.
Lander ES (1996). The new genomics: global views of biology. *Science* 274: 536–539.
Lander ES, Linton LM, Birren B, Nusbaum C, Zody MC, Baldwin J, Devon K, Dewar K, Doyle M, Fitzhugh W, *et al.* (2001). Initial sequencing and analysis of the human genome. *Nature* 409: 860–921.
Lifschytz E, Lindsley DL (1972). The role of X-inactivation during spermatogenesis. *Proc Natl Acad Sci* 69: 182–186.
Lin MS, Davidson RL (1974). Centric fusion, satellite DNA and DNA polarity in mouse chromosomes. *Science* 185: 1,179–1,181.
Maleszka R, de Couet HG, Miklos GLC (1998). Data transferability from model organisms to human beings: Insights from the functional genomics of the flightless region of *Drosophila*. *Proc Natl Acad Sci* 95: 3,731–3,736.
Manuelidis L (1997). Interphase chromosome positions and structure during silencing, transcription and replication. In: Van Driel R and Otte AP, eds. *Nuclear Organization, Chromatin Structure, and Gene Expression*. Oxford: Oxford University Press, pp. 145–168.
Meyne J, Goodwin EH, Moyzis RK (1994). Chromosome localization and orientation of the simple sequence repeat of human satellite I DNA. *Chromosoma* 103: 99–103.
Miklos GLC (1974). Sex-chromosome pairing and male sterility. *Cytogenet Cell Genet* 13: 558–577.
Miller OJ, Therman E (2001). *Human Chromosomes*. New York: Springer-Verlag.
Misteli T, Spector DL (1998). The cellular organization of gene expression. *Curr Opin Cell Biol* 10: 323–331.
Mittwoch U, Mahadevaiah S, Setterfield LA (1984). Chromosome anomalies that cause male sterility in the mouse also reduce ovary size. *Genet Res* 44: 219–224.
Monaghan P, Metcalfe N (2000). Genome size and longevity. *Trends Genet* 16: 331–332.
Nagele RG, Freeman T, Fazekas J, Lee KM, Thomson Z, Lee HY (1998). Chromosome spatial order in human cells: evidence for early origin and faithful propagation. *Chromosoma* 107: 330–338.
van der Ploeg M (2000). Cytochemical nucleic acid research during the twentieth century. *Eur J Histochem* 44: 7–42.
Redi CA, Hilscher B, Winking H (1983). Stage-dependent enzymatic activities in spermatogenesis of mice with the standard karyotype and of chromosomal variants with impaired fertility. *Andrologia* 15: 322–330.
Redi CA, Garagna S, Pellicciari C, Manfredi Romanini MG, Capanna E, Winking H, Gropp A (1984). Spermatozoa of chromosomally heterozygous mice and their fate in male and female genital tracts. *Gamete Res* 9: 273–286.
Redi CA, Garagna S, Hilscher B, Winking H (1985). The effects of some Robertsonian chromosome combinations on the seminiferous epithelium of the mouse. *J Embryol Experim Morphol (Development)* 85: 1–19.
Redi CA, Capanna E (1988). Robertsonian heterozygotes in the house mouse and the fate of their germ cells. In: Daniel A, ed., *The Cytogenetics of Mammalian Autosomal Rearrangements*. New York: Alan R. Liss, pp. 315–359.
Redi CA, Garagna S, Della Valle G, Bottiroli G, Dell'Orto P, Viale G, Peverali F, Raimondi E, Forejt J (1990a). Differences in the organisation and chromosomal allocation of satellite DNA between the European long-tailed house mice of the *domesticus* and *musculus* groups. *Chromosoma* 99: 11–17.

Redi CA, Garagna S, Zuccotti M (1990b). Robertsonian chromosome formation and fixation: the genomic scenario. *Biol J Linn Soc* 41: 235–255.
Redi CA, Garagna S, Zacharias H, Zuccotti M, Capanna E (2001). The other chromatin. *Chromosoma* 110: 136–147.
Rieseberg LH (2001). Chromosomal rearrangements and speciation. *Trends Ecol Evol* 16: 351–358.
Rose D, Holm C (1993). Meiosis-specific arrest revealed in DNA topoisomerase II mutants. *Mol Cell Biol* 13: 3,445–3,455.
Shabalina SA, Ogurtsov AY, Kondrashov VA, Kondrashov AS (2001). Selective constraint in intergenic regions of human and mouse genomes. *Trends Genet* 17: 373–376.
Strouboulis J, Wolffe AP (1996). Functional compartmentalization of the nucleus. *J Cell Sci* 109: 1,991–2,000.
Tiersch TR, Wachtel SS (1991). On the evolution of genome size in birds. *J Hered* 82: 363–368.
Vinogradov AE (1995). Nucleotypic effect in homeotherms: body mass-corrected basal metabolic rate of mammals is related to genome size. *Evolution* 49: 1,249–1,259.
Wallace B, Searle JB, Everett CA (1992). Male meiosis and gametogenesis in wild house mice (*Mus musculus domesticus*) from a chromosomal hybrid zone: a comparison between "simple" Robertsonian heterozygotes and homozygotes. *Cytogenet Cell Genet* 61: 211–220.
Weibel ER, Taylor CR, Hoppeler H (1991). The concept of symmorphosis: a testable hypothesis of structure–function relationship. *Proc Natl Acad Sci* 88: 10,357–10,361.
White MJD (1978). *Modes of Speciation.* San Francisco: Freeman.
Wong AKC, Rattner JB (1988). Sequence organization and cytological localization of the minor satellite of mouse. *Nucl Acids Res* 16: 11,645–11,661.
Zijlmans JM, Martens UM, Poon SSS, Raap AK, Tanke HJ, Ward RK, Lansdorp PM (1997). Telomeres in the mouse have large inter-chromosomal variations in the number of T2AG3 repeats. *Proc Natl Acad Sci* 94: 7,423–7,428.

4.2. New Developments in Multicolour Fluorescence *in situ* Hybridization

Christine Fauth, Jürgen Kraus, Sabine Langer and Michael R. Speicher
Institut für Humangenetik, Technische Universität München, Trogerstr. 32, D-81675 München, Germany

Introduction

In recent years various fluorescence *in situ* hybridization (FISH)-based multicolour technologies have been developed (for review see Fauth and Speicher, 2001). This multitude of different methods was fuelled by efforts to use FISH for the high-resolution analysis of metaphases and the automated generation of karyotypes. The initial multicolour karyotyping technologies include multiplex-FISH (M-FISH/Speicher *et al.*, 1996) and spectral karyotyping (SKY/Schröck *et al.*, 1996). Later combined binary ratio labelling (COBRA/Tanke *et al.*, 1999) and colour-changing karyotyping (CCK/Henegariu *et al.*, 1999) were added for FISH-based karyotype analysis. Up to date these technologies were supplemented by a variety of additional methods. Some of these applications aim at improving the resolution for intrachromosomal aberrations, others are tailored for a detailed analysis of specific subregions within the genome (for a summary see Table 1).

The multicolour karyotyping technologies are increasingly being used in clinical and research applications. The elucidation of previously hidden abnormalities or the unequivocal deciphering of complex rearrangements indicated unprecedented resolution limits. However, exact resolution limits have not yet been defined for any multicolour-system. Here, we focus on how both sensitivity and specificity for the detection and correct classification of small (~3 Mb or less) interchromosomal rearrangements can be determined. Furthermore, we present a strategy how classification accuracy of existing systems can be optimized. Finally, we provide an outlook on future possible multicolour-FISH developments.

M. Schmid and I. Nanda (eds.), Chromosomes Today, Volume 14, 187–196.

Table 1. *Currently existing multicolour-FISH technologies and their main applications*

Method	Preferential application	Reference
Multiplex-FISH (M-FISH)	Karyotyping	Speicher *et al.* (1996)
Spectral karyotyping (SKY)	Analysis of inter-chromosomal aberrations	Schröck *et al.* (1996)
COmbined Binary RAtio labelling (COBRA)		Tanke *et al.* (1999)
Colour-changing karyotyping (CCK)	Identification of euchromatin in marker-chromosomes	Henegariu *et al.* (1999)
Chromosome arm COBRA		Wiegant *et al.* (2000)
Arm-specific multicolour-FISH		Karhu *et al.* (2000)
M-FISH telomere integrity assay (8 chromosomes simultaneously)	Screen for cryptic translocations	Granzow *et al.* (2000)
M-TEL (12 chromosomes sim.)		Brown *et al.* (2001)
TM-FISH (24 chromosoms sim.)		Henegariu *et al.* (2001a)
cenM-FISH	Origin of marker-chromosomes	Nietzel *et al.* (2001)
CM-FISH		Henegariu *et al.* (2001b)
AcroM-FISH	Complete analysis of marker-chromosomes derived from acrocentric chromosomes	Langer *et al.* (2001a)
Multicolour chromosome bar code	Karyotyping	Müller *et al.* (1997)
Cross-species colour segmenting	Analysis of inter- and intrachromosomal aberrations	Müller *et al.* (1998)
Multicolour-banding (Mband)	Intrachromosomal rearrangements	Chudoba *et al.* (1999)
Simultaneous hybridization of mouse and human M-FISH mix	Analysis of mouse–human hybrid cell lines	Langer *et al.* (2001b)

Sensitivity and specificity of multicolour karyotyping

A large number of publications demonstrated the ability of M-FISH, SKY and COBRA to identify subtle rearrangements suggesting a superior resolution as compared to standard banding technologies. However, the true performance of multicolour karyotyping has never been assessed in detail. So far approaches for an evaluation have included the comparison of different systems based on experimental data (Strefford *et al.*, 2001; Rens *et al.*, 2001) or on mathematical models for classification accuracy (Garini *et al.*, 1999; Castleman *et al.*, 2000). In addition, some classification inaccuracies were reported (Zhang *et al.*, 2000; Lee *et al.*, 2001). However, these efforts provide no reliable estimation of the sensitivity and specificity for the detection and classification of small (~3 Mb or less) interchromosomal rearrangements. As a consequence the true extent of classification artifacts and the frequency of small rearrangements, which may even be overlooked, has remained unknown.

Both sensitivity and specificity depend on a number of factors such as the sampling resolution of the scanning device, the intensity of fluorescence, the type and quality of the probes, the properties of metaphase spreads, the condensation of chromosomes, and so on. Using "sensitivity" we refer to the chance for the correct *detection* and using "specificity" we mean the chance for the correct *classification* of small interchromosomal rearrangements. Assuming that appropriate imaging systems and probe sets are used together with optimized protocols for probe labelling and metaphase spread preparation (which should be the case in most laboratories applying multicolour karyotyping), sensitivity and specificity depend critically on the labelling strategy, which determines the "fluorochrome labelling complexity". The necessity to understand the impact of the labelling strategy on classification accuracy is documented by a recent paper (Lee *et al.*, 2001), which describes a number of classification inaccuracies for various multicolour systems such as M-FISH, COBRA and SKY. Although these observations are an important confirmation of previously reported classification limitations (Azofeifa *et al.*, 2000), the underlying reasons for the classification artifacts are not identified. In fact, a systematic analysis of the causes of classification artifacts has never been provided although such an analysis should be mandatory considering the widespread use of multicolour karyotyping in clinical cytogenetic applications and tumourcytogenetics.

Here, we present such a systematic approach, which was refined and updated compared to our previously published data (Azofeifa *et al.*, 2000). This was used for precise estimates of sensitivity and specificity and in addition for the development of a new strategy to optimize multicolour karyotyping.

As illustrated in Figure 1, colour blending due to fluorescence flaring (Figure 1a,d) or fluorescence flaring alone (Figure 1b,c) are major causes for classification errors. The extent of possible misclassifications is dictated by the fluorochrome-combinations generated by an interchromosomal rearrangement on a derivative chromosome. The fluorochrome-combination types identified for 5-fluorochrome probe mixes are shown on the left-hand side (Figure 1a–c), for the 8-fluorochrome probe mix on the right-hand side (Figure 1d). "A" refers to the chromosome, which forms the centric, "B" to the chromosome, which forms the translocated segment after a translocation. The bars in the scheme indicate the position of the breakpoints. The resulting derivative chromosome is shown in the columns "visual" and "auto" because analysis can be based either on visual inspection of individual colour-channels or by automated classification using sophisticated computer algorithms. For simplicity only one of the involved chromosomes is discussed. Using a 5-fluorochrome probe mix three different types of fluorochrome-combinations can be identified, which determine critically sensitivity and specificity (Figure 1a–c).

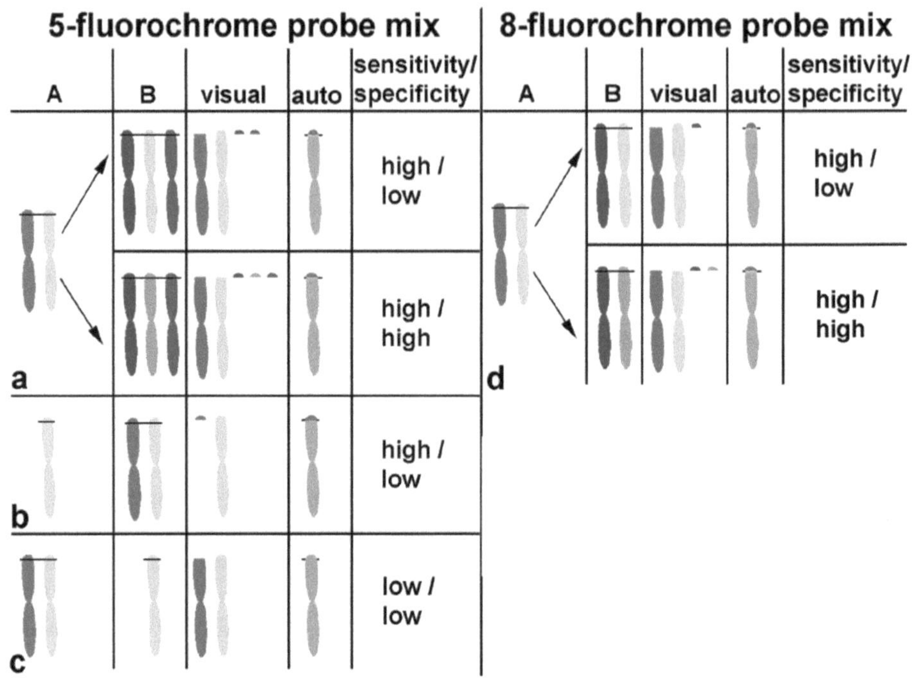

Figure 1. *This subdivision is based on the fluorochrome-combinations on derivative chromosomes as they determine critically the sensitivity and specificity in multicolour karyotyping. The categorization shown in Figure 1 is based on our previously published experience with misclassifications (Azofeifa* et al., *2000). The detection (sensitivity) and classification (specificity) accuracies are exemplarily shown for small (~3 Mb or less) translocations, but the same considerations apply to insertions as well. The fluorochrome-combination types which can occur if a 5-fluorochrome probe mix is used are shown in (a–c), for a 8-fluorochrome probe mix in (d). Column "A" shows the chromosome, which forms the centric, "B" the chromosome which forms the translocated segment after a translocation. The resulting derivative chromosome is shown in the columns "visual" (analysis based on visual inspection of individual colour-channels) and "auto" (analysis based on automated classification). For details see text. (a) Type-1 fluorochrome combination. (b) Type-2 fluorochrome combination. (c) Type-3 fluorochrome combination. (d) Using an 8-fluorochrome probe mix all colour-combinations observed on derivative chromosomes correspond to the "type-1" as shown in (a).*

Type-1 (Figure 1a)

The main principle of the type-1 fluorochrome-combination is that both chromosomes have each at least one different fluorochrome. In the example shown, chromosome A is labelled with two, chromosome B with three fluorochromes. Colour blending through fluorescence flaring is visible as an additional colour at the site of the translocation breakpoint. This "extra band" is the hallmark of type-1 fluorochrome-combinations (Azofeifa *et al.*, 2000). In the first example (upper row) both chromosomes have one fluorochrome in common (yellow). Such a translocation will be detected with high sensitivity as it results in additional fluorescence signals easily visible in two colour channels (blue and magenta). However, accurate classification of the translocated segment may be impossible as fluorescence flaring may hamper to distinguish between the possible colour combinations blue/magenta, red/blue/magenta or yellow/blue/magenta. In contrast, the lower row shows a translocation, which can be detected with both, high sensitivity and high specificity. The small translocated segment is labelled with three fluorochromes, the maximum fluorochrome number for labelling of a chromosome in a 5-fluorochrome probe mix of the first kind. As none of these three fluorochromes occurs in the centric segment the translocated segment can be assigned unequivocally to the chromosome with the corresponding three-colour combination. Thus, despite fluorescence flaring the classification is possible by visual inspection and robust computer algorithms.

Type-2 (Figure 1b)

The main principle of type-2 fluorochrome combinations is that the centric segment is labelled with a subset of fluorochromes that occurs in the translocated segment. An additional band at the breakpoint caused by blending of colours is not observed because both chromosomes involved in the translocation have only a single fluorochrome difference. The additional signal caused by the translocation ensures that it will be identified with high sensitivity. However, specificity may be low as it may not be possible to distinguish between possible colour-combinations red only versus red and yellow.

Type-3 (Figure 1c)

The main principle is that the translocated segment is labelled with a subset of fluorochromes that occur in the centric segment. As there is again only a single fluorochrome difference there is no additional band at the translocation breakpoint. However, as no new fluorochrome is added to the derivative chromosome fluorescence flaring may completely obscure the translocation. This is the most challenging situation in 24-colour karyotyping as both sensitivity and specificity are low. Currently there are two different kinds of 5-fluorochrome probe mixes available. The first kind has a minimum fluorochrome labelling complexity and utilizes one, two or a maximum number of three fluorochromes for the labelling of individual chromosomes. This labelling strategy was used in the first description of M-FISH (Speicher *et al.*, 1996) and corresponds to the majority of commercially available probe kits. The second kind is used for the SKY-system and is characterized by a higher fluorochrome labelling complexity as four chromosomes are labelled with four fluorochromes.

The 8-fluorochrome probe mix reduces the number of possible fluorochrome-combination types to one (Figure 1d). In the 8-fluorochrome probe mix all chromosomes are labelled each with a double combination. Thus, the occurrence of rearrangements, which are detected with a low sensitivity, is completely avoided. As a result all colour-combinations observed on derivative chromosomes correspond to the "type-1" as shown in Figure 1a.

In order to calculate the frequency of interchromosomal rearrangements, which by using a 5-fluorochrome probe mix can be misclassified or overlooked, we used the subdivision of all possible aberrations into three categories as shown in Figure 1a–c. The frequency of aberrations in the respective categories determines the specificity

Table 2. *Sensitivity and specificity for the detection of small (~3 Mb or less) interchromosomal rearrangements with different probe mixes*

	SKY-probe mix[a] (%)	5-fluorochrome probe mix[b] (%)	7-fluorochrome probe mix[c] (%)	8-fluorochrome probe mix[d] (%)
Sensitivity	82.0	86.5	96.7	100.0
Specificity	3.4	4.9	38.0	56.5

[a]The SKY-probe mix uses 5 fluorochromes for probe labelling. As 4 chromosomes are labelled with a combination of 4 fluorochromes it has a higher probe-complexity as compared to other 5-fluorochrome probe mixes.
[b]Most 5-fluorochrome probe mixes have a minimum probe-complexity as chromosomes are labelled with a maximum number of three different fluorochromes.
[c]7-fluorochrome probe mix describes a probe set with three chromosomes each labelled with a single fluorochrome and 21 chromosomes each labelled with a double combination (Azofeifa *et al.*, 2000).
[d]8-fluorochrome probe mix in which all chromosomes are labelled each with a double combination.

and sensitivity as listed in Table 2. For example, in a normal human male metaphase spread there are 552 possible two-way translocations. With the minimum-complexity 5-fluorochrome probe mixes 74 (13.5%) of these 552 two-way translocations correspond to type-3 fluorochrome-combinations (Figure 1c). This sets the sensitivity for the detection of small interchromosomal changes to 86.5%. On the other hand, there are only 27 translocations resulting in an unequivocal classification (Figure 1a, lower row), which defines a specificity of merely 4.9%. The higher complexity of the SKY-probe kit reduces both sensitivity and specificity even further as it increases the number of type-3 colour-combinations to 98 (18%) yielding a sensitivity of 82% and decreases the number of explicit rearrangements to 19 (3.4%).

Several important conclusions can be drawn: Firstly, resolution limits should never be given in absolute numbers. Although any 24-colour karyotyping system is capable of identifying small interchromosomal aberrations the detection of for example one 1.5 Mb interchromosomal rearrangement does not allow the conclusion that all possible 1.5 Mb exchanges will be detected. Resolution may vary tremendously depending on the fluorochrome composition. As a result some small rearrangements may be detected with ease while others with the same size may be missed. Secondly, often one- or two-colour chromosome painting experiments will be needed to corroborate the multicolour karyotyping interpretation. Filter-based multicolour karyotyping systems, such as M-FISH, offer the option to control hybridization results by visual inspection of individual fluorochrome channels. The visual inspection corresponds in principle to additional control-hybridizations of painting probes. Thus, a filter based system allows an interactive control to double-check ambiguous classification results. Third, considering the low sensitivity and specificity of the currently used multicolour karyotyping systems there is an urgent need for improvement.

A strategy for the optimization of classification accuracy

The classification problems described by us (Azofeifa *et al.*, 2000) and confirmed by Lee *et al.* (2001) are determined by the fluorochrome labelling complexity and can be avoided by a conceptual change in probe labelling. The theory behind this concept is that interchromosomal rearrangements are in general much easier identified if the translocated or inserted segment adds a new fluorochrome to the derivative chromosome. This can be achieved if all chromosomes are labelled with the same number of fluorochromes. Thus, the aim is not to reduce the fluorochrome labelling complexity but rather to ensure that no chromosome is labelled with a subset of fluorochromes used in the fluorescence profiles of other chromosomes. We suggest the use of eight different fluorochromes with a two-fluorochrome labelling of each chromosome. As shown in Figure 1d and in Table 2 a theoretical sensitivity of 100% is thus possible in multicolour karyotyping, provided the painting probes have a good quality and cover the entire chromosome including the subtelomeric regions. In addition, the specificity is substantially increased because 56.5% of all possible translocations result in unambiguous colour-combinations.

As an intermediate towards this goal we have already presented a 7-fluorochrome probe mix (Azofeifa *et al.*, 2000), which is now being routinely used in our laboratory. Currently we are testing several fluorochrome-candidates to be incorporated as an eighth fluorochrome in our probe mix as we have no doubt that the concepts discussed here are absolutely necessary if multicolour karyotyping should evolve to a robust standard technology in routine applications.

Outlook

There is still a demand for new multicolour FISH developments. These developments may include a further increase in probe number. The differential painting of all chromosome arms (42-colour experiment) was already accomplished with both the COBRA (Wiegant *et al.*, 2000) and the M-FISH (Karhu *et al.*, 2001) system. The multicolour techniques facilitate the karyotyping of other species. The usefulness of multicolour mouse karyotyping is well documented in a number of publications (reviewed in Fauth and Speicher, 2001). By combining probe mixes of different species, various genomes can be analysed simultaneously in hybrid cell systems, as shown for human–mouse hybrid cells (45-colour experiment/Langer *et al.*, 2001b).

Furthermore we anticipate the development of specific diagnostic probe sets, designed to facilitate identification of chromosomal aberrations frequently observed e.g. in leukemia and lymphomas. The completion of the human genome project and the identification of new DNA-probes (The international human genome mapping consortium 2001; Cheung *et al.*, 2001) will assist the design of specific probe sets.

Multicolour-FISH to interphase cytogenetics has yet to be explored in detail. The application of appropriate 3-D laser scanning imaging systems or algorithms for deconvoluting optical section images recorded by CCD cameras has provided first promising results. Such studies could provide important new insights into a broad spectrum of specific biological and clinical questions.

An unsolved problem remains the efficient screening for unknown microdeletions with cytogenetic tools. The array/chip-CGH technologies (Solinas-Toldo *et al.*, 1997; Pinkel *et al.*, 1998; Pollack *et al.*, 1999) may fill this gap. Furthermore, innovative microscope generations, new fluorescent dyes or novel imaging systems may again change the way we will analyse chromosomes.

Acknowledgements

Research in our laboratory is supported by the Deutsche Forschungsgemeinschaft (Sp460/4-1), the Deutsche Krebshilfe (10-1392-Pa1), the Wilhelm Sander-Stiftung (2000.015.1) and the BMBF.

References

Azofeifa J, Fauth C, Kraus J, Maierhofer C, Langer S, Bolzer A, Reichman J, Schuffenhauer S, Speicher MR (2000). An optimized probe set for the detection of small interchromosomal aberrations by 24-color FISH. *Am J Hum Genet* 66: 1,684–1,688.

Brown J, Saracoglu K, Uhrig S, Speicher MR, Eils R, Kearney L (2001). Subtelomeric chromosome rearrangements are detected using an innovative 12-colour FISH assay (M-TEL). *Nature Med* 7: 497–501.

Castleman KR, Eils R, Morrison L, Piper J, Saracoglu K, Schulze MA, Speicher MR (2000). Classification Accuracy in Multiple Color Fluorescence Imaging Microscopy. *Cytometry* 41: 139–147.

Cheung VG, Nowak N, Jang W, Kirsch IR, Zhao S, Chen XN, Furey TS, Kim UJ, Kuo WL, Olivier M, Conroy J, Kasprzyk A, Massa H, Yonescu R, Sait S, Thoreen C, Snijders A, Lemyre E, Bailey JA, Bruzel A, Burrill WD, Clegg SM, Collins S, Dhami P, Friedman C, Han CS, Herrick S, Lee J, Ligon AH, Lowry S, Morley M, Narasimhan S, Osoegawa K, Peng Z, Plazjer-Frick I, Quade BJ, Scott D, Sirotkin K, Thorpe AA, Gray JW, Hudson J, Pinkel D, Ried T, Rowen L, Shen-Ong GL, Strausberg RL, Birney E, Callen DF, Cheng JF, Cox DR, Doggett NA, Carter NP, Eichler EE, Haussler D, Korenberg JR, Morton CC, Albertson D, Schuler G, De Jong PJ, Trask BJ (2001). Integration of cytogenetic landmarks into the draft sequence of the human genome. *Nature* 409: 953–958.

Chudoba I, Plesch A, Lörch T, Lemke J, Claussen U, Senger G (1999). High resolution multi-color-banding: a new technique for refined FISH analysis of human chromosomes. *Cytogenet Cell Genet* 84: 156–160.

Fauth C, Speicher MR (2001). Classifying by colors: FISH-based genome analysis. *Cytogenet Cell Genet* 93: 1–10.

Garini Y, Gil A, Bar-Am I, Cadib D, Katzir N (1999). Signal to noise analysis of multiple color fluorescence imaging microscopy. *Cytometry* 35: 214–226.

Granzow M, Popp S, Keller M, Holtgreve-Grez H, Brough M, Schoell B, Rauterberg-Ruland I, Hager HD, Tariverdian G, Jauch A (2000). Multiplex FISH telomere integrity assay identifies an unbalanced cryptic translocation der (5)t(3;5)(q27;p15.3) in a family with three mentally retarted individuals. *Hum Genet* 107: 51–57.

Henegariu O, Heerema NA, Bray-Ward P, Ward DC (1999). Colour-changing karyotyping: an alternative to M-FISH/SKY. *Nature Genet* 23: 263–264.

Henegariu O, Artan S, Greally JM, Chen XN, Korenberg JR, Vance GH, Stubbs L, Bray-Ward P, Ward DC (2001a). Cryptic translocation identification in human and mouse using several telomeric multiplex FISH (TM-FISH) strategies. *Lab Invest* 81: 483–491.

Henegariu O, Bray-Ward P, Artan S, Vance GH, Qumsyieh M, Ward DC (2001b). Small Marker Chromosome Identification in Metaphase and Interphase Using Centromeric Multiplex FISH (CM-FISH). *Lab Invest* 81: 475–482.

Karhu R, Ahlstedt-Soini M, Bittner M, Meltzer P, Trent JM, Isola JJ (2001). Chromosome arm-specific multicolor-FISH. *Genes Chrom Cancer* 30: 105–109.

Knight SJL, Regan R, Nicod A, Horsley SW, Kearney L, Homfray T, Winter RM, Bolton P, Flint J (1999). Subtle chromosomal rearrangements in children with unexplained mental retardation. *Lancet* 354: 1,676–1,681.

Knight SJL, Lese CM, Precht KS, Kuc J, Ning Y, Lucas S, Regan R, Brenan M, Nicod A, Lawrie NM, Cardy DLN, Nguyen H, Hudson TJ, Riethman HC, Ledbetter DH, Flint J (2000). An optimized set of human telomere clones for studying telomere integrity and architecture. *Am J Hum Genet* 67: 320–333.

Langer S, Fauth C, Rocchi M, Murken J, Speicher MR (2001a). AcroM-FISH analyses of marker chromosomes. *Hum Genet* 109: 152–158.

Langer S, Jentsch I, Gangnus R, Yan H, Lengauer C, Speicher MR (2001b). Facilitating haplotype analysis by fully automated analysis of all chromosomes in human-mouse hybrid cell lines. *Cytogenet Cell Genet* 93: 11–15.

Lee C, Gisselsson D, Jin C, Nordgren A, Ferguson DO, Blennow E, Fletcher JA, Morton CC (2001). Limitations of chromosome classification by multicolor karyotyping. *Am J Hum Genet* 68: 1,043–1,048.

Müller S, Rocchi M, Ferguson-Smith MA, Wienberg J (1997). Toward a multicolor chromosome bar code for the entire human karyotype by fluorescence *in situ* hybridization. *Hum Genet* 100: 271–278.

Müller S, O'Brien PC, Ferguson-Smith MA, Wienberg J (1998). Cross-species color segmenting: a novel tool in human karyotype analysis. *Cytometry* 33: 445–452.

Nietzel A, Rocchi M, Starke H, Heller A, Fiedler W, Wlodarska I, Loncarevic IF, Beensen V, Claussen U, Liehr T (2001). A new multicolor-FISH approach for the characterization of marker chromosomes: centromere-specific multicolor-FISH (cenM-FISH). *Hum Genet* 108: 199–204.

Pinkel D, Segraves R, Sudar D, Clark S, Poole I, Kowbel D, Collins C, Kuo WL, Chen C, Zhai Y, Dairkee SH, Ljung BM, Gray JW, Albertson DG (1998). High resolution analysis of DNA copy number variations using comparative genomic hybridization to microarrays. *Nat Genet* 20: 207–211.

Pollack JR, Perou CM, Alizadeh AA, Eisen MB, Pergamenschikov A, Williams CF, Jeffrey SS, Botstein D, Brown PO (1999). Genome-wide analysis of DNA copy-number changes using cDNA microarrays. *Nat Genet* 23: 41–46.

Rens W, Yang F, O'Brien PCM, Solanky N, Ferguson-Smith MA (2001). A classification efficiency test of spectral karyotyping and multiplex fluorescence *in situ* hybridization: identification of chromosome homologies between Homo sapiens and Hylobates leucogenys. *Genes Chrom Cancer* 31: 65–74.

Schröck E, du Manoir S, Veldman T, Schoell B, Wienberg J, Ferguson-Smith MA, Ning Y, Ledbetter DH, Bar-Am I, Soenksen D, Garini Y, Ried T (1996). Multicolor spectral karyotyping of human chromosomes. *Science* 273: 494–497.

Solinas-Toldo S, Lampel S, Stilgenbauer S, Nickolenko J, Benner A, Döhner H, Cremer T, Lichter P (1997). Matrix-based comparative genomic hybridization: biochips to screen for genomic imbalances. *Genes Chrom Cancer* 20: 399–407.

Speicher MR, Ballard SG, Ward DC (1996). Karyotyping human chromosomes by combinatorial multi-fluor FISH. *Nat Genet* 12: 368–375.

Strefford JC, Lillington DM, Young BD, Oliver RTD (2001). The use of multicolor fluorescence technologies in the characterization of prostate carcinoma cell lines a comparison of

multiplex fluorescence *in situ* hybridization and spectral karyotyping data. *Cancer Genet Cytogenet* 124: 112–121.
Tanke HJ, Wiegant J, van Gijlswijk RPM, Bezrookove V, Pattenier H, Heetebrij RJ, Talman EG, Raap AK, Vrolijk (1999). New strategy for multi-colour fluorescence *in situ* hybridisation: COBRA: COmbined Binary RAtio labelling. *Eur J Hum Genet* 7: 2–11.
The international human genome mapping consortium (2001). A physical map of the human genome. *Nature* 409: 934–941.
Wiegant J, Bezrookove V, Rosenberg C, Tanke HJ, Raap AK, Zhang H, Bittner M, Trent JM, Meltzer P (2000). Differentially painting human chromosome arms with combined binary ratio-labelling fluorescence *in situ* hybridization. *Genome Res* 10: 861–865.
Zhang FF, Murata-Collins JL, Gaytan P, Forman SJ, Kopecky KJ, Willman CL, Appelbaum FR, Slovak ML (2000). Twenty-four-color spectral karyotyping reveals chromosome aberrations in cytogenetically normal acute myeloid leukemia. *Genes Chrom Cancer* 28: 318–328.

4.3. Artificial Chromosomes – Past, Present and Future

Howard J. Cooke
MRC Human Genetics Unit, Crewe Road, Edinburgh EH4 2XU, United Kingdom

Artificial chromosomes have long been touted as the ideal vector for gene therapy and biotechnology purposes based on the idea that such a chromosome would mimic the natural state of DNA in the cell. This, it is argued, would mean that essentially unlimited amounts of DNA could be incorporated into such a vector enabling either large genes or whole metabolic pathways to be provided to the recipient cell or organism. Additionally such a vector would not integrate into the genome of the host cell and so would not cause mutagenesis by insertion and could perhaps be withdrawn from the cell or organism when no longer required. A number of preconditions are implicit in these claims. Firstly, the chromosome should have a segregation efficiency approaching 100% in order to be useful in a cell population undergoing multiple rounds of cell divisions. Secondly, the chromosome should have a defined structure for regulatory and practical reasons. A defined structure is needed to maximise the control of expression of the genes that it contains. Third the chromosome should not be so large that delivery becomes a problem. Finally chromosomal effects such as centromeric or telomeric silencing should not dominate the expression of genes contained in an artificial chromosome.

In this chapter we discuss our own and others efforts to achieve these aims using a variety of non viral approaches to the problem.

Past

Two contrasting approaches, based on the demonstration that repeats of the vertebrate telomere sequence TTAGGG can provide telomere function in mammalian cells, have been used to generate artificial chromosomes. The initial experiments demonstrating telomeric DNA function showed that transformation with constructs containing these sequences resulted in chromosome fragmentation (Farr *et al.*, 1991; Itzhaki *et al.*, 1993). The logical extension of this process was then to apply this methodology,

M. Schmid and I. Nanda (eds.), Chromosomes Today, Volume 14, 197–205.

initially using random non homologous recombination approaches, to fragment chromosomes – most notably the human X (Bayne *et al.*, 1993) and Y (Heller *et al.*, 1996) {Brown, Barnett, *et al.*, 1994 350/id} chromosomes to produce progressively smaller products. This use of non homologous recombination results in a limitation. As the fragmentation target becomes smaller then the difficulty of obtaining the desired chromosome decreases.

This approach has the significant advantage that the chromosome remains a chromosome within a cell, or at least a nucleus, throughout the process of manipulation and so the possibility of a requirement to form or reform an particular epigenetic state does not arise. This process of fusion can also be used to transfer chromosomes from cell line to cell line. It also makes no assumptions about the nature of centomeric DNA but in principle can be used to define the minimum region capable of function as a centromere. Chromosomes manipulated in this manner have been transmitted through the mouse germ line (Shen *et al.*, 2000).

The second approach also relies on the proven ability of telomeric DNA to provide telomere function but requires two further chromosomal components, origins of replication and centromere competent DNA. In this context centromere competent DNA means DNA which is capable of nucleating the formation of a functional kinetochore on the artificial chromosome. This has been a controversial subject with views ranging from the purist *Saccharomyces cerevisiae* stand point of a critical "magic" sequence being necessary and sufficient (in the case of *S. cerevisiae* about 125 bp) (Carbon and Clarke, 1984) to the epigenetic viewpoint of Karpen and Allshire (1997) with no sequence requirement explicit. The approach of building a chromosome from these three naked DNA components (telomeres, origins of replication and centromeric sequences) then requires assumptions about origins and centromeric DNA which must in part be guesses. Replication origins were largely ignored. The first reason being that some reports did not suggest that there was a stringent definition in mammalian cells (Heinzel *et al.*, 1991; Krysan *et al.*, 1989). Secondly on the basis that the best guess for a centromeric candidate in human cells – alphoid DNA – was reported in some assays to function as a replication origin. Alphoid DNA was chosen as the candidate centromeric sequence because of its ubiquitous presence at human kinetochores (Sullivan *et al.*, 1996) despite uncertainty about the significance of recognition sites for the CENPB protein (Goldberg *et al.*, 1996; Yoda *et al.*, 1996). The two approaches used – illustrated in Figure 1 – both use a combination of telomeric and alphoid sequences to generate a new chromosome in the cell (Ikeno *et al.*, 1998; Harrington *et al.*, 1997). In one case this is generated by recombination within the host human cell and in the other in a preceding host yeast strain in which the construct is propogated. The similarities in the chromosomes produced are in the controlled copy number, approximate size and stability and an increase in the size of the input DNA by multimerisation to generate the visible chromosomes seen. The kinetochores of the chromosomes formed contain the same components as the active kinetochores of other human chromosomes. Both initial experiments used HT1080 a fibrosarcoma cell with a stable karyotype as the DNA recipient. Although the number of alphoid sequences used in these experiments was limited controls suggest that regularity of repeat may be important in conferring the ability of alphoid DNA to nucleate a kinetochore. To

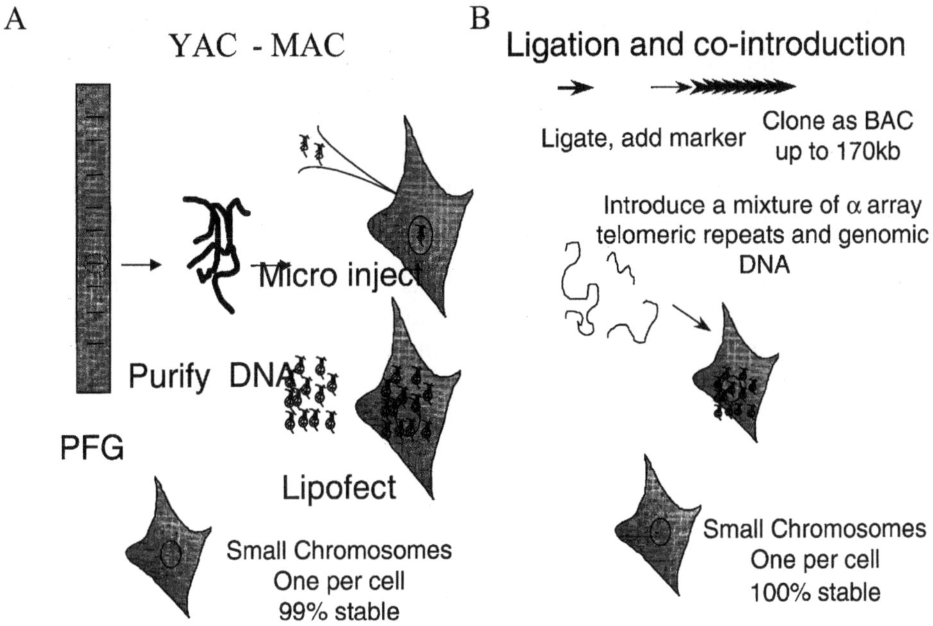

Figure 1. *Two "bottom up" approaches to MAC formation. In A yeast artificial chromosomes retrofitted with mammalian telomeres in transiently recombination proficient yeast cells are purified from pulse field gels as DNA molecules of 100 kb and introduced into mammalian cells by microinjectin or lipofection. In B ligation* ex vivo *followed by recombination in the host cell is used to generate artificial chromosomes.*

date neither of these "bottom up" approaches have produced a chromosome which has been transmitted through the germ line.

In addition to these attempts to be systematic in constructing or deconstructing chromosomes a number of naturally occurring small chromosomes have been engineered to contain selectable markers and to form the basis for future manipulations (Guiducci *et al.*, 1999). Mammalian cells have a propensity to produce amplified segments of DNA when provoked by selection and uptake of exogenous DNA (Hadlaczky *et al.*, 1991). Subsequently these HSR regions can fragment to small chromosomes and this process has been commercialised to produce these chromosomes as biotechnology vectors. Some chromosomes produced by these methods have been transmitted through the mouse germ line (Co *et al.*, 2000).

Present

Since the first successful attempts to construct chromosomes effort has concentrated in a number of areas. We and others have moved to use vectors other than YACs as the host in which DNA for bottom up construction of MACs is produced (Mejia and

Larin, 2000). The incentive for this is the relative ease of producing useable amounts of DNA in bacterial systems, the similarity to procedures used for production of small plasmids in significant quantities to GMP standards for clinical trials and most recently the development of recombination based methods for manipulation of these large molecules. We have carried this simplification one stage further by omitting telomeres and have been able to generate constructs which efficiently form macs (Ebersole *et al.*, 2000). These chromosomes are likely to be circular based on the absence of telomeric sequences in the starting constructs and in the chromosomes produced.

One of the motivations for all of this work is to develop vectors based on these chromosomes. It is therefore important to demonstrate that genes incorporated into MACs can be expressed and this has now been demonstrated in MACs, constructed either in bacteria or mammalian cells, to contain the human HPRT locus (Mejia *et al.*, 2001). Both of these produce chromosomes consist of multimeric forms of the input DNA and both chromosomes are capable of expressing sufficient HPRT protein to complement the HPRT deficiency in the cell line used. Transcription of the gene is maintained throughout many cell divisions in the absence of selection for the retention of the artificial chromosome. This is a critical proof of principle for this application since these chromosomes have a substantial load of sequences likely to be in a heterochromatic state. On this basis silencing of the genes contained might have been an inevitable consequence of kinetochore formation.

Other approaches have been based on the use of neocentromeres. Neocentromeres arise as a result of chromosome rearrangements and have been reported in a range of species. The best defined is the mardel(10) (du *et al.*, 1997) where the underlying DNA has been sequenced and methylation mapped in both centric and non centric states (Barry *et al.*, 1999). No differences were found underlining the importance of epigenetic effects in kinetochore function. Attempts to use the 80 kb of DNA from this centromere in bottom up experiments were unsuccessful (Saffery *et al.*, 2001). This probably reflects the rarity of neocentromere formation which, in turn, is likely to be due to a hierarchy in efficiency of kinetochore formation between different DNA sequences. Viewed most simply this scheme suggests that although once formed a kinetochore can inhabit any DNA sequence its probability of forming *de novo* is very much higher on repetitive sequences than elsewhere in the genome. This idea is supported by Saffery *et al.'s* ability to use telomere directed fragmentation to construct a deleted chromosome utilising the mardel(10) neocentromere. These chromosomes segregated efficiently in the HT1080 host cell used. Interestingly several of the chromosomes produced were likely to be circular. A systematic analysis of the sequence requirements for kinetochore formation has not yet been attempted.

We have observed that neocentromere forming on a chromosome generated by telomere fragmentation of the human Y chromosome and transfer into mouse ES cells. In the ES cells various rearrangements occurred and a number of derivative chromosomes were formed. When these in turn were moved to chicken DT40 cells one was characterised as lacking alphoid DNA and mouse minor satellite sequences. Subsequent analysis using fibre FISH to detect DNA sequences in combination with immunocytochemistry to detect kinetochore proteins showed that the DNA underlying the kinetochore was derived from interval 5 of the Y chromosome. This is a region previously associated with the formation of a neocentromere (Floridia *et al.*, 2000) but the chromosome does not appear to contain the repeated DAZ genes which are found in this region.

Requirements for telomere and centromere function are likely to be different in mitosis and meiosis and this is and area which artificial chromosomes can be used to investigate in terms of DNA sequence requirements. A major shortfall in the current technology is that non of the chromosomes constructed by the bottom up approaches have been transmitted through the germ line to date. The reasons for this are unclear but not surprising since there have been no reports of "bottom up" chromosomes which function in mouse cells. This is a major gap in our capabilities since mouse provides the only accessible mammalian system in which to examine the meiotic functions of artificial chromosomes. Size is one factor which seems unlikely to be important as some of the chromosomes created by *in vivo* manipulation, which have been transmitted through the mouse germ line have been of approximately the same size as those created by the bottom up approach. A listing of such chromosomes includes those made by fragmentation of amplified regions of chromosomes associated with DNA uptake (Co *et al.*, 2000), chromosomes produced by telomere fragmentation (Shen *et al.*, 2000) and passage through a range of cell types and those generated using Cre Lox systems (Voet *et al.*, 2001).

Future

Despite being relatively recent developments, artificial chromosomes have been the subject of a considerable amount of "hype". Willard (2000) reports that they are presented in some text books as a simple tool for gene therapy. Statements are frequently made along the lines that artificial chromosomes will provide vectors which will be free of position effects, have a capacity for large genes with no side effects from integration. Not surprisingly the field is some way from providing these features and developments are necessary in a wide range of areas before the practical application of this technology will develop. What is needed and what are the prospects for development?

Perhaps first and foremost we do not understand the requirements for kinetochore formation and the difference in the requirements for formation versus replication. When using naked DNA there is a requirement for the recipient cell to overcome any initial barriers to centromere formation. This seems likely to be a combination of the DNA sequence and the host cell machinery. For example the majority of host cell lines used for this type of approach have been HT1080 and derivatives. As mentioned above DNA from a neocentromere has not been shown to function in this type of approach, even in HT1080. HeLa and other cell types have been shown to form MACs but at significantly lower frequencies. The reasons for this restriction are unclear and need to be established since practical applications will depend on being able to deliver DNA to a primary cell population with a high probability of the formation of stable chromosomes. The prospects here seem good. Advances in the understanding of kinetochore and chromosome function in yeast (Pidoux and

Allshire, 2000) are being increasingly applied to mammalian systems as the basic conservation of these systems becomes apparent. For example homologues of the histone like CENP A protein are now known in many organisms (Saitoh *et al.*, 1997; Keith *et al.*, 1999; Malik and Henikoff, 2001). Similarly many of the basic processes of cell cycle control and DNA repair now emerging from cancer related research will have a bearing on the cellular basis for chromosome formation. A concrete example of this might come from the telomerase field where it is now possible to immortalise some primary cells by expressing the catalytic component of telomerase (Counter *et al.*, 1998). Transient expression of this enzyme in non transformed cells might extend the range of cell types in which MAC formation might occur efficiently with an input of naked DNA.

Efficient methods of constructing large DNA molecules are also needed to assemble the input DNA from multiple different fragments of genomic DNA if the use of very large genes or assembly of pathways is envisaged. Site specific recombinases such as Cre and Flp and others (Groth *et al.*, 2000) (for review see Gorman and Bullock, 2000) provide one way of achieving this, which – because of the tight definition of the process – may be acceptable to regulatory bodies. A limitation of this, which may no longer apply, is the need to introduce the sites recognised by the recombinases. When dealing with large DNA molecules this was a non trivial task, especially if several genes needed to be assembled. Recent developments such as ET cloning (Muyrers *et al.*, 2000, 2001) make it relatively trivial to use recombination in *E. coli* to carry out sequential manipulations. One potential problem, which is general to the manipulation of these types of construct, is the tendency of tandem repeats to rearrange in all of the available hosts and vectors. Manipulation of these constructs may be limited by this property of one of the basic components needed for chromosome formation.

These advances will also apply to the *in vivo* manipulation of chromosomes constructed from the top down route. Constructs for manipulation will be easier to make. One of the major problems remaining for this approach are the development of efficient ways of transferring these chromosomes from one cell type to another, for example, from the homologous recombination competent chicken DT40 cell line to a human cell. Rearrangement often seems to accompany this process (Shen *et al.*, 2000) and since the starting chromosome is also not fully sequence defined this will pose regulatory problems. This lack of sequence definition may also pose problems for gene expression from both up and down chromosomes. Development is needed in delivery systems for bottom up approaches also. Both up and down systems require the delivery of molecules which are large in terms of current methods of introducing molecules to cells and the methods used only work in a limited range of cells with low efficiency. In the case of the down methods, the requirement for fusion or injection based approaches implies a limitation to *ex vivo* approaches and subsequent selection and expansion of cell populations. Reagents for DNA/Chromatin transfer are improving but this may remain the limiting step in the practical application of artificial chromosome for some time to come. Although it is frequently claimed that position effects will be avoided by these systems experience is currently limited to a few cases. The basis for this claim is that large regions of DNA will contain endogenous signals such as locus control regions and insulators but until more is known about the role and spacing of these elements in individual genes this is unsubstantiated. Prospects for improvement here are again good as the complete genomic sequence of several mammals should make it possible to deduce the location of conserved but non coding elements by comparison.

These limitations suggest that the primary use for artificial chromosomes will for some time be in the research field. Whole genome sequence availability opens up the prospect of understanding genomic organisation at a level not previously possible and raises the necessity of bridging the gap between DNA sequence and nuclear organisation and function. Mammalian artificial chromosomes, by providing an independent test bed for analysing the role of megabase regions in these processes, will be one of the tools used to build the bridge.

References

Barry AE, Howman EV, Cancilla MR, Saffery R, Choo KA (1999). Sequence analysis of an 80 kb human neocentromere. *Hum Mol Genet* 8: 217–227.

Bayne R, Taggart MT, Farr C, Petit C, Guilford P, Toniolo D, Sala C, Cooke HJ (1993). An analysis of xq breakpoints created by telomere-associated chromosome fragmentation. *Cytogenet Cell Genet* 64: 175–175.

Carbon J, Clarke L (1984). Structural and functional analysis of a yeast centromere (CEN3). *J Cell Sci* Suppl 1: 43–58.

Co DO, Borowski AH, Leung JD, van der KJ, Hengst S, Platenburg GJ, Pieper FR, Perez CF, Jirik FR, Drayer JI (2000). Generation of transgenic mice and germline transmission of a mammalian artificial chromosome introduced into embryos by pronuclear microinjection. *Chromos Res* 8: 183–191.

Counter CM, Meyerson M, Eaton EN, Ellisen LW, Caddle SD, Haber DA, Weinberg RA (1998). Telomerase activity is restored in human cells by ectopic expression of hTERT (hEST2), the catalytic subunit of telomerase. *Oncogene* 16: 1,217–1,222.

du SD, Cancilla MR, Earle E, Mao JI, Saffery R, Tainton KM, Kalitsis P, Martyn J, Barry AE, Choo KH (1997). A functional neo-centromere formed through activation of a latent human centromere and consisting of non-alpha-satellite DNA [see comments]. *Nat Genet* 16: 144–153.

Ebersole TA, Ross A, Clark E, Mcgill N, Schindelhauer D, Cooke H, Grimes B (2000). Mammalian artificial chromosome formation from circular alphoid input DNA does not require telomere repeats. *Hum Mol Genet* 9: 1,623–1,631.

Farr C, Fantes J, Goodfellow P, Cooke H (1991). Functional reintroduction of human telomeres into mammalian cells. *Proc Natl Acad Sci USA* 88: 7,006–7,010.

Floridia G, Gimelli G, Zuffardi O, Earnshaw WC, Warburton PE, Tyler-Smith C (2000). A neocentromere in the DAZ region of the human Y chromosome. *Chromosoma* 109: 318–327.

Goldberg IG, Sawhney H, Pluta AF, Warburton PE, Earnshaw WC (1996). Surprising deficiency of CENP-B binding sites in African green monkey alpha-satellite DNA: implications for CENP-B function at centromeres. *Mol Cell Biol* 16: 5,156–5,168.

Gorman C, Bullock C (2000). Site-specific gene targeting for gene expression in eukaryotes. *Curr Opin Biotechnol* 11: 455–460.

Groth AC, Olivares EC, Thyagarajan B, Calos MP (2000). A phage integrase directs efficient site-specific integration in human cells. *Proc Natl Acad Sci USA* 97: 5,995–6,000.

Guiducci C, Ascenzioni F, Auriche C, Piccolella E, Guerrini AM, Donini P (1999). Use of a human minichromosome as a cloning and expression vector for mammalian cells. *Hum Mol Genet* 8: 1,417–1,424.

Hadlaczky G, Praznovszky T, Cserpan I, Kereso J, Peterfy M, Kelemen I, Atalay E, Szeles A, Szelei J, Tubak V, Burg K (1991). Centromere formation in mouse cells cotransformed with human dna and a dominant marker gene. *Proc Natl Acad Sci USA* 88: 8,106–8,110.

Harrington JJ, VanBokkelen G, Mays RW, Gustashaw K, Willard HF (1997). Formation of *de novo* centromeres and construction of first-generation human artificial microchromosomes. *Nat Genet* 15: 345–355.

Heinzel SS, Krysan PJ, Tran CT, Calos MP (1991). Autonomous DNA replication in human cells is affected by the size and the source of the DNA. *Mol Cell Biol* 11: 2,263–2,272.

Heller R, Brown KE, Burgtorf C, Brown WRA (1996). Minichromosomes derived from the human Y-chromosome by telomere directed chromosome breakage. *Proc Natl Acad Sci USA* 93: 7,125–7,130.

Ikeno M, Grimes B, Okazaki T, Nakano M, Saitoh K, Hoshino H, Mcgill NI, Cooke H, Masumoto H (1998). Construction of YAC-based mammalian artificial chromosomes. *Nat Biotechnol* 16: 431–439.

Itzhaki JE, Barnett MA, Buckle VJ, Brown WRA, Porter ACG (1993). Targeted breakage of a human chromosome mediated by cloned human telomeric DNA. *Nat Genet* 2: 283–287.

Karpen GH, Allshire RC (1997). The case for epigenetic effects on centromere identity and function. *Trends Genet* 13: 489–496.

Keith KC, Baker RE, Chen Y, Harris K, Stoler S, Fitzgerald-Hayes M (1999). Analysis of primary structural determinants that distinguish the centromere-specific function of histone variant Cse4p from histone H3. *Mol Cell Biol* 19: 6,130–6,139.

Krysan PJ, Haase SB, Calos MP (1989). Isolation of human sequences that replicate autonomously in human cells. *Mol Cell Biol* 9: 1,026–1,033.

Malik HS, Henikoff S (2001). Adaptive evolution of Cid, a centromere-specific histone in Drosophila. *Genetics* 157: 1,293–1,298.

Mejia JE, Larin Z (2000). The assembly of large BACs by *in vivo* recombination. *Genomics* 70: 165–170.

Mejia JE, Willmott A, Levy E, Earnshaw WC, Larin Z (2001). Functional complementation of a genetic deficiency with human artificial chromosomes. *Am J Hum Genet* 69: 315–326.

Muyrers JP, Zhang Y, Benes V, Testa G, Ansorge W, Stewart AF (2000). Point mutation of bacterial artificial chromosomes by ET recombination. *EMBO Rep* 1: 239–243.

Muyrers JP, Zhang Y, Stewart AF (2001). Techniques: recombinogenic engineering – new options for cloning and manipulating DNA. *Trends Biochem Sci* 26: 325–331.

Pidoux AL, Allshire RC (2000). Centromeres: getting a grip of chromosomes. *Curr Opin Cell Biol* 12: 308–319.

Saffery R, Wong LH, Irvine DV, Bateman MA, Griffiths B, Cutts SM, Cancilla MR, Cendron AC, Stafford AJ, Choo KH (2001). Construction of neocentromere-based human minichromosomes by telomere-associated chromosomal truncation. *Proc Natl Acad Sci USA* 98: 5,705–5,710.

Saitoh S, Takahashi K, Yanagida M (1997). Mis6, a fission yeast inner centromere protein, acts during G1/S and forms specialized chromatin required for equal segregation. *Cell* 90: 131–143.

Shen MH, Mee PJ, Nichols J, Yang J, Brook F, Gardner RL, Smith AG, Brown WR (2000). A structurally defined mini-chromosome vector for the mouse germ line. *Curr Biol* 10: 31–34.

Sullivan BA, Schwartz S, Willard HF (1996). Centromeres of human-chromosomes. *Environ Mol Mutagen* 28: 182–191.

Voet T, Vermeesch J, Carens A, Durr J, Labaere C, Duhamel H, David G, Marynen P (2001). Efficient male and female germline transmission of a human chromosomal vector in mice. *Genome Res* 11: 124–136.

Willard HF (2000). Genomics and gene therapy. Artificial chromosomes coming to life. *Science* 290: 1,308–1,309.

Yoda K, Nakamura T, Masumoto H, Suzuki N, Kitagawa K, Nakano M, Shinjo A, Okazaki T (1996). Centromere protein-b of african-green monkey cells – gene structure, cellular expression, and centromeric localization. *Mol Cell Biol* 16: 5,169–5,177.

4.4. Differential Demethylation of Paternal and Maternal Genomes in the Preimplantation Mouse Embryo: Implications for Mammalian Development

Thomas Haaf[1], Wei Shi[1], Reinald Fundele[1], Katharine L. Arney[2], M. Azim Surani[2] and Sheila C. Barton[2]
[1]*Max Planck Institute of Molecular Genetics, Berlin, Germany;* [2]*Wellcome CRC Institute, Cambridge, UK*

A paternal and a maternal genome are required for normal mammalian development

Higher eukaryotes are endowed with a diploid somatic genome. At fertilization the paternal and maternal genomes are combined, forming the new diploid organism. Nuclear transfer experiments in the 1980s (McGrath and Solter, 1984; Surani *et al.*, 1986) demonstrated for the first time the functional non-equivalence of the two parental genomes in mammals. In contrast to many lower vertebrates, mammalian uniparental embryos cannot develop beyond mid-gestation. Androgenetic embryos, with two paternal genomes, are usually very stage-retarded but often have well developed mural trophoblast and yolk sac. Gynogenetic embryos, with two maternal genomes, can develop relatively normally to mid-term but are small and have very restricted trophoblast and yolk sac. In both cases there is no normal placental development. Because of opposing patterns of gene expression from maternal and paternal alleles of imprinted genes (Fundele and Surani, 1994; Tilghman, 1999), normal mammalian development requires the participation of both a maternal and a paternal genome. Genomic imprinting is thought to be a by-product of the unique fetal-maternal relationship in mammals to control the different parental interests on embryo development. Consistent with the parental conflict hypothesis (Moore and Haig, 1991), many paternally expressed genes enhance fetal growth, whereas maternally expressed genes inhibit growth.

M. Schmid and I. Nanda (eds.), Chromosomes Today, Volume 14, 207–214.

Topological genome separation in the early embryo

In mammals, the one-cell embryo is formed from two very different sets of chromatin. The sperm chromatin is highly compacted by protamines into an almost crystalline nucleoprotamine structure (Balhorn, 1982). The oocyte chromosomes are arrested at the second meiotic metaphase and have the normal nucleosomal structure. Sperm-egg fusion activates the oocyte which then remodels the two parental genomes into their functional forms (Perreault, 1992). This very dramatic process of converting the two complementary sets of chromosomes into a specialized diploid somatic genome occurs in separate nuclear compartments in the early mammalian embryo. Two independent experimental approaches, BrdU labeling of mouse sperm DNA followed by immunofluorescent detection of BrdU in embryos of the next generation and differential fluorescence *in situ* hybridization labeling of paternal and maternal heterochromatin in interspecific (*Mus musculus* × *M. spretus*) F1 hybrid embryos, have demonstrated topological separation of chromatin according to parental origin during preimplantation development (Mayer *et al.*, 2000a). This genome separation is preserved up to the four-cell embryo stage and then gradually disappears.

Genome-wide demethylation during preimplantation development

The separate nuclear compartments are likely to be functionally important for reprogramming the sperm and egg genomes for somatic development rather than being a passive consequence of their different developmental histories before fertilization. Genome-wide opposing patterns in paternal and maternal DNA methylation are fundamental to this process. Molecular studies quantifying unmethylated *Hpa*II restriction sites in digested genomic DNA suggested that ovulated mouse oocytes are globally undermethylated, compared with the highy methylated sperm genome (Monk *et al.*, 1987; Sanford *et al.*, 1987). The intermediate methylation levels in the early embryo have been explained as a combination of undermethylated maternal and methylated paternal DNA. A genome-wide gradual demethylation during preimplantation development then leads to indistinguishable alleles at most gene loci, but not at those that are imprinted (Howlett and Reik, 1991; Razin and Shemer, 1995). Differentially methylated regions, which are maintained during development, could serve as primary imprinting signals (Reik and Walter, 1998). DNA methylation is essential for the control of gene activity in a number of developmental processes. Knockout mice that are deficient in DNA methylation activity die at mid-gestation (Li *et al.*, 1992; Okano *et al.*, 1999).

The primary function of CpG methylation is an epigenetic DNA modification, which establishes or maintains an inactive chromatin structure (Ng and Bird, 1999). The methyl-CpG-binding protein MeCP2 binds to DNA in a methylation-dependent manner and then recruits a corepressor complex including histone deacetylase. Histone deacetylation renders the chromatin inaccessible to the transcriptional machinery (Jenuwein and Allis, 2001). It is plausible that genome-wide demethylation in the early mammalian embryo is necessary for the restoration of pluripotency.

Differential demethylation of the two parental genomes in the early mouse embryo

Immunofluorescent staining with an antibody against 5-methylcytosine (MeC) demonstrated that the two parental genomes are demethylated by different mechanisms and at different times during preimplantation development (Rougier *et al.*, 1998; Mayer *et al.*, 2000b). In contrast to previous molecular analyses (Monk *et al.*, 1987; Sanford *et al.*, 1987), the maternal chromosomes of the unfertilized mouse egg, which were arrested at metaphase II of meiosis, stained very intensely with anti-MeC antibody (Figure 1A). Upon fertilization, the oocyte genome completed its meiotic maturation by extrusion of the second polar body. At 1–2 h after fertilization (before the male pronucleus became visible), the "bulge stage" embryos showed equally high methylation levels of sperm nucleus, egg chromosomes and polar body (Figure 1B). Decondensation of sperm chromatin was followed by formation of the male pronucleus. At 3–6 h after fertilization both parental chromosome sets were heavily methylated (Figure 1C). However, the paternal genome was then drastically and rapidly demethylated before the onset of the first DNA replication. At 7–8 h after fertilization, the paternal pronucleus was usually larger than the maternal pronucleus and showed very little, if any, detectable methylation (Figure 1D). Demethylation of the zygotic paternal genome even occurred in aphidicolin-treated embryos, which did not undergo DNA replication. It therefore requires the presence of a demethylating enzyme(s) in the fertilized egg. Although several biochemical activities have been reported, including a repair enzyme that excises MeC, a processive enzyme that cleaves the methyl group from MeC, and a ribozyme-like enzyme, the existence and identification of the elusive demethylase remain in doubt (Hsieh, 2000). Demethylation and activation of the paternal zygotic genome may depend on transient histone H4 hyperacetylation of the male pronucleus (Adenot *et al.*, 1997). In the mouse, the paternal pronucleus replicates earlier and exhibits greater transcriptional activity than the female pronucleus (Aoki *et al.*, 1997; Bouniol-Baly *et al.*, 1997).

Even the dramatic chromosome movements during the first mitosis did not lead to an intermingling of the demethylated paternal and methylated maternal chromosome sets (Figure 1E). It is important to emphasize that at first metaphase both chromatids on the maternal chromosomes remained methylated. The global methylation of the maternal genome was largely maintained from the oocyte to the two-cell embryo stage. The differentially methylated parental genomes each occupied one half of the nucleus in two-cell embryos (Figure 1F). At the second metaphase, the maternal chromosomes showed methylation on only one of the two sister chromatids (Figure 1G). This differential sister chromatid staining is caused by a replication-dependent demethylation mechanism (Rougier *et al.*, 1998). MeC cannot be incorporated directly into replicating DNA. The maintenance methyltransferase DNMT1, which has a high affinity for hemimethylated sites, detects methylated CpG sites in the parental DNA strand and adds methyl groups to the corresponding sites in the newly synthesized strand. Since DNMT1 is actively retained in the cytoplasm from the oocyte to the blastocyst stage (Cardoso and Leonhardt, 1999), it cannot bind to target sites in the nucleus during early development. When half of the methyl groups are lost with every

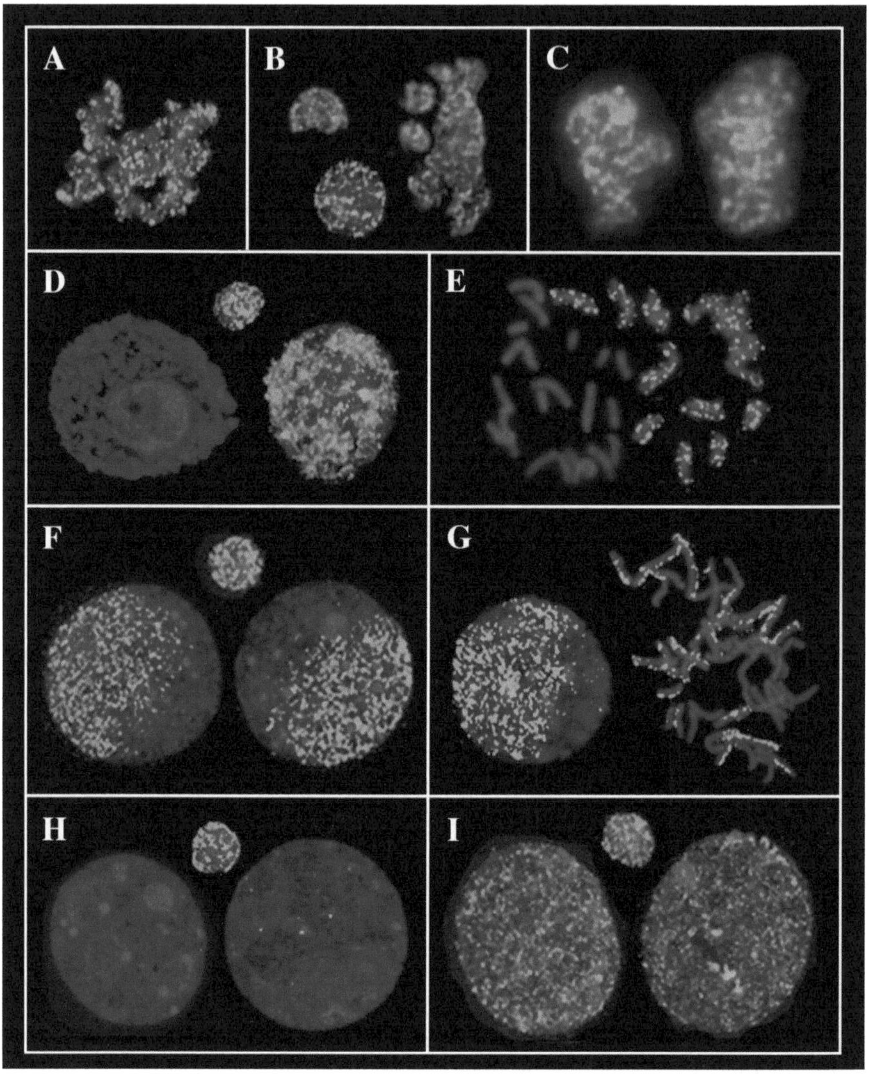

Figure 1. *Dynamic and opposing changes in global DNA methylation of parental genomes in early mouse embryos. Nuclei were stained with FITC-conjugated anti-MeC antibody (green) and counterstained with DAPI (blue). (A) Unfertilized egg. (B) Very early (bulge stage) embryo at 1–2 h after fertilization, showing equivalently high methylation levels of sperm nucleus, oocyte chromosomes and second polar body. (C) Highly methylated paternal and maternal chromosome sets at 3 h after fertilization. (D) Active demethylation of the paternal pronucleus at 8 h after fertilization. The somewhat smaller maternal pronucleus and the second polar body remain fully methylated. (E) Spatially separated and differentially methylated chromosome sets at first metaphase. (F) Two-cell embryo showing a demethylated paternal half and a methylated maternal half of the nucleus. (G) Late two-cell embryo. The mitotic nucleus reveals differential sister chromatid methylation of the maternal chromosomes. (H) Androgenetic two-cell embryo showing complete demethylation of the two paternal genomes in each nucleus. (I) Gynogenetic two-cell embryo with two methylated maternal genomes.*

round of replication, full double-stranded demethylation in one chromatid occurs after two cell cycles. Consequently, four-cell embryos had a much weaker MeC density over the maternal half of the nucleus. After the 8-cell embryo stage, paternal and maternal chromosomes showed equivalently low methylation levels. The second polar body was always highly methylated throughout preimplantation development. Evidently, genome-wide demethylation of the maternal genome occurs step-by-step from lack of maintenance methylation.

Because most CpGs are found in repetitive DNA sequences, which make up 30–40% of the mammalian genome, the intensity of MeC staining reflects the methylation of various repeat DNA families (Yoder *et al.*, 1997). However, consistent with the immunofluorescent results, bisulphite sequencing of single copy sequences, including an imprinted gene, *Igf2*, a house-keeping gene, *α-actin*, a tissue-specific gene, *myl-C*, and a transgene, *TKZ*, has also demonstrated rapid demethylation at sperm-derived loci in the zygote (Olek and Walter, 1998; Oswald *et al.*, 2000). Thus, active zygotic demethylation of paternal DNA is a general phenomenon, affecting widely different classes of repetitive and single-copy sequences. This argues in favor of the notion that the striking asymmetry of demethylation is important for biparental epigenetic reprogramming in the early embryo. It is likely to be associated with parent-specific genomic imprinting which is a hallmark of mammalian development.

Implications for genomic imprinting

Because there is currently no evidence that one genome is protected more than the other from replication-dependent demethylation (by sequestration of DNMT1 into the cytoplasm), the paternal genome is thought to become actively and passively demethylated in the early embryo. In contrast, the maternal genome has evolved a protection system against the active zygotic demethylation wave and undergoes only passive demethylation. Because methylation imprints imposed on the paternal chromosomes may be much more difficult to protect against demethylation during preimplantation development than maternal germline imprints, it is not surprising that most currently known methylation imprints are derived from the maternal germline (Reik and Walter, 2001). Thus, the battle of the sexes starts at the genomic level in the one-cell embryo, when the paternal genome decondenses and is totally unprotected (Haaf, 2001). Then the fertilized egg that is the maternal genome can actively erase most paternal methylation imprints. Consistent with this hypothesis, most maternally repressed genes are silenced epigenetically by maternal methylation, whereas paternally repressed genes have evolved other (genetic) silencing mechanisms, i.e., maternally repressed antisense RNAs.

It is generally assumed that sperm- and oocyte-derived imprints are transmitted throughout further development (Constancia *et al.*, 1998; Brannan and Bartolomei, 1999). However, the fact that the two parental genomes are spatially separated and differentially demethylated during preimplantation development suggest the possibility that parent-specific imprinting may also occur in the early embryo. The analysis of Prader–Willi imprinting center deletion patients indicated that the maternal methylation imprints of chromosome 15q11-13 are established during or after fertilization (El-Maarri *et al.*, 2001). Considering that the preimplantation embryo is much less protected than the germline, this may be the time when environmental factors have the greatest impact on the epigenotype of the individual.

Abnormal methylation patterns in uniparental and normal fertilized mouse embryos

The analysis of methylation reprogramming in uniparental embryos suggests that the cellular machinery of the fertilized egg is not capable of changing the methylation patterns of transplanted pronuclei. Diploid androgenetic embryos, whose two genomes are wholly of paternal origin, showed completely demethylated nuclei at the two-cell stage (Figure 1H). The paternal pronucleus was already demethylated at the time of manipulation, since only eggs where the pronuclei were well formed and distinguishable by size could be used for transplantation (Barton and Surani, 1993). In contrast, the two female genomes in gynogenetic (Figure 1I) and parthenogenetic two-cell embryos remained fully methylated. Evidently, the demethylation activity in the fertilized egg is restricted to paternal DNA, regardless of the number of genomes present. We conclude that demethylation of the paternal zygotic genome is associated with sperm chromatin decondensation and remodelling before pronucleus formation. During the period of protamin-histone exchange, the paternal DNA is unusually loosely packaged or even naked (Rodman *et al.*, 1981; Nonchev and Tsanev, 1990) and, therefore, may outrival the maternal chromatin for the binding of demethylating enzymes. Following restoration of the nucleosomal configuration, the paternal chromatin physically resembles the maternal chromatin and may no longer be susceptible to active demethylation. This has important implications for mammalian cloning. Since a somatic cell nucleus that has been introduced into an oocyte during cloning does not decondense, its DNA may not be able to bind demethylases and, similar to a maternal pronucleus, may undergo only gradual replication-dependent demethylation. Indeed, incomplete methylation reprogramming of donor genomic DNA is believed to cause developmental failure and anomalies in cloned bovine embryos (Kang *et al.*, 2001).

In the normal diploid mammalian embryo, differential demethylation of the paternal and maternal genomes occurs during preimplantation development in a spatially and temporally highly coordinated fashion. Nevertheless, a significant proportion (10–30%) of one- and two-cell embryos derived from normal matings display abnormal methylation patterns, some of which are indistinguishable from those in androgenetic or gynogenetic embryos. These embryos may be expected to fail early in development. It is likely that defects in nuclear compartmentalization and parent-specific epigenetic reprogramming contribute to the high incidence of early pregnancy failure in mammals.

References

Adenot PG, Mercier Y, Renard J-P, Thompson EM (1997). Differential H4 acetylation of paternal and maternal chromatin precedes DNA replication and differential transcriptional activity in pronuclei of 1-cell mouse embryos. *Development* 124: 4,615–4,625.

Aoki F, Worrad DM, Schultz RM (1997). Regulation of transcriptional activity during the first and second cell cycles in the preimplantation mouse embryo. *Dev Biol* 181: 296–307.
Balhorn R (1982). A model for the structure of chromatin in mammalian sperm. *J Cell Biol* 93: 298–305.
Barton SC, Surani MA (1993). Manipulations of genetic constitution by nuclear transplantation. *Meth Enzymol* 225: 732–744.
Bouniol-Baly C, Nguyen E, Besombes D, Debey P (1997). Dynamic organization of DNA replication in one-cell mouse embryos: relationship to transcriptional activation. *Exp Cell Res* 236: 201–211.
Brannan CI, Bartolomei MS (1999). Mechanisms of genomic imprinting. *Curr Opin Genet Dev* 9: 164–170.
Cardoso MC, Leonhardt H (1999). DNA methyltransferase is actively retained in the cytoplasm during early development. *J Cell Biol* 147: 25–32.
Constancia M, Pickard P, Kelsey G, Reik W (1998). Imprinting mechanisms. *Genome Res* 8: 881–900.
El-Maarri O, Buiting K, Peery EG, Kroisel PM, Balaban B, Wagner K, Urman B, Heyd J, Lich C, Brannan CI, Walter J, Horsthemke B (2001). Maternal methylation imprints on human chromosome 15 are established during or after fertilization. *Nature Genet* 27: 341–344.
Fundele R, Surani MA (1994). Experimental embryological analysis of genetic imprinting in mouse development. *Dev Genet* 15: 515–522.
Haaf T (2001). The battle of the sexes after fertilization: behaviour of paternal and maternal chromosomes in the early mammalian embryo. *Chrom Res* 9: 263–271.
Howlett SK, Reik W (1991). Methylation levels of maternal and paternal genomes during preimplantation development. *Development* 113: 119–127.
Hsieh CL (2000). Dynamics of DNA methylation pattern. *Curr Opin Genet Dev* 10: 224–228.
Jenuwein T, Allis CD (2001). Translating the histone code. *Science* 293: 1,074–1,080.
Kang Y-K, Koo D-B, Park J-S, Choi Y-H, Chung A-S, Lee K-K, Han Y-M (2001). Aberrant methylation of donor genome in cloned bovine embryos. *Nat Genet* 28: 173–177.
Li E, Bestor TH, Jaenisch R (1992). Targeted mutation of the DNA methyltransferase gene results in embryonic lethality. *Cell* 69: 915–926.
Mayer W, Smith A, Fundele R, Haaf T (2000a). Spatial separation of parental genomes in preimplantation mouse embryos. *J Cell Biol* 148: 629–634.
Mayer W, Niveleau A, Walter J, Fundele R, Haaf T (2000b). Active demethylation of the zygotic paternal genome. *Nature* 403: 501–502.
McGrath J, Solter D (1984). Completion of mouse embryogenesis requires both the maternal and paternal genomes. *Cell* 37: 179–183.
Monk M, Boubelik M, Lehnert S (1987). Temporal and regional changes in DNA methylation in the embryonic, extraembryonic and germ cell lineages during mouse embryo development. *Development* 99: 371–382.
Moore T, Haig D (1991). Genomic imprinting in mammalian development: a parental tug-of-war. *Trends Genet* 7: 45–49.
Ng HH, Bird A (1999). DNA methylation and chromatin modification. *Curr Opin Genet Dev* 9: 158–163.
Nonchev S, Tsanev R (1990). Protamine-histone replacement and DNA replication in the male pronucleus. *Mol Reprod Dev* 25: 72–76.
Okano M, Bell DW, Haber DA, Li E (1999). DNA methyltransferases Dnmt3a and Dnmt3b are essential for de novo methylation and mammalian development. *Cell* 99: 247–257.
Olek A, Walter J (1998). The pre-implantation ontogeny of the H19 methylation imprint. *Nat Genet* 17: 275–276.

Oswald J, Engemann S, Lane N, Mayer W, Olek A, Fundele R, Dean W, Reik W, Walter J (2000). Active demethylation of the paternal genome in the mouse zygote. *Curr Biol* 10: 475–478.

Perreault SD (1992). Chromatin remodeling in mammalian zygotes. *Mut Res* 296: 43–55.

Razin A, Shemer R (1995). DNA methylation in early development. *Hum Mol Genet* 4: 1,751–1,755.

Reik W, Walter J (1998). Imprinting mechanisms in mammals. *Curr Opin Genet Dev* 8: 154–164.

Reik W, Walter J (2001). Evolution of imprinting mechanisms: the battle of the sexes begins in the zygote. *Nat Genet* 27: 255–256.

Rodman TC, Pruslin FH, Hoffmann HP, Allfrey VG (1981). Turnover of basic chromosomal proteins in fertilized eggs: a cytoimmunochemical study of events *in vivo. J Cell Biol* 90: 351–361.

Rougier D, Bourc'his D, Gomes DM, Niveleau A, Plachot M, Pàldi A, Viegas-Péquignot E (1998). Chromosome methylation patterns during mammalian development. *Genes Dev* 12: 2,108–2,113.

Sanford JP, Clark HJ, Chapman VM, Rossant J (1987). Differences in DNA methylation during oogenesis and spermatogenesis and their persistence during early embryogenesis in the mouse. *Genes Dev* 1: 1,039–1,046.

Surani MAH, Barton SC, Norris ML (1986). Nuclear transplantation in the mouse: heritable differences between parental genomes after activation of the embryonic genome. *Cell* 45: 127–136.

Tilghman SM (1999). The sins of the fathers and mothers: genomic imprinting in mammalian development. *Cell* 96: 185–193.

Yoder JA, Walsh CP, Bestor TH (1997). Cytosine methylation and the ecology of intragenomic parasites. *Trends Genet* 13: 335–340.

4.5. Integrated Comparative Genome Maps and Their Implications for Karyotype Evolution of Carnivores

Fengtang Yang[1, 2] and Alexander S. Graphodatsky[3]

[1]*Centre for Veterinary Medicine, University of Cambridge, Madingley Road, Cambridge CB3 0ES, UK;* [2]*Kunming Institute of Zoology, Kunming, Yunnan 650223, People's Republic of China;* [3]*Institute of Cytology and Genetics, Novosibirsk, Russia*

Introduction

Comparative chromosome maps record the history of karyotype changes that have occurred during speciation. The rates, types and directions of chromosomal rearrangements as well as phylogenetic relationships between species being compared can be inferred by comparative analysis of the distribution patterns of conserved segments in different phylogenetic lineages (Nadeau and Sankoff, 1998; Yang *et al.*, 2000a). The successful identification of chromosomal homology between species is fundamental for comparative cytogenetic and genomic analysis. Cross-species chromosome painting, being accurate, efficient, and suitable for genome-wide comparison, has become the method of choice for comparative cytogenetics, particularly for comparing distantly related species or species with highly rearranged karyotypes (Scherthan *et al.*, 1994; Yang *et al.*, 1995; Wienberg and Stanyon, 1997; Ferguson-Smith *et al.*, 1998). When combined with chromosome banding and gene mapping, comparative chromosome painting can provide the most accurate comparative chromosome maps for species being compared. Most recently we have been applying this approach to the study of chromosome evolution of carnivores, in particular, the species of the Canidae family, which have the most highly rearranged karyotypes relative to the other carnivores. A series of genome-wide comparative chromosome maps have been established (Yang *et al.*, 1999, 2000a,b; Graphodatsky *et al.*, 2000a,b, 2001; Sargan *et al.*, 2000). The aim of this chapter is to provide an overview of these recent advances in our endeavour to decipher the karyotype evolution of carnivores by integrating the published results together with some of our latest unpublished results.

M. Schmid and I. Nanda (eds.), Chromosomes Today, Volume 14, 215–224.

High-resolution comparative chromosome map of four canid species

The domestic dog (*Canis familiaris*, $2n = 78$), raccoon dog (*Nectereutes procyonoides viverrinus*, $2n = 38–39 + 2–4$ B), Arctic fox (*Alopex lagopus*, $2n = 48–50$) and red fox (*Vulpes vulpes*, $2n = 34 + 0–8$ Bs), are members of the same family (Canidae) and diverged from a common ancestor some 10 Myr ago (Wayne, 1993). Recent comparative chromosome painting with a complete set of dog chromosome-specific paint probes enabled, for the first time, the entire genomes of four canid species to be compared accurately on the basis of both molecular and cytogenetic homologies (Yang *et al.*, 1999, 2000a,b; Graphodatsky *et al.*, 2000a,b, 2001). Integrating these previously published comparative maps between dog and red fox, dog and Arctic fox and dog and raccoon dog produces a map showing the genome-wide correspondence of all G-banded chromosomes of these four canid species with diverged karyotypes. This map is based solely on homologies defined by chromosome painting, while G-banding comparison is used to define the orientation of homologous chromosomes or chromosomal segments. In total, 38 dog autosomal paints reveal 42 pSCEUS (i.e., the **s**mallest **c**onserved **e**volutionary **s**egment **u**nits defined by painting, Nash *et al.*, 1998) with conserved banding patterns across the four canid species. Dog chromosomes 1, 13, 18 and 19 are each composed of two pSCEUS, while the remaining 34 autosomes each correspond to one pSCEUS. The karyotypes of the four canid species represent different combinations of 42 pSCEUS. All bi-armed autosomes of the red fox, Arctic fox and raccoon dog display only chromosomal arm homologies (i.e., monobrachial homologies), and in contradiction of previous reports (Wayne *et al.*, 1987), not a single bi-armed autosome is shared among the Arctic fox, red fox and raccoon dog (Figure 1). Further comparative analysis of the distribution patterns of the 42 pSCEUS in the canids and outgroup species such as the cat and human reveals 38 ancestral autosomal segments. These segments could represent the ancestral chromosome arms in the karyotype of the most recent Canidae ancestor, which we suggest could have a low diploid number (Graphodatsky *et al.*, 2001).

An integrated comparative chromosome map of human, cat and mink defined by painting with dog probes

It is well established that cat, American mink and human have highly conserved karyotypes. The karyotypes of the human and cat have been intensively investigated for about two decades by the various approaches available, from comparative banding and gene mapping to reciprocal chromosome painting (Nash and O'Brien, 1982; Dutrillaux and Couturier, 1983; Rettenberger *et al.*, 1995; Wienberg *et al.*, 1997). The karyotype relationships among cats, humans and minks have been further elucidated by our recent comparative painting with probes derived from the highly rearranged dog chromosomes, which generate comparative maps of resolution higher than the human–cat (Rettenberger *et al.*, 1995; Wienberg *et al.*, 1997) and human–mink maps (Hameister *et al.*, 1997) based on chromosome painting with human probes. Chromosome-specific paints from the 38 dog autosomes identified 68, 71 and 90 conserved segments in cat, mink and human, respectively. In contrast, 22 human autosomal paints have highlighted 34 conserved segments only in the cat and mink (Murphy *et al.*, 2000; Yang *et al.*, 2000a).

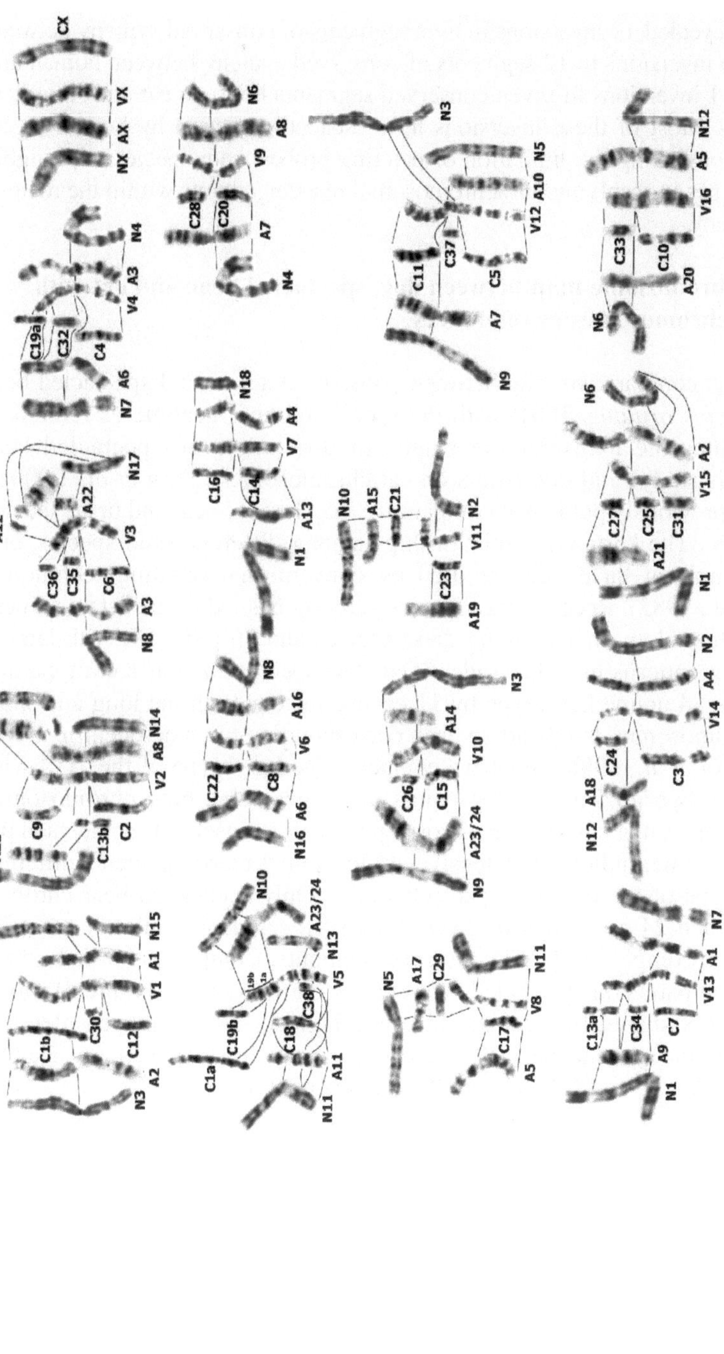

Figure 1. *Genome-wide comparative chromosome maps among domestic dog (C), red fox (V), Arctic fox (A) and Japanese raccoon dog (N), based on homologies revealed by dog paint probes and high resolution G-banding comparisons. The red fox chromosomes were used as references. Note the remarkable conservation in banding patterns among homologous chromosomal arms and extensive chromosomal homologies between chromosomes of the domestic dog, Japanese raccoon dog, Arctic fox and red fox. The four canid species do not share a single bi-armed autosome.*

In addition to confirming the previously identified interchromosomal rearrangements, the comparative map generated using dog paints has revealed extensive intrachromosomal rearrangements (inversions) in conserved syntenies delineated by human and cat paints. Furthermore, the use of dog paints also helps to define the orientation of homologous segments on chromosomes of these species (Figure 2). As summarized in Table 1, dog paint probes revealed 19 inversions in nine segments of conserved synteny between human and cat, 26 inversions in 12 segments of conserved synteny between human and American mink; 11 inversions in seven conserved segments between cat and American mink, respectively. Most of these inversions have escaped detection by human or cat paint probes, demonstrating the limitation of painting probes from species with highly conserved karyotypes in resolving intrachromosomal rearrangements within the regions of conserved synteny.

A comparative chromosome map between dog, spectacled bear and cat with spectacled bear chromosomes as references

Figure 3 shows the comparative map between domestic dog, cat and spectacled bear ($2n = 52$, *Tremarctos ornatus*, TOR) with the spectacled bear genome as reference. This map integrates the hybridization results of dog paints on spectacled bear chromosomes (this study) and dog paints on cat chromosomes (Yang *et al.*, 2000a), together with the previously published comparative map between cat and ursid species (Nash *et al.*, 1998). The karyotypic relationships among different ursid species, and between ursids and cat have been studied by conventional banding techniques (Wurster-Hill *et al.*, 1982). Recent cross-species painting has led to the establishment of genome-wide homology maps among these species and 39 pSCEUS with largely conserved banding patterns have been identified. The spectacled bear karyotype differs from the $2n = 74$ ursine karyotype by 11 centric fusions with the long and short arms of TOR chromosomes 1–10 and 25 each corresponding to an acrocentric ursine chromosome (Nash *et al.*, 1998). We have compared the karyorype of the TOR with that of the dog by hybridizing the dog paints onto spectacled bear chromosomes (Figure 3). Thirty-eight dog chromosomes correspond to 72 conserved segments in the spectacled bear genome, indicating extensive chromosomal rearrangements differentiating the karyotype of the dog and spectacled bear. Only spectacled bear chromosomes 15, 21, 22 and 24 each consist of one conserved segment, while the other 21 autosomes are each composed of 2–7 conserved segments. Comparison of hybridization patterns of dog paints on spectacled bear and cat chromosomes (Figure 4) shows that the 39 pSCEUS shared between Ursidae and Felidae have displayed identical painting patterns in the cat and spectacled bear. Only two inversions were revealed by dog paints (see FCA C1q and E2 and their TOR equivalents).

An integrated comparative map between cat, dog, spectacled bear, American mink and human with cat chromosomes as references

Comparative banding, gene mapping and chromosome painting have established that the karyotype of the domestic cat closely resembles the ancestral carnivore karyotype

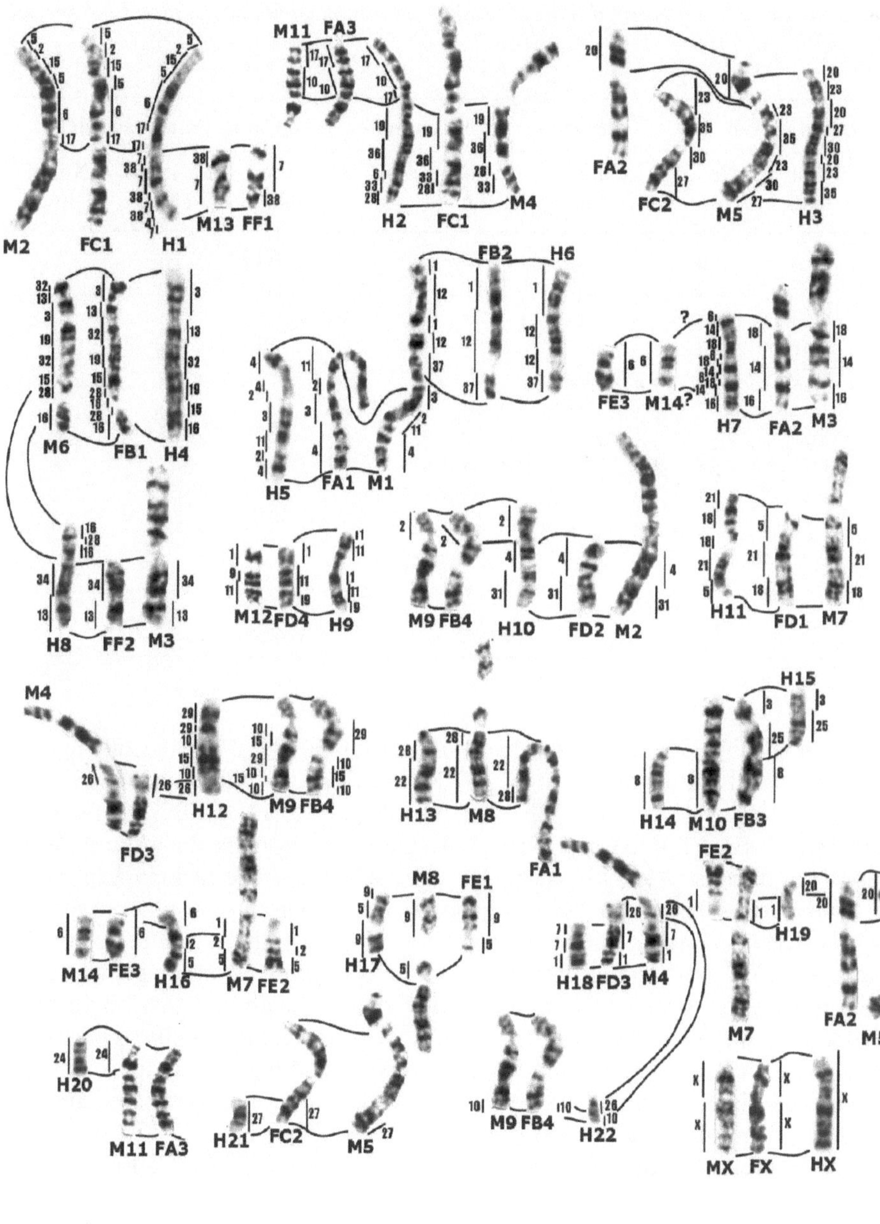

Figure 2. *A genome-wide comparative chromosome map of human (H), cat (F) and American mink (M), generated by integrating the hybridization results of dog paint probes onto G-banded chromosomes of human (Yang* et al., *1999; Sargan* et al., *2000), domestic cat (Yang* et al., *2000a) and American mink (Graphodatsky* et al., *2000b). Human chromosomes are used as reference. The use of paints from the highly rearranged dog chromosomes has enabled the generation of a high-resolution comparative map, which illuminates both the inter- and intra-chromosomal rearrangements.*

Table 1. *Inversions revealed by dog paint probes in chromosomal segments of conserved synteny*

Human *vs.* Cat			Human *vs.* American mink			Cat *vs.* American mink		
Human chr.	Cat chr.	No. of inv.	Human chr.	Mink chr.	No. of inv.	Cat chr.	Mink chr.	No. of inv.
1q	F1	4	1q	13	4	A1q	1q	1
2p-q	A3	2	2p-q	11q	2	B1	6	3
3	C2	3	2q	4q	1	B2	1p	1
5	A1q	3	3	5	3	B4q	9q	3
7	A2q	3	4q	6q	2	C1q	4q	1
8p	B1 prox.	1	5q	1q	3	C2	5q	1
9	D2	1	6	1p	1	D4	12	1
11	D1	1	7	3q	3			
17	E1	1	8p	6 prox.	1			
			9q	12	2			
			12	9q	3			
			17	8p	1			
Total	9	19		12	26		7	11

(ACK) (Nash and O'Brien, 1982; Dutrillaux and Couturier, 1983). The recently revised consensus ancestral carnivore karyotype (ACK, Murphy *et al.*, 2001) has a $2n = 42$, composed of the equivalents of cat A2p + C2, A1q, A1p, A2q, A3, B1, B2, B3, B4, C1p, C1q, D1–D4, E1–E3, F1, F2, X and Y. The ACK thus differs from the domestic cat karyotype by three fissions (A1p–A1q, A2p–A2q, C1p–C1q) and one fusion (A2p + C2). Figure 4 summarizes the genome wide correspondence among chromosomes of the cat, dog, spectacled bear, American mink, human and the hypothetical ancestor of the Carnivora order. The map demonstrates the three levels of genomic rearrangement, each of which represents an independent trend of karyotype evolution that could have occurred during species radiation of the carnivores. The Canidae family, represented by the dog, has the most highly rearranged karyotypes and highest rate of karyotype evolution in carnivores. The 42 pSCEUS conserved among the four canid species correspond to 68 segments in the cat genome. At least 47 fissions and 23 fusions are required to "convert" the current 19 cat autosomes, into the 42 pSCEUS found in the four canids so far studied. Compared with the dog, the spectacled bear, however, has a rather "conserved" karyotype. The 39 pSCEUS shared by the cat and bears appear to have been derived from the cat-like ancestral chromosomes through 20 fissions with 10 breakpoints mapped to the centromeric regions and 10 breakpoints localised inside chromosomal arms. Only a few inversions appear to have been involved in this process. The extent of chromosomal rearrangements found between the spectacled bear and cat karyotypes is comparable to that seen among different canid species. Chromosome fission and fusions are the dominant trend of karyotype evolution in

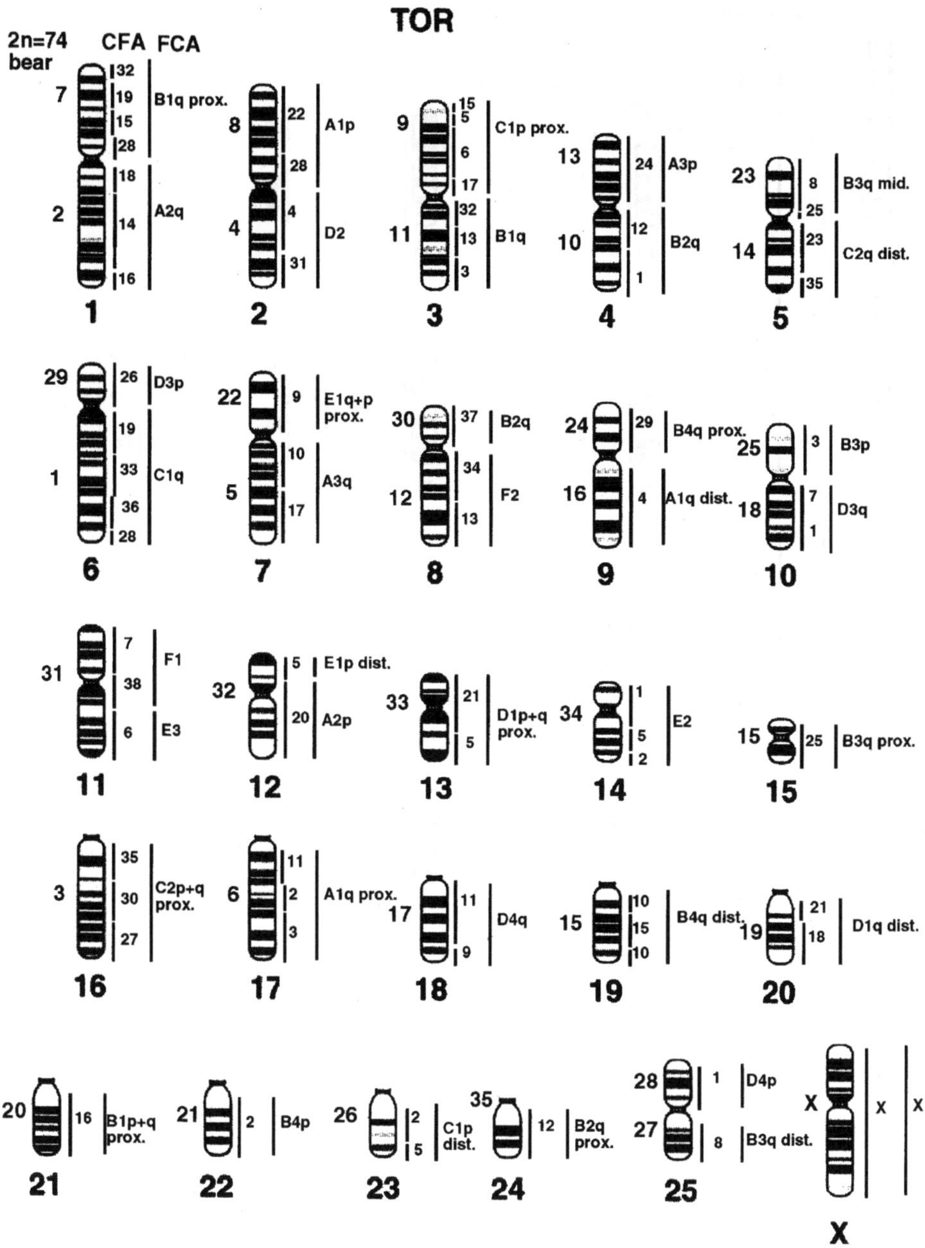

Figure 3. *A comparative map of the dog (CFA), cat (FCA) and spectacled bear (TOR) generated by summarising the hybridization patterns of the dog paints onto TOR chromosomes. The correspondence between cat and 2n = 74 bear chromosomes are inferred by integrating the hybridization results of dog paints on cat (Yang* et al., *2000a) and TOR chromosomes (this study) as well as the comparative maps reported in Nash* et al. *(1998).*

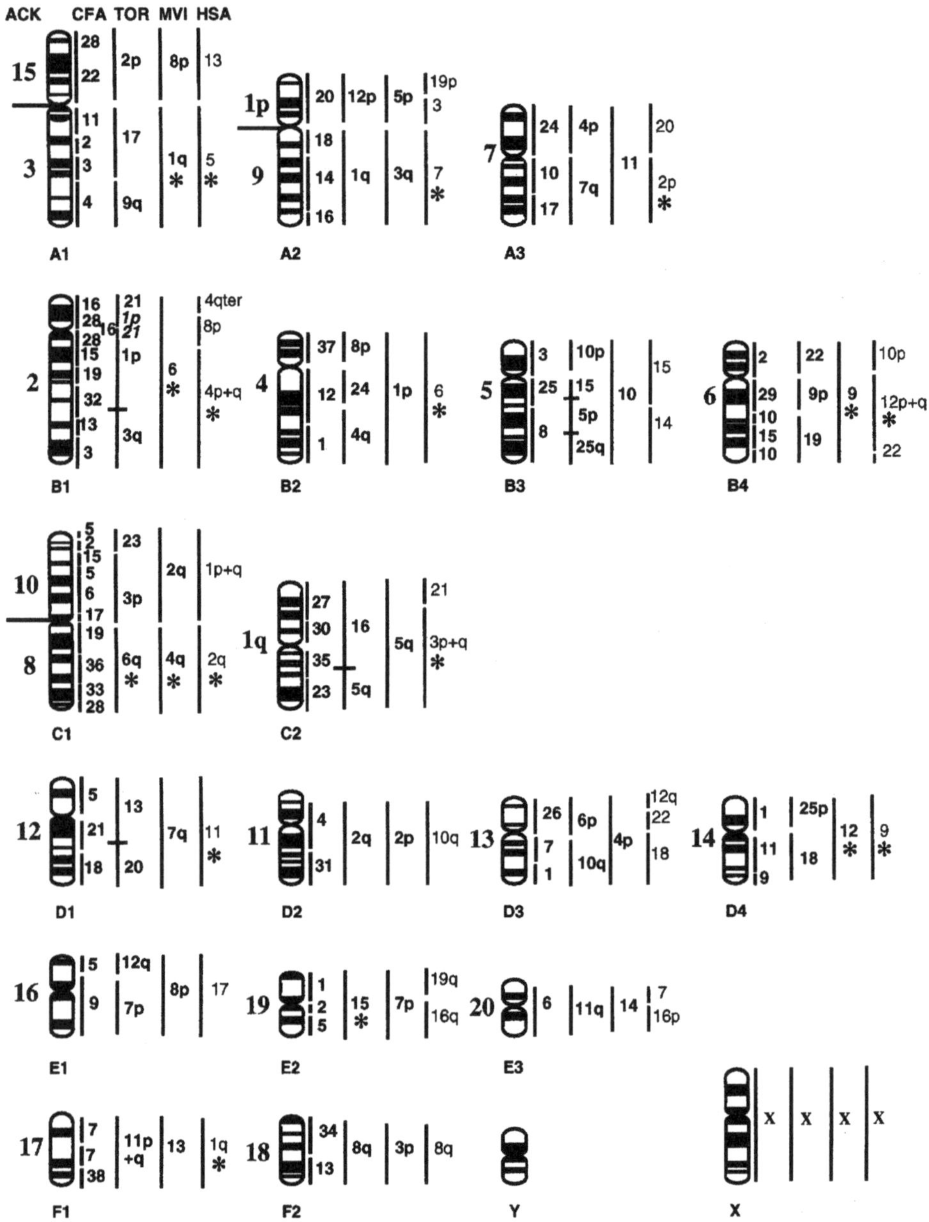

Figure 4. *A comparative chromosome map of cat (FCA), dog (CFA), spectacled bear (TOR), American mink (MVI), and human (HSA) inferred from comparative painting with paints from the dog, cat, mink and human. Asterisks (*) note where dog paints have revealed inversions in the segments of conserved synteny between cat and the species being compared. The corresponding chromosome numbers of the hypothetical carnivore ancestor (ACK, Murphy* et al., *2001) are given to the left of each cat chromosome.*

Ursidae (Nash *et al.*, 1998). The American mink and cat shared 22 conserved segments, each of which correspond to one entire chromosome or chromosomal arm of the ACK. This could represent the third trend of karyotypic evolution: a high degree of conservation in chromosomal synteny accompanied by inversions. As shown in Table 1, at least 7 of the 22 segments of conserved synteny between cat and mink have undergone 11 inversions. In conclusion, comparative chromosome maps between the cat and other carnivores reveal at least three sets of evolutionary conserved chromosome segments, each of which is shared by a group of canivore species, indicating three levels of karyotype conservation. The canids have the most rearranged karyotypes, the Ursidae have moderately rearranged karyotypes, while the mustelids and the felids have a highly conserved karyotype.

Summary

We have constructed a series of genome-wide, high resolution, comparative maps by integrating the hybridization results of dog paint probes on the chromosomes of human, cat, American mink, spectacled bear and three canid species. These maps reveal three levels of genomic reorganization, with canids having the most highly rearranged karyotypes, felids and mustelids having highly conserved karyotypes, and the ursids having moderately rearranged karyotypes. The use of paint probes derived from the highly rearranged dog chromosomes has enabled the detection of many cryptic intrachromosomal rearrangements that have occurred during the genome evolution of species with conserved karyotypes and has provided further insights into the karyotype evolution of carnivores.

Acknowledgements

We thank Mrs. Patricia C.M. O'Brien and Professor Malcolm A. Ferguson-Smith for help in preparing this manuscript, Ms. Beiyuan Fu for technical assistance, and Dr. Steve J. O'Brien for the provision of spectacled bear idiogram.

References

Dutrillaux B, Couturier J (1983). The ancestral karyotype of Carnivora: comparison with that of platyrrhine monkeys. *Cytogenet Cell Genet* 35: 200–208.

Ferguson-Smith MA, Yang F, O'Brien PCM (1998). Comparative mapping using chromosome sorting and painting. *ILAR J* 39: 68–76.

Graphodatsky AS, Yang F, O'Brien PCM, Serdukova N, Milne BS, Trifonov V, Ferguson-Smith MA (2000a). A comparative chromosome map of the Arctic fox, red fox and dog defined by chromosome painting and G-banding. *Chromosome Res* 8: 253–263.

Graphodatsky AS, Yang F, Serdukova N, Perelman P, Zhdanova N, Ferguson-Smith MA (2000b). Dog chromosome-specific paints reveal evolutionary inter- and intra-chromosomal rearrangements in the American mink and human. *Cytogenet Cell Genet* 90: 275–278.

Graphodatsky AS, Yang F, O'Brien PCM, Perelman P, Milne BS, Serdukova N, Kawada S-I, Ferguson-Smith MA (2001). Phylogenetic implications of the 38 putative ancestral chromosomal segments for four canid species. *Cytogenet Cell Genet* 92: 243–247.

Hameister H, Klett CH, Bruch J, Dixkens CH, Vogel W, Klott C (1997). ZOO-FISH analysis: the American mink (*Mustela vison*) closely resembles the cat karyotype. *Chromosome Res* 5: 5–11.

Murphy WJ, Sun S, Chen Z, Yuhki N, Hirschmann D, Menotti-Raymond M, O'Brien SJ (2000). A radiation hybrid map of the cat genome: implications for comparative mapping. *Genome Res* 10: 691–702.

Murphy WJ, Stanyon R, O'Brien SJ (2001). Evolution of mammalian genome organisation inferred from comparative gene mapping. *Genome Biol* 2: 1–8.

Nadeau JH, Sankoff D (1998). Counting on comparative maps. *Trends Genet* 14: 495–500.

Nash WG, O'Brien SJ (1982). Conserved regions of homologous G-banding chromosomes between orders in mammalian evolution: Carnivore and primates. *Proc Natl Acad Sci USA* 79: 6,631–6,635.

Nash WG, O'Brien SJ (1987). A comparative chromosome banding analysis of the Ursidae and their relationship to other carnivores. *Cytogenet Cell Genet* 45: 206–212.

Nash WG, Wienberg J, Ferguson-Smith MA, Menninger JC, O'Brien SJ (1998). Comparative genomics: tracking chromosome evolution in the family Ursidae using reciprocal chromosome painting. *Cytogenet Cell Genet* 83: 182–192.

Rettenberger G, Klett C, Zechner U, Bruch J, Just W, Vogel W, Hameister H (1996). ZOO-FISH analysis: cat and human karyotypes closely resemble the putative ancestral mammalian karyotype. *Chromosome Res* 3: 479–486.

Sargan DR, Yang F, Squire M, Milne SB, O'Brien PCM, Ferguson-Smith MA (2000). Use of flow-sorted canine chromosomes in the assignment of canine linkage, radiation hybrid and syntenic groups to chromosomes: Refinement and verification of the comparative chromosome map for dogs and humans. *Genomics* 69: 182–195.

Scherthan H, Cremer T, Arnason U, Weier HU, Lima-de-Faria A, Fronicke L (1994). Comparative chromosome painting discloses homologous segments in distantly related mammals. *Nat Genet* 6: 342–347.

Wayne RK (1993). Molecular evolution of the dog family. *Trends Genet* 9: 218–224.

Wayne RK, Nash WG, O'Brien SJ (1987). Chromosome evolution of the Canidae II: Divergence from the primitive carnivore karyotype. *Cytogenet Cell Genet* 44: 134–141.

Wienberg J, Stanyon R, Nash WG, O'Brien PCM, Yang F, O'Brien SJ, Ferguson-Smith MA (1997). Conservation of human vs. feline genome organization revealed by reciprocal chromosome painting. *Cytogenet Cell Genet* 77: 211–217.

Wurster-Hill DH, Bush M (1980). The interrelationship of chromosome banding patterns in the giant panda (*Ailuropoda melanoleuca*), hybrid bear (*Ursus middendorfi* × *Thalarctos maritimus*), and other carnivores. *Cytogenet Cell Genet* 27: 147–154.

Yang F, Carter NP, Shi L, Ferguson-Smith MA (1995). A comparative study of karyotypes of muntjacs by chromosome painting. *Chromosoma* 103: 642–652.

Yang F, O'Brien PCM, Milne BS, Graphodatsky AS, Solanky N, Trifonov V, Rens W, Sargan D, Ferguson-Smith MA (1999). A complete comparative chromosome map for the dog, red fox and human and its integration with canine genetic maps. *Genomics* 62: 189–202.

Yang F, Graphodatsky AS, O'Brien PCM, Colabella A, Solanky N, Squire M, Sargan DR, Ferguson-Smith MA (2000a). Reciprocal chromosome painting illuminates the history of genome evolution of the domestic cat, dog and human. *Chromosome Res* 8: 393–404.

Yang F, Milne BS, Schelling C, Dolf G, Schläpfer J, Switonski M, Ladon D, Pienkowska A, Bosma AA, Sargan DR, Ferguson-Smith MA (2000b). Chromosome identification and assignment of DNA clones in the dog using a red fox and dog comparative map. *Chromosome Res* 8: 93–100.

4.6. Complex Relationships Between DNA Methylation Status and Chromosome Compaction and Cohesion

J. Bernardino-Sgherri[1], D. Flagiello[2] and B. Dutrillaux[1,2]
[1]*CEA, Department of Radiobiology and Radiopathology, Fontenay-aux-Roses, France;*
[2]*UMR147 CNRS, Institut Curie, 26 rue d'Ulm, 75248 Paris Cedex 05, France*

Introduction

Mitotic chromosomes are generally studied on cultured cells. For instance, almost all data collected in human cytogenetics since the early 1960s were obtained from fibroblast, lymphocyte, amniotic and trophoblastic cell cultures. Even for cancer cells, which are spontaneously proliferating, cultures are almost always performed before metaphase harvesting. Consequently, the chromosomes of spontaneously dividing cells are rarely studied. When such studies are successfully performed, chromosomes of "poor quality" are invariably obtained. Indeed, the notion of quality for mitotic chromosomes is purely subjective, the goal of cytogeneticists being the obtention of sharply delineated chromosomes, which can resist to treatments inducing banding or allowing "*in situ*" hybridization. The "poor quality" of chromosomes from non-cultured cells may have several explanations. It is more difficult to standardize the cytological techniques. Mitotic indexes are generally low, which reduces the choice of mitotic cells. It is also possible that mitotic chromosome structure is different.

In cultured cells, a variety of treatments are able to induce alterations of chromosome appearance. It is noteworthy that the most extreme alterations are induced by the DNA demethylating agents, 5-aza-cytidine (5-aza-C) and 5-aza-deoxycytidine (5-aza-dC). They can induce drastic alterations of chromosome compaction and appearance, such as fuzzyness, recalling the "poor quality" of chromosomes from non-cultured cells.

Variations of DNA methylation constitute the most important epigenetic change. DNA methylation may vary from tissue to tissue, but also within a same tissue with time or physiopathological conditions (Reik *et al.*, 2001). If these changes are important enough, they may alter mitotic chromosome structure, as does 5-aza-dC in

M. Schmid and I. Nanda (eds.), Chromosomes Today, Volume 14, 225–234

cultured cells (Viegas-Pequignot and Dutrillaux, 1976; Coffigny *et al.*, 1999; Haaf and Schmid, 2000). We analyzed a number of situations in which DNA methylation is studied at the whole chromosome level, using anti-5-mC antibody. This allows us to propose that variations of chromosome appearance such as chromatid compaction and cohesion may depend on their DNA methylation status.

Material and methods

Human cells

Lymphocyte from voluntary donors and fibroblasts from our tissue bank were cultured according to our usual conditions (Dutrillaux and Couturier, 1981). In addition, several breast cancer cell lines, among which CAL51 and MDA-MB231 were used for long-term cultures.

Mouse and rat cells

Fibroblast cultures were performed in the same conditions as for humans. Spontaneously dividing gonadic cells were studied for both animals in addition to cells from various organs of mouse embryos. No colchicine treatment was applied, fixation procedures were uniformized, but conditions of hypotonic shock were adapted to testicular cells.

Treatments by 5-aza-dC

Human cells were treated by 5-aza-dC (Sigma, Aldrich, France) added at various culture times. When the treatment was prolonged, the addition of 5-aza-dC was repeated each 24 h. A stock solution of 5-aza-dC (10^{-2} M) in bi-distilled water was used to prepare final fresh solutions for each experiment (Kokalj-Vokac *et al.*, 1993). The final concentration of 5-aza-dC was 10^{-5} M. For obtaining independent data on cell cycle, BrdU treatments (10 μg/ml) were performed in some parallel cultures, or in addition to 5-aza-dC treatments. Preparations were stained by Giemsa, analysed and distained before immunochemical procedures. Monoclonal antibody against 5-methylcytosine (5-mC) and immunostaining revelation methods were described in Barbin *et al.*, 1994). Chromosomes were counterstained with propidium iodide (PI). BrdU incorporation was visualized by FITC-labelled anti BrdU IgG (Boehringer, Mannheim, Germany). When cultures were treated by both BrdU and 5-aza-dC, antibody against 5-mC was revealed by Texas Red labelled anti-mouse IgG (Boehringer, Mannheim, Germany). Images were captured using a cooled CCD camera and a capture software (Quips-Smart, Vysis, Downers Grove, IL, USA). Colchicine (Sigma, St-Louis, MO), at a concentration of 0.04 μg/mL, was added in some cultures for 3 h.

Results and discussion

Chromosome compaction alteration and DNA demethylation induced by 5-aza-dC on human lymphocytes

5-aza-C or 5-aza-dC was used as demethylating agent at various times of the last cell cycle (Viegas-Pequignot and Dutrillaux, 1976), and during two (Haaf *et al.*, 1986) or more (Flagiello *et al.*, 2001) cell cycles. The relationships between the number of cell cycles under treatment and chromosome appearance was checked by using BrdU incorporation. Various alterations of chromosome structure were induced in relation with the timing of treatment. They can be summarized as follows:

(1) *Last late S-phase*: Strong lack of longitudinal compaction of constitutive heterochromatin and G-bands, i.e., late replication chromosome segments (Figure 1B).
(2) *Last whole S-phase*: Intense shortening, but fuzzy and spiralized appearance of chromosomes. Treatments during the last early S-phase gave similar results (Figure 1A).
(3) *Last two cell cycles*: Uneven compaction of sister chromatids giving chromosomes a curved (asymmetrical) appearance. The shorter chromatids were always the demethylated ones (unlabelled by anti-5-mC antibody) (Figures 1C and 2A).
(4) *Last three cell cycles*: In each metaphase, half of chromosomes were as described above. The other half of chromosomes were symmetrical, with both chromatids demethylated. They were generally shorter than the shorter chromatid of their asymmetrical homologs. Similar results were obtained when 5-aza-dC was removed during the last cell cycle (Figures 1E, F and 2B).
(5) *Penultimate cell cycle*: Chromosomes were very similar to those obtained in experiment 3 (Figure 1D).
(6) *Antepenultimate cell cycle*: All chromosomes were symmetrical. Half of them were methylated and longitudinally compacted. The other half was less compacted and demethylated (Figure 1G).
(7) *Antepenultimate and last cell cycles*: All chromosomes were symmetrical but the relationships between DNA methylation and compaction were opposite: the methylated chromosomes were longer than their demethylated homologs (Figure 1H).

In these experiments, the relationships between DNA methylation and chromosome–chromatid compaction are neither direct nor univocal. The most intense effect on chromosome compaction was achieved when the methylation status varied along a given chromatid or chromosome. This suggests the existence of *cis*-acting factors. The effect was less intense, although detectable, between homologous chromosomes with different DNA methylation status, suggesting *trans*-acting factors also exist.

For a large part, apparently contradictory results (compare experiments 6 and 7) may be explained by a desynchronization of chromosome compaction when the cells reach mitosis with differently modified DNA methylation status. Such desynchronization would induce premature chromosome condensation (PCC). Another important parameter seems to be the hemi-methylation of the pre-replicative DNA. The chromatid inheriting the methylated DNA strand is always the less longitudinally compacted, independently of the methylation status of its other DNA strand. This suggests that DNA–protein interactions may depend on the DNA methylation status. Much work has been published about the multisubunit proteic complexes called condensin, which comprises SMC family members (review in Heck, 1997). To the best of our knowledge, their possible relationship with the DNA methylation status, suggested by our results, was never envisaged.

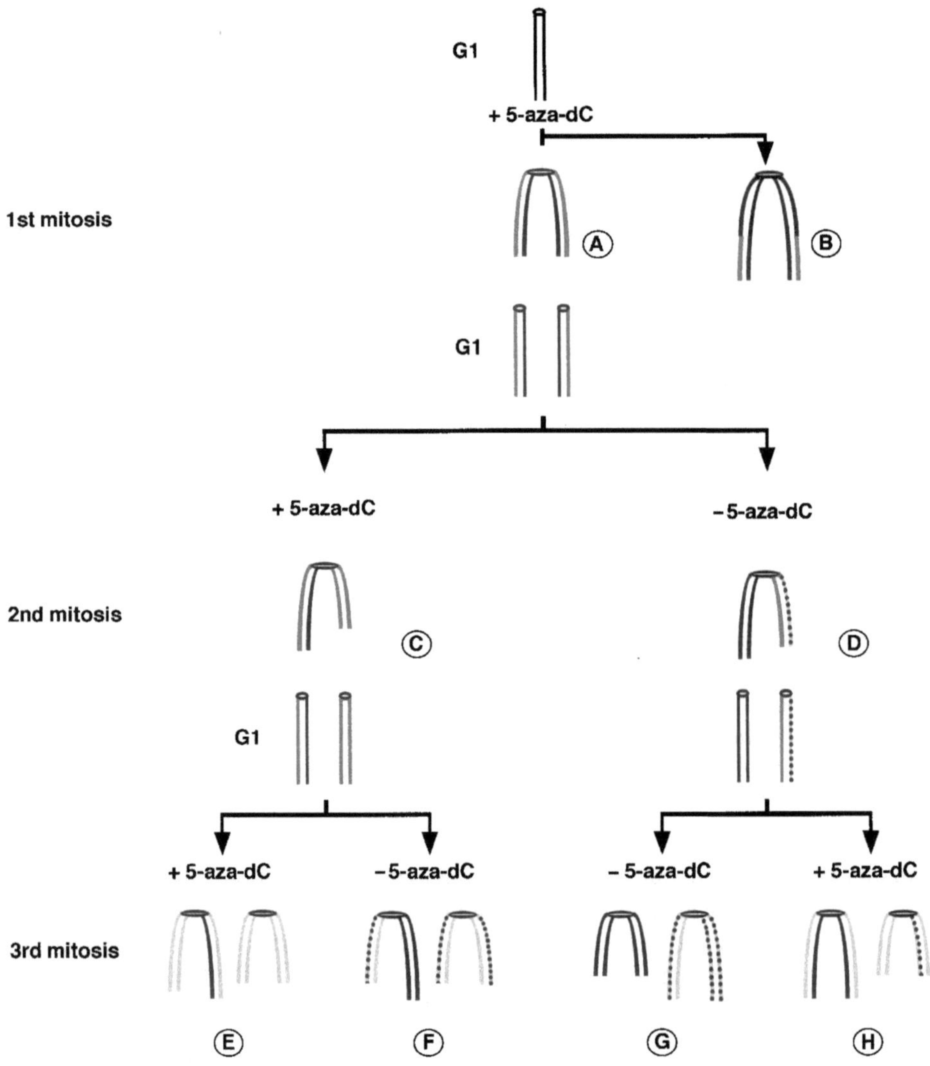

Figure 1. *Variations of DNA strand composition and mitotic chromosome compaction in the various conditions of 5-aza-dC treatments studied. Black lines: normal DNA; grey lines: DNA strand containing 5-aza-dC; dotted lines: demethylated DNA without 5-aza-dC; +5-aza-dC: presence of 5-aza-dC in the culture medium; −5-aza-dC: absence of 5-aza-dC in the culture medium. When two chromosomes are represented, each corresponds to about half of the whole set in a cell, as a result of chromosome segregation. A to H indicate DNA methylation status at the various moments when metaphases were analysed.*

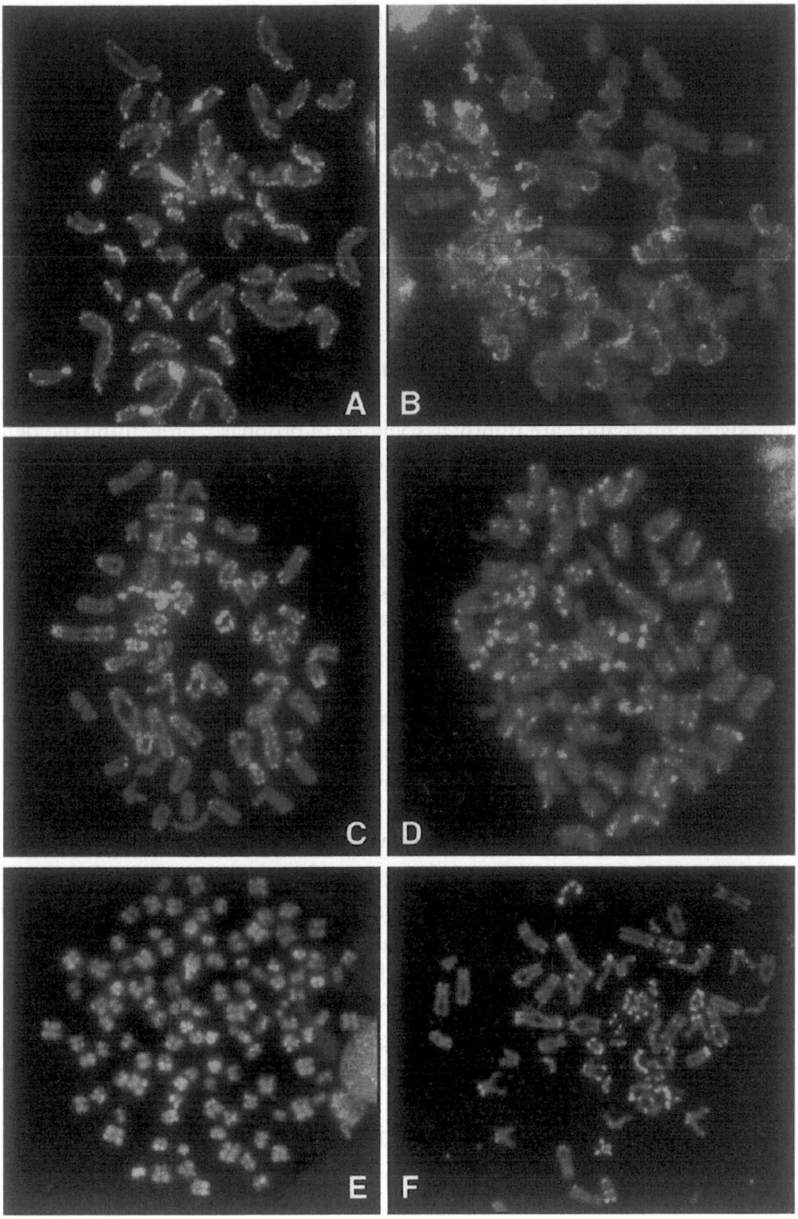

Figure 2. *Mitotic figures observed after immunolabelling by anti-5-mC antibody (green) and counterstaining by propidium iodide (red). Treatments of human lymphocytes by 5-aza-dC were performed during two (A) or three (B) consecutive cell cycles. Treatments by 5-aza-dC of MDA MB 231 cell line for one (C) or two (D) cell cycles induce premature centromere division. Culture of the same cell line in a medium with methionine deprivation, induces strong chromosome compaction (E) and premature centromere division (F).*

Chromosome compaction in other culture conditions inducing DNA hypomethylation of human cells

To avoid possible additional effects which could be caused by 5-aza-dC incorporation into DNA, we used other conditions for reducing DNA methylation. We used a RPMI medium without methionine, in which we added ethionine (a methionine competitor) and homocysteine to allow cells to survive. We could check, by using anti-5-mC antibody that DNA methylation of mitotic chromosomes was reduced. After 24 h treatment on the breast cancer cell line MDA-MB231, most metaphases exhibited highly compacted, spiralized and fuzzy chromosomes, even in the absence of colchicine block (Figure 2E).

Centromere and sister chromatid cohesion in cultured human cells

The morphology of mitotic chromosomes varies from prophase to anaphase. Classically, sister chromatids are tightly associated in prophase, preventing the detection of centromeres and become more separated at prometaphase. They separate at metaphase, remaining linked by centromeres, whose cleavage initiates anaphase. The cohesion between chromatids is ensured by cohesin, another multisubunit protein complex composed of Smc1, Smc3, Scc1 and Scc3 proteins (review in Nasmyth *et al.*, 2000). This complex can be indirectly destroyed by the ubiquitin protein ligase anaphase-promoting complex (APC). Cohesin is assumed to start binding specific chromosomal loci during S-phase. Interestingly, both condensin and cohesin contain proteins of the Smc family.

In most mammalian species, large heterochromatic segments surround centromeres. In the few species we studied, many (as in man and rat) or all (as in the mouse) these regions are rich in 5-mC and labelled with anti 5-mC antibody.

Treatments of lymphocytes in culture failed to recurrently induce detectable changes of centromere or chromosome arm cohesion. We further used the breast cancer cell line MDA MB 231, which spontaneously exhibits about 3% of metaphases with premature centromere division (PCD), as defined by (Fitzgerald, 1975). After 24–96 h treatments by 5-aza-dC, PCD was observed in 35% of metaphases (Figure 2C and D). When using the culture medium depleted in methionine described above, most metaphases exhibited PCD associated with a strong chromosome compaction (Figure 2E and F). The use of anti-5-mC antibody indicated the efficiency of the treatment, by a decrease of the labelling. Thus, a link seems to exist between hypo- or demethylation of DNA located in juxtacentromeric regions and centromere lack of cohesion, but it remains unclear why it is not observed in all cell types. Possibly, hypomethylation of DNA is not sufficient *per se* to induce PCD, but it may induce or amplify it when the maintenance of centromere cohesion is fragilized, as in some cancer cells.

Chromosome arm cohesion is also assumed to be ensured by cohesin. Its disruption at metaphase may largely be artefactual, induced by colchicine treatments usually applied by cytogeneticists. Treatments by 5-aza-dC were applied on the breast cancer cell line CAL51, which is apparently euploid. After long-term treatment, even in metaphases with condensed chromosomes, centromere cohesion was maintained, while sister chromatid cohesion was disrupted in several chromosomes from all metaphases. Compared to the cells from the same cell line at the same passage but untreated by 5-aza-dC, sister chromatid cleavage were about 10-fold more frequent after treatment. A systematic analysis of 13 karyotyped metaphases did not allow us to draw any rule about this disruption which seemed to be completely at random. In particular, homologous chromosomes or chromosome arms behave independently, as well as both arms of a given chromosome. The use of anti-5-mC antibody indicated that this long-term treatment by 5-aza-dC did not totally inhibited DNA methylation. In particular, usually 5-mC rich juxtacentromeric heterochromatin remained weakly labelled, which may explain the lack of PCD. Nevertheless, in this cell line, the overall decrease of DNA methylation correlates with a defect in sister chromatid cohesion. Thus, in addition to condensation, it is likely that DNA methylation interfers with chromatid cohesion, by altering the interactions between condensin, cohesin and DNA.

"*In vivo*" studies in mouse and rat

During the perinatal period, in male gonades, gonocytes undergo a cell cycle arrest while Sertoli cells actively proliferate (Coffigny *et al.*, 1999). Thus, with the possible exception of rare other somatic cells, only Sertoli cells reach metaphase. At this moment, all mitoses share the same typical chromosomes, independently of their contraction status (Figure 3B):

- chromatids are fuzzy, weakly stained by Giemsa, and not cohesive;
- centromeric heterochromatin regions are strongly cohesive and intensely stained by Giemsa. This and the pale staining of chromosome arms give a C-banding appearance;
- DAPI staining reveals the same characteristics, as well as large halos surrounding chromosomes: thus chromatin fibres are dispersed around the core of chromatids;
- with anti-5-mC antibody, chromatids are weakly and centromeric heterochromatin intensely labelled;
- although prometaphases exist, most mitoses ressemble late metaphases, with short chromosomes.

When gonocytes resume their cycle, i.e., 1 and 4 days after birth for mouse and rat respectively, a completely new type of mitoses appears, coexisting with that described above (Figure 3A):

- chromatids are long, sharply stained by either Giemsa or DAPI staining, and tightly cohesive;
- centromeric heterochromatin is weakly stained, thus hardly visible in the same conditions of staining;
- no DNA background surrounds chromosomes after DAPI staining;
- with anti-5-mC antibody, chromosome arms are strongly labelled while centromeric heterochromatin is not labelled at all.

The hypermethylation of euchromatin is further semi-conservatively lost, while heterochromatin becomes progressively more methylated (Coffigny *et al.*, 1999).

Thus, in male gonads, two types of spontaneously dividing cells exhibit opposite characteristics of chromosome staining, compaction and cohesion, and this correlates with

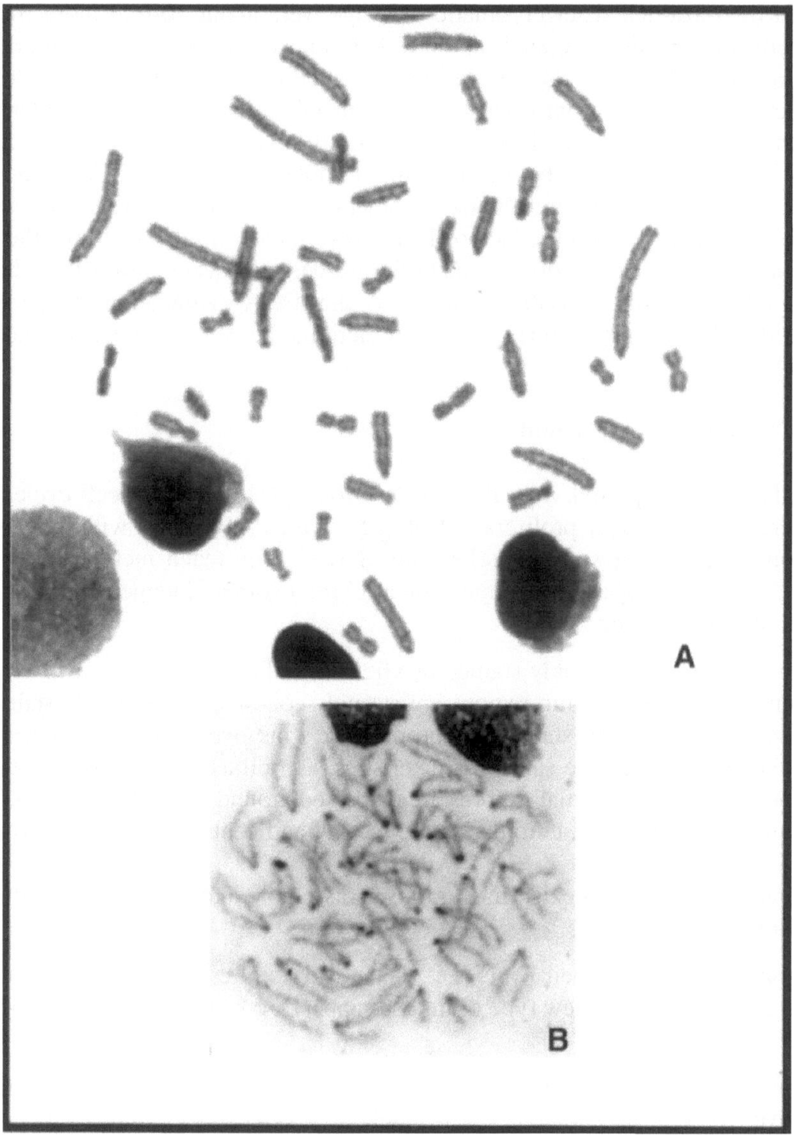

Figure 3. *Giemsa staining of rat gonocyte (A) and mouse Sertoli cell (B). Chromatids from gonocyte chromosomes have strongly methylated DNA. They are long, well delineated and strongly cohesive. Chromatids from Sertoli cell chromosomes have weakly methylated DNA. They are pale, thin, fuzzy, and not cohesive. At contrast, their centromeric regions have strongly methylated DNA and are intensely stained and cohesive.*

opposite DNA methylation patterns. The absence of colchicine use and the obtention of these two aspects on the same preparations avoid the possibility of artefacts. Longitudinal compaction (chromatid shortening) is associated with a low DNA methylation status, while lateral compaction (sharply delineated and thin chromatids without surrounding halos) is associated with a high DNA methylation status. In parallel, cohesion of sister chromatids also seems to be determined or modulated by the DNA methylation status.

DNA methylation, chromatin compaction and chromosome banding

The strong differences between euchromatin and heterochromatin are easily visualized and were discussed, above. Within euchromatin, the two main components, R- and G-bands also differ in several respects. It was shown that DNA is globally more methylated in R-bands than in G-bands (Barbin *et al.*, 1994). In Indian Muntjac cells, it was also proposed that R- and G-bands are differently compacted (Saitoh and Laemmli, 1994). Is the difference of chromatid compaction between Sertoli and germ cells sufficient to induce a spontaneous banding in one or the other cell type, and eventually in both with opposite patterns?

The staining of Sertoli cell chromomes is not homogeneous. It is also the case of spontaneously dividing cells from other fetal tissues, whose euchromatin is not strongly methylated. Their karyotype can be established after a simple Giemsa staining: chromosomes exhibit an imperfect but indisputable G-banding pattern. Detection of R-band material can be done using both YOYO and methylgreen stains. Preliminary assays performed on Sertoli cell metaphases exhibited a staining of surrounding halos, suggesting they are composed of chromatin usually located in R-bands. Thus, when euchromatin has a low DNA methylation content, chromatin from R-bands has a low lateral compaction and forms halos giving chromosomes their fuzzy appearance. It is less easy to answer for hypermethylated germ cells, which are intensely stained by Giemsa. They obviously do not spontaneously exhibit a G-banding pattern. By decreasing the concentration of the stain, it is possible to obtain heterogeneously stained chromatids. Then, an incomplete R-banding can be occasionally observed. Altogether, our results suggest that relationships between condensin, cohesin and DNA methylation status of the chromosome exist. The opposite behaviour of euchromatin and heterochromatin indicates this relation is complexe. It may result from their differential methylation status, interfering with the same proteins, or with proteins specific for each chromosome structure, such as CENP proteins at centromeres.

The hierarchy of these relationships, between primary and higher levels of chromatin structure is not well understood, and probably all the proteins that may interact are not yet known. Finally, data about proteins involved in chromosome condensation and cohesion largely come from studies on Yeast and Xenopus, which may not be totally extrapolated to mammals. This dictates much caution about mechanistic interpretations.

References

Barbin A, Montpellier C, Kokalj-Vokac N, Gibaud A, Niveleau A, Malfoy B, Dutrillaux B, Bourgeois CA (1994). New sites of methylcytosine-rich DNA detected on metaphase chromosomes. *Hum Genet* 94: 684–692.

Coffigny H, Bourgeois C, Ricoul M, Bernardino J, Vilain A, Niveleau A, Malfoy B, Dutrillaux B (1999). Alterations of DNA methylation patterns in germ cells and Sertoli cells from developing mouse testis. *Cytogenet Cell Genet* 87: 175–181.

Dutrillaux B, Couturier J (1981) La pratique de l'analyse chromosomique. Paris: Masson.

Fitzgerald PH (1975). A mechanism of X chromosome aneuploidy in lymphocytes of aging women. *Humangenetik* 28: 153–158.

Flagiello D, Bernardino-Sgherri J, Dutrillaux B (2003). Complex relationships between 5-aza-dC induced DNA demethylation and chromosome compaction at mitosis, submitted.

Haaf T, Ott G, Schmid M (1986). Differential inhibition of sister chromatid condensation induced by 5-azadeoxycytidine in human chromosomes. *Chromosoma* 94: 389–394.

Haaf T, Schmid M (2000). Experimental condensation inhibition in constitutive and facultative heterochromatin of mammalian chromosomes. *Cytogenet Cell Genet* 91: 113–123.

Heck MM (1997). Condensins, cohesins, and chromosome architecture: how to make and break a mitotic chromosome. *Cell* 91: 5–8.

Kokalj-Vokac N, Almeida A, Viegas-Pequignot E, Jeanpierre M, Malfoy B, Dutrillaux B (1993). Specific induction of uncoiling and recombination by azacytidine in classical satellite-containing constitutive heterochromatin. *Cytogenet Cell Genet* 63: 11–15.

Nasmyth K, Peters JM, Uhlmann F (2000). Splitting the chromosome: cutting the ties that bind sister chromatids. *Science* 288: 1,379–1,385.

Reik W, Dean W, Walter J (2001). Epigenetic reprogramming in mammalian development. *Science* 293: 1,089–1,093.

Saitoh Y, Laemmli UK (1994). Metaphase chromosome structure: bands arise from a differential folding path of the highly AT-rich scaffold. *Cell* 76: 609–622.

Viegas-Pequignot E, Dutrillaux B (1976). Segmentation of human chromosomes induced by 5-ACR (5-azacytidine). *Hum Genet* 34: 247–254.

4.7. Constitutive Heterochromatin of *Microtus agrestis*: Molecular Organization and Genetic Activity in Mitotic and Meiotic Cells

Karl Sperling, Sabine Henschel, Ilka Schulze and Heidemarie Neitzel
Institut fuer Humangenetik, Augustenburger Platz 1, 13353 Berlin, Germany

Introduction

The descriptive term "heterochromatin" was introduced by the botanist Emil Heitz in 1928 for chromosomal regions that remain condensed during interphase and stain intensively in contrast to euchromatin. Later a distinction was made between facultative heterochromatin (a transient state) and constitutive heterochromatin (Brown, 1966). The constitutive heterochromatin is a permanent entity of a chromosome and according to text book opinion considered "to be DNA that is never transcribed in any cell" (Alberts *et al.*, 1983). Even more, the possibility was suggested "that constitutive heterochromatin *per se* has no function in either development or evolution" (John, 1988). It is a constituent of both plant and animal genomes. For mammals, the gonosomal c-heterochromatin of the European field vole *Microtus agrestis* has been considered typical for constitutive heterochromatin in general: it is C-band positive, late replicating, and transcriptionally inactive in various tissues (Sieger *et al.*, 1970).

With the advent of methods to induce premature chromosome condensation, it became possible to visualize mammalian interphase chromosomes which are still genetically active. The sites of activity can be mapped after pulse-labelling with [^{3}H] uridine (Sperling, 1982; Marcus and Sperling, 1984). Using this approach, it was possible to illustrate that constitutive heterochromatin is indeed transcribed in mammals in its decondensed state (Sperling *et al.*, 1987) – as had been shown earlier already for insects and amphibians. The transcriptional activity is paralleled by its sensitivity to digestion with DNase I (Sperling *et al.*, 1985, 1989). Combined with the new data and the molecular organization of constitutive heterochromatin it is now evident that the historical term "heterochromatin" does not adequately describe its complex nature

M. Schmid and I. Nanda (eds.), Chromosomes Today, Volume 14, 235–246

(Redi *et al.*, 2001). Here, we briefly summarize our findings on the molecular organization of constitutive heterochromatin in *M. agrestis* and its behaviour during mitosis and meiosis.

Molecular characterization

M. agrestis has a diploid number of $2n = 50$. Most of the constitutive heterochromatin (about 20% of total DNA) is accumulated in the extremely large sex chromosomes (Figure 1). We have established a series of *M. agrestis* × mouse somatic cell hybrids which differed with respect to the number and activity of the *M. agrestis* X chromosomes (Bat-Sheva *et al.*, 1988) (Figure 2). These enabled us to identify *M. agrestis* DNA sequences and transcripts that are derived from the X chromosome. We could show that simple GAT/CA repeat clusters are localized predominantly in the X chromosomal heterochromatin which have been thought to be involved in the initial heterochromatization of the incipient vertebrate sex chromosomes (Nanda *et al.*, 1988). In addition, we have cloned several interspersed middle repetitive elements, which represent most of the noncentromeric X-heterochromatin. One element, pMAHAE2, is confined to the constitutive heterochromatin, shows a tissue-specific methylation pattern, and is transcriptionally inactive. It is a component of a large, multimeric, higher-order repeat unit with a length of 20 kb, that accounts for at least 15–30% of the gonosomal heterochromatin (Kalscheuer *et al.*, 1996) (Table 1). Furthermore, the gonosomal heterochromatin is highly enriched for an L1 element, pMAECO14, which is also interspersed at autosomal locations (Neitzel *et al.*, 1998) (Table 1). In addition, a novel non-L1 retroposon, pMA11/3, has been isolated which has almost the same

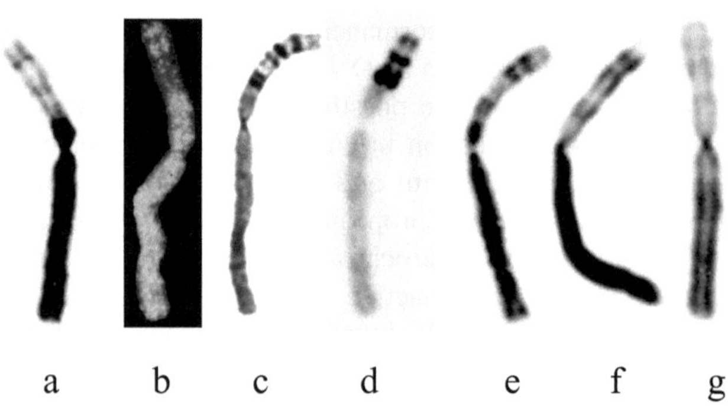

Figure 1. *Metaphase of a male* Microtus agrestis *X chromosome after C-banding (a), FISH with the biotin-labeled clone pMA11/31 (b), GTG-banding (c), replication banding using BrdU incorporation during late S phase (d), and nuclease banding with Alu I (e), Hae III (f), and Hinf I (g). Note that the insertion sites of the retroposon correspond with the C bands and the dark, late replicating G band in the distal part of Xp. The Alu I bands correspond with the C-bands, while the Hae III and Hinf I bands correspond with the labelling pattern of pMAHAE2 depicted in Figure 4.*

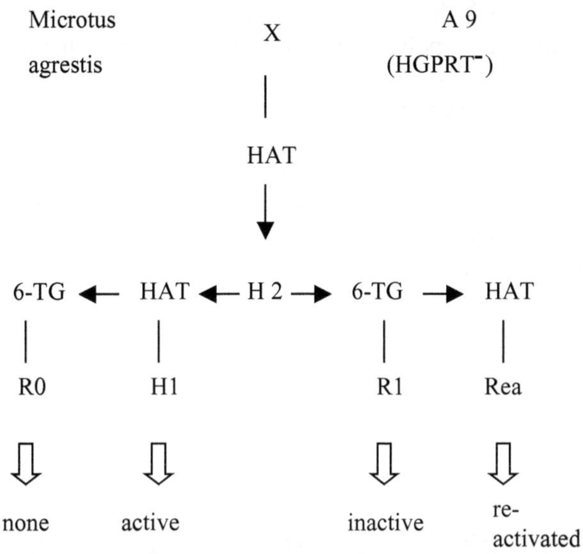

Figure 2. *Scheme for the selection and isolation of A9* mouse x M. agrestis *cell hybrids with one functionally active, one inactive, and one reactivated* M. agrestis *X chromosome, or none* M. agrestis *X. Selection media: HAT (* Hypoxanthin + Aminopterin + Thymidine*); 6-TG (6-Thioguanine).*

Table 1. *Characterization of cloned middle repetitive elements from* M. agrestis

	pMAHAE2	pMAECO14	pMA11/3
AT content	0.65	0.61	0.58
Length (kb)	2.999	1.771	2.293
CpGs (observed)	9	12	21
CpGs (expected)[1]	64	65	101
CpGs obs/exp.	0.141	0.185	0.207
No. of homologous genomic copies	$1–2 \times 10^4$	1×10^4	$1–2 \times 10^5$
Length of higher order repeat unit	> 20 kb	> 5.5 kb	—
Percentage of the gonosomal heterochromatin	15–30%	~4%	8–16%
Homologies to known sequences	$(GATA)_n$	L1	—

[1]The average ratio of observed to expected CpGs is 0.493 for housekeeping genes, 0.209 for processed pseudogenes, and 0.159 for LINEs (Kricker *et al.* (1992). *Proc Natl Acad Sci* 89: 1,075–1,079).

chromosomal distribution as the L1 element (Table 1). It contains a number of inverted repeats along its length and a poly A tail. There is no evidence that it is translated into a protein (Neitzel *et al.*, 2002).

These middle repetitive elements are not organized in tandem arrays because they do not show the typical ladder of multiplies in partial digests (Kalscheuer *et al.*, 1996; Neitzel *et al.*, 1998). Their interspersed organization can be directly seen in fiber fish preparations (Figure 3). After *in situ* hybridization on metaphase chromosomes, the pMAHAE2 element is confined to the noncentromeric heterochromatin of both X and Y, but excluded from the telomeric region of the Y (Figure 4a; Nanda *et al.*, 1988; Kalscheuer *et al.*, 1996). In contrast, the L1 and the non-L1 retroposons are highly enriched in the centromeric and non-centromeric heterochromatin of the gonosomes, including a faint signal in the short arm of the X, and interspersed at autosomal locations (Figure 4b; Neitzel *et al.*, 1998, 2002).

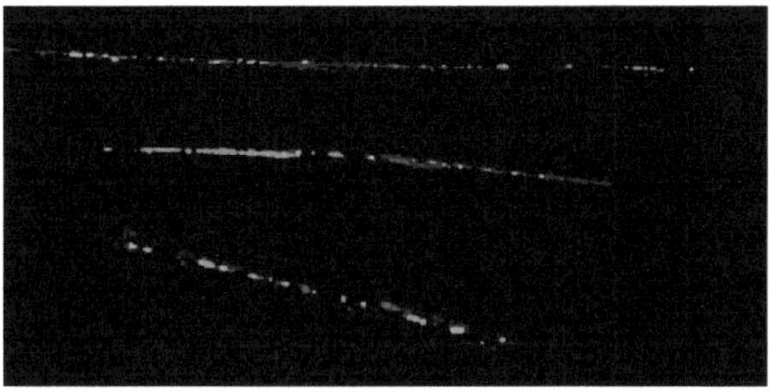

Figure 3. *Fiber FISH analysis of* M. agrestis *DNA with pMAHAE2 (green) and the retroposon pMA11/31 (red). Note the interspersed organization (Original C. Ramel, Berlin).*

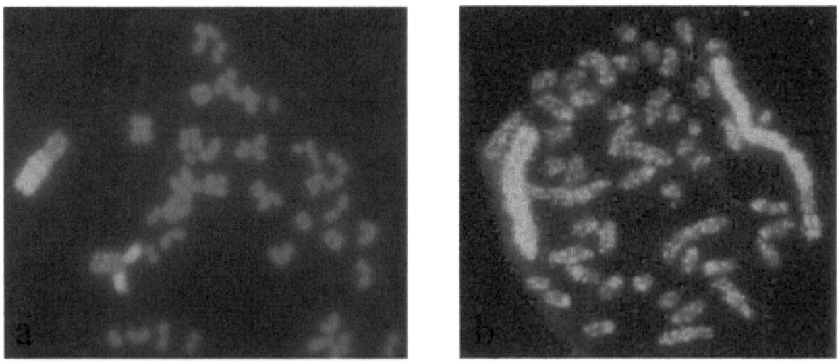

Figure 4. *Localization of the repetitive sequences pMAHAE2 (a) and pMAECO14 (b) after FISH analysis with biotinylated probes on male metaphases of* Microtus agrestis. *The pMAHAE2 signal is restricted to the non-centromeric heterochromatin of the gonosomes, while the pMAECO14 probe binds to the centromeric and non-centromeric heterochromatin of the gonosomes and is also interspersed at autosomal locations.*

After *Eco*RI digestion of genomic DNA from *M. agrestis* and *M. agrestis* × A9 somatic cell hybrids and hybridization with pMA11/31, three distinct bands of 1.0, 1.5 and 1.7 kb are observed (Figure 5). Since these bands are absent in the hybrids without *M. agrestis* X or with the iso Xp, these repeats must have been accumulated in the constitutive heterochromatin of Xq. In addition, a smear of higher molecular weight is observed in all cell lines indicating that numerous elements are also interspersed in the genome. Thus, the genomic organization of this sequence is typical for a mobile element.

Evolutionary origin

Sequences homologous to pMAHAE2 are absent in the mouse genome and abundant in the genomes of three closely related *Microtinae* (*M. guentheri, M. arvalis*, and

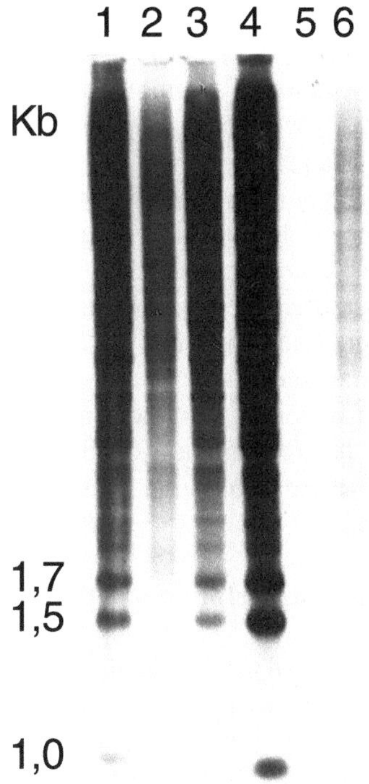

Figure 5. *Genomic organization of the non-L1-retroposon, pMA11/31: Southern blot of* Eco*RI digested DNA of (1)* M. agrestis *female, (2) somatic cell hybrid R0 without* M. agrestis *X chromosome, (3) somatic cell hybrid R1 with one* M. agrestis *X chromosome, (4) somatic cell hybrid H1 with five* M. agrestis *X chromosomes, (5) the parental cell line A9 of the somatic hybrids, and (6) somatic cell hybrid with an isochromosome Xp demonstrating that the distinct bands of 1.7, 1.5 and 1.0 kb are confined to the heterochromatin of Xq (from Neitzel* et al., *2002).*

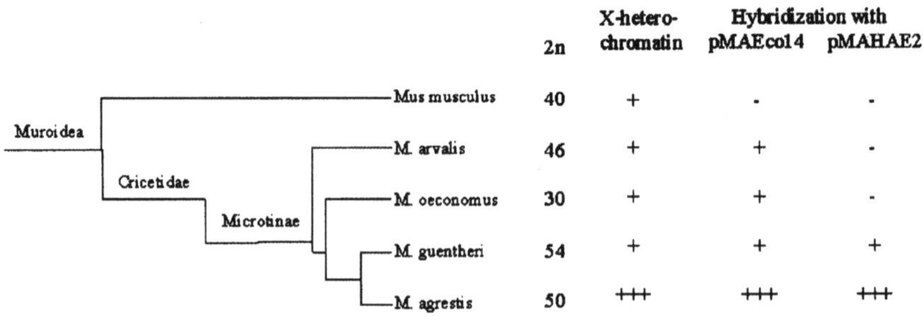

Figure 6. *Prevalence of three middle repetitive elements of* Microtus agrestis *in* Mus musculus *and four related Microtinae.*

M. oeconomus) which have retained the small original X chromosome (Figure 6). The copy number varies from ~100 per diploid genome in *M. arvalis and M. oeconomus* to ~500 in *M. guentheri* and up to $1–2 \times 10^4$ in *M. agrestis*. Thus, this sequence family probably arose during the evolution of a common ancestor of *Microtus* and has subsequently been amplified extensively in the gonosomes of *M. agrestis* in a short period of less than 1 million years (Kalscheuer *et al.*, 1996).We estimated that there are $1–2 \times 10^4$ copies of the L1 element (pMAECO14) and $1–2 \times 10^5$ copies of the novel retroposon per diploid genome of *M. agrestis*. Interestingly, these elements are not found in the closely related *Microtinae*. Thus, the rate of retrotransposition in the *M. agrestis* genome is higher than that of the majority of invertebrate retroposons and also that of murine L1 elements (Neitzel *et al.*, 2002). One consequence of its high-transposon activity might be an increased rate of spontaneous mutations that are caused by insertions of these elements into existing genes (Kazazian and Moran, 1998). DNA methylation and heterochromatin condensation are considered defence mechanism against the spread of these genomic invaders (Henikoff, 2000).

Behaviour during meiosis and transcriptional activity

The pMAHAE2 clone is ideally suited to map the heterochromatin in interphase nuclei and to study the meiotic behaviour of the gonosomes during spermatogenesis and oogenesis. The latter studies can be performed on neonatal ovaries which show almost all stages of prophase I (Singh *et al.*, 1993). As can be seen from Figure 7a, on silver-stained synaptonemal complexes, the X and Y chromosomes do not pair but show occasional telomeric associations. They form a compact sex body at pachytene. During oogenesis the two X chromosomes exhibit sequential pairing with crossing-over restricted to the euchromatic short arm. In the course of pachytene the heterochromatin becomes gradually condensed and shows positive heteropycnosis (Figure 7b; Singh *et al.*, 2000).

One main question was, whether the heterochromatin is transcribed during meiosis. Figure 8 illustrates that the sex body in spermatocytes is almost free of labelling after

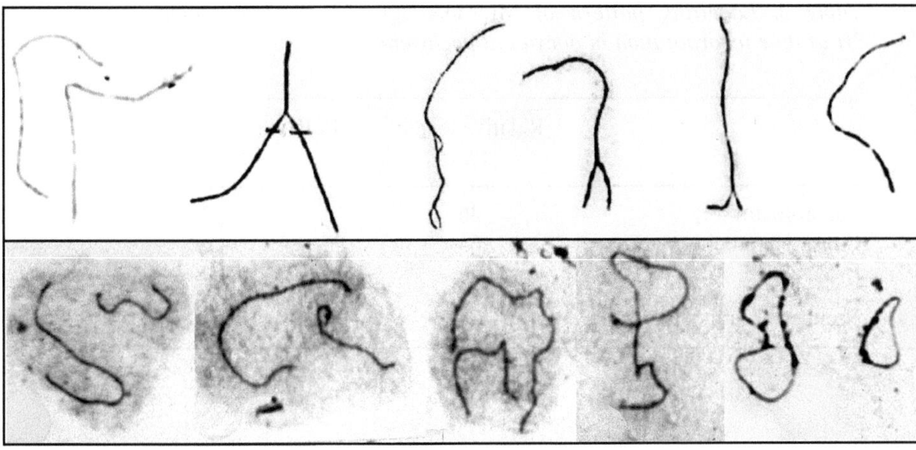

Figure 7. *Silver-stained synaptonemal complexes of oocytes and spermatocytes of* Microtus agrestis. *(Top) Both X chromosomes align through their euchromatic terminals during early pachytene and the alignment progresses unidirectionally till both had paired completely. (Bottom) In contrast, there is no pairing of the axial elements of X and Y in the sex body, except some telomeric associations (after Singh* et al., *2000).*

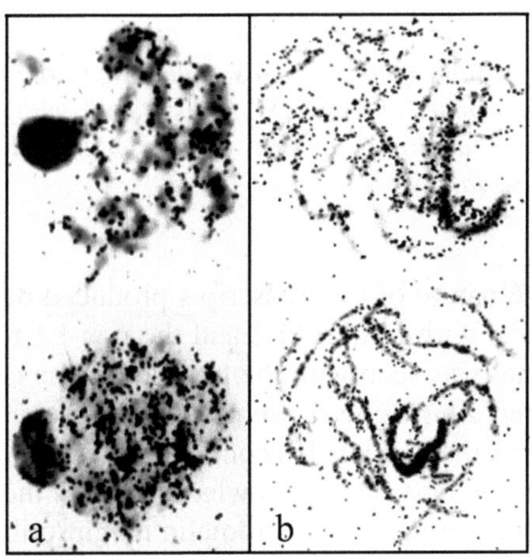

Figure 8. *The sites of transcription on pachytene chromosomes of male (a) and female (b)* M. agrestis *are mapped after ^{3}H uridine incorporation. Note that the sex body is almost free of labelling while the X-bivalent is heavily labelled throughout its length (a. Neitzel, unpubl.; b. from Neitzel* et al., *2002).*

[^{3}H] uridine incorporation. However, during oogenesis the labelling intensity of the constitutive heterochromatin at pachytene in not different from that of euchromatin as has been shown after combined [^{3}H] uridine/FISH analysis, using the pMAHAE2 clone to identify the gonosomal heterochromatin (Figure 8b; Table 2).

Table 2. Labelling pattern of Microtus agrestis *X chromosomes after* ^{3}H *uridine incorporation in oocytes at pachytene*

	Relative length (%)	Relative labelling pattern (%)
Euchromatin	36	38
Const. Heterochromatin	64	62
P		0.86
Number of grains mapped		630

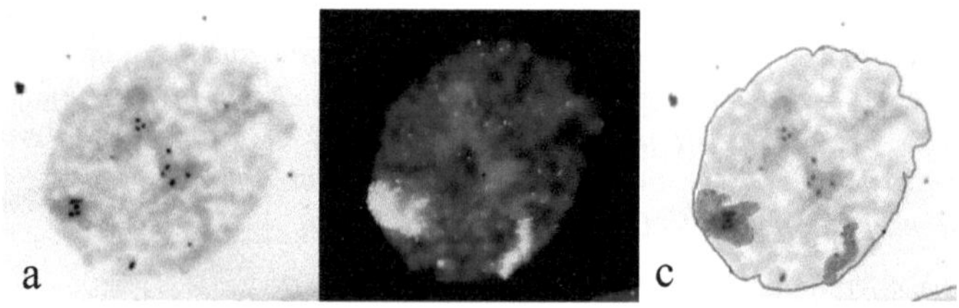

Figure 9. *Combined autoradiography after pulse labelling with* ^{3}H *uridine (a) and FISH analysis on prophase chromosomes of female* M. agrestis *cells with the biotin-labelled clone pMAHAE2 (b). Note the labelling pattern of the pMAHAE2 positive gonosomal heterochromatin (c) (see also Table 3).*

In order to identify the nature of the transcripts produced during oogenesis, DNA-RNA FISH was performed with pMAHAE2 and the non-L1 retroposon, pMA11/31, as probes. While no signal was seen with the first probe, the constitutive heterochromatin was heavily labelled after using the retroposon as probe. Furthermore, it could be shown in tissue sections that the label is confined almost exclusively to the oocytes (Neitzel *et al.*, 2002). To the best of our knowledge, this is the first clear demonstration that a distinct sequence of the heterochromatin is transcribed in mammals.

Using the combined [^{3}H] uridine/FISH analysis with the pMAHAE2 clone (Figure 9) in somatic cells, we could also demonstrate that the labelling intensity of the constitutive heterochromatin is not different from euchromatin (Table 3), confirming our earlier findings in prematurely condensed chromosomes (Sperling *et al.*, 1987). There is convincing, albeit indirect evidence from Northern blot analysis that at least part of this transcriptional activity is due to the transcription of the retroposon. When poly $(A)^{+}$ RNA was extracted from a transformed *M. agrestis* cell line and probed with pMA11/31, a smear of transcripts varying in length from 1 to 5 kb could be identified. This finding has been confirmed *in vivo*. Here, dot blots were performed on total cellular RNA from various tissues and the somatic cell hybrids and probed with *in vitro*

Table 3. *Labelling pattern of* Microtus agrestis *constitutive X-heterochromatin in comparison to total euchromatin after ^{3}H uridine incorporation in somatic cells at prophase*

	Relative length (%)	Relative labelling pattern (%)
Euchromatin (total)	89	87
Const. heterochromatin	11	13
P		0.78
Number of grains mapped		118

synthesized sense and anti-sense RNA of pMA11/31. The results indicate that the retroposon is expressed in the cell hybrids, including those without the *M. agrestis* X chromosome, both in sense and anti-sense orientation. It is most heavily expressed in the ovary and to a lesser extent in heart, lung, brain, and testes. Generally, this RNA is more abundant than that of the housekeeping gene GAPDH (Neitzel *et al.*, 2002).

The retroposon resembles members of the LINE family, however, it has neither an internal promoter nor an open reading frame. It contains a polyadenylation signal, a poly A tail, and typical pol II promoter motifs in the 5'UTR. Therefore, it is likely to be transcribed by RNA polymerase II. The heterogeneous size of the transcripts can be explained by a readthrough mechanism, which is characteristic for many L1 and SINE elements in mammals (Soares *et al.*, 1985; Daniels and Deininger, 1985).

Conclusions

A characteristic of the mammalian genome is its compartmentalization, which finds its expression in the light microscope as G-, R-, and C-bands. These chromatin domains differ both in their gene content and with respect to their repetitive elements, especially the non-viral retroposons, which can be classified as either autonomous or non-autonomous elements. (Holmquist, 1992). While the most abundant members of the autonomous retroposons, the L1 elements, are preferentially located in the dark G-bands, the most prominent members of the non-autonomous elements, the SINE elements, are clustered in the R-bands (Ostertag and Kazazian, 2001). The C-bands of the pericentromeric heterochromatin in mammals are preferentially composed of simple repetitive elements, organized as tandem arrays, and are extremely depleted of retroposons (Korenberg and Rykowski, 1988). This chromatin fraction can be considered equivalent to the α-heterochromatin in *Drosophila*. The gonosomal heterochromatin of the field vole, *M. agrestis*, has a much more complex organization. It is composed predominantly of middle repetitive elements and has accumulated retroposons which are transcribed. In general, it strongly resembles the β-heterochromatin of *Drosophila* (Neitzel *et al.*, 1998). With respect to the abundance of L1 elements, it also shows resemblance to G-bands. This is illustrated in Figure 1, in which a small region at the distal end of Xp is both C- and G-band positive. As demonstrated after *in situ* hybridization, this band harbours the middle repetitive elements pMAHAE2 and pMAECO14 as well as the retroposon pMA11/3 and is obviously identical to the DNase I sensitive band detected by the *in situ* nick translation technique (Sperling *et al.*, 1985).

The cytological properties of (hetero)chromatin depend on the number of repetitive elements (Kunze *et al.*, 1996) and DNA-specific proteins (Csink *et al.*, 1997) and, thus, cannot be predicted from the DNA sequences alone. Clearly, heterochromatin is not always heterochromatic (i.e. condensed), but rather a state of chromatin (Redi *et al.*, 2001).

Here is not the space to discuss the functional relevance of constitutive heterochromatin. However, it is remarkable that the enormous increase in the amount of gonosomal heterochromatin and the high transposition activity observed in *M. agrestis* is paralleled by a wide geographical expansion which argues more in favour of an evolutionary advantage than against it. It could play an active role in the speciation process by establishing a fertility barrier or by contributing regulatory elements and even new genes (Dimitri and Junakovic, 1999; Hughes, 2000). However, the recent observation in fission yeast and the ciliate, *Tetrahymena*, that short RNAs complementary to (centromeric) repeats play an important role in the regulation of gene expression could also shed new light on the role of RNA transcripts from repetitive DNA in higher organisms. This new concept explains the conversion of non-specific, repetitive sequence information into condensed chromatin and thus the epigenetic control of gene expression. It involves bidirectional transcription of DNA repeats, the formation of double-stranded RNAs (dsRNA) that trigger the production of short heterochromatic RNAs (shRNAs) which accumulate at these nuclear sites and initiate heterochromatin formation (reviewed in Jenuwein, 2002). The sense- and anti-sense transcripts of the *M. agrestis* retroposon fulfil the criteria for the formation of dsRNA, not only on gonosomal but also on autosomal sites and thus might induce chromatin alterations (epigenetic changes) as part of cellular differentiation. Moreover, the enormous rate of transcription and transposition during oogenesis should result in a high rate of new (epigenetic) modifications which are prone to the selection process and are propagated in case of improved Darwinian (biological) fitness. Clearly, this is still pure speculation, but *M. agrestis* is the ideal mammal to test the relevance of this novel concept.

References

Csink AK, Sass GL, Henikoff S (1997). *Drosophila* heterochromatin: retreats for repeats. In: van Driel R, Otte AP, eds. *Nuclear Organization, Chromatin Structure, and Gene Expression.* Oxford University Press, pp. 223–235.

Daniels GR, Deininger PL (1985). Repeat sequence families derived from mammalian tRNA genes. *Nature* 317: 819–822.

Dimitri P, Junakovic N (1999). Revising the selfish hypothesis – new evidence on accumulation of transposable elements in heterochromatin. *Trends Genet* 15: 1123–1124.

Henikoff S (2000). Heterochromatin function in complex genomes. *Biochimica et Biophysika Acta* 1,470: 1–8.

Holmquist GP (1992). Chromosome bands, their chromatin flavors, and their functional features. *Am J Hum Genet* 51: 17–37.

Hughes DC (2000). MIRs as agents of mammalian gene evolution. *Trends Genet* 16: 60–62.

Jenuwein T (2002). An RNA-guided pathway for the epigenome. *Science* 297: 2215–2218.

John B (1988). The biology of heterochromatin. In: Verma RS, ed., *Heterochromatin: Molecular and Structural Aspects*, Cambridge University Press, pp. 1–147.

Kalscheuer V, Singh AP, Nanda I, Sperling K, Neitzel H (1996). Evolution of the gonosomal heterochromatin of *Microtus agrestis*: rapid amplification of a large, multimeric, repeat unit containing a 3.0-kb (GATA)11-positive, middle repetitive element. *Cytogenet Cell Genet* 73: 171–178.

Kazazian HH, Moran JV (1998). The impact of L1 retrotransposons on the human genome. *Nat Genet* 19: 19–24.

Kerem B, Kottusch-Geiseler V, Kalscheuer V, Goitein R, Sperling K, Marcus M (1988). DNase I sensitivity of *Microtus agrestis* active, inactive and reactivated X chromosomes in mouse-*Microtus* cell hybrids. *Chromosoma* 96: 277–230.

Korenberg JR, Rykowski MC (1988). Human genome organization: Alu, Lines, and the molecular structure of metaphase chromosome bands. *Cell* 53: 391–400.

Kunze B, Weichenhan D, Virks P, Traut W, Winking H (1996) Copy number of a clustered long-range repeat determine C-band staining. *Cytogenet Cell Genet* 73: 86–91.

Nanda I, Neitzel H, Sperling, Studer R, Epplen JT (1988). Simple GATA/GACA repeats characterize the X chromosomal heterochromatin of *Microtus agrestis*, European field vole (*Rodentia, Cricetidae*). *Chromosoma* 96: 213–219.

Neitzel H, Kalscheuer V, Henschel S, Digweed M, Sperling K (1998). Beta-heterochromatin in mammals: evidence from studies in *Microtus agrestis* based on the extensive accumulation of L1 and non-L1 retroposons in the heterochromatin. *Cytogenet Cell Genet* 80: 165–172.

Neitzel H, Kalscheuer V, Singh AP, Henschel S, Sperling K (2002). Copy and paste: the impact of a new non-L1 retroposon on the gonosomal heterochromatin of *Microtus agrestis*. Cytogenet Genome Res 96: 179–185.

Ostertag EM, Kazazian HH Jr (2001). Biology of mammalian L1 retrotransposons. *Annu Rev Genet* 35: 501–538.

Redi CA, Garagna S, Zacharias H, Zuccotti M, Capanna E (2001). The other chromatin. *Chromosoma* 110: 136–147.

Singh AP, Maerzke S, von Deimling O, Raman R, Sperling K, Neitzel H (1993). Oocytes from pachytene to dictyotene can easily be analysed in neonatal rodents. *Chromosome Res* 1: 209–312.

Schulze I (2000). Untersuchung der Transkriptionsaktivität und des Replikationsverhaltens des konstitutiven Heterochromatins and des Tiermodells *Microtus agrestis*. Inaug. Dissertation, Humboldt-Universität zu Berlin.

Sieger M, Pera F, Schwarzacher WG (1970). Genetic inactivity of heterochromatin and heteropycnosis in *Microtus agrestis. Chromosoma* 29: 349–364.

Singh A, Henschel S, Sperling K, Kalscheuer V, Neitzel H (2000). Differences in the meiotic pairing behavior of gonosomal heterochromatin between female and male *Microtus agrestis*: implications for the mechanism of heterochromatin amplification on the X and Y. *Cytogenet Cell Genet* 9: 253–260.

Soares MB, Schon E, Efstratiadis (1985). Rat LINE1: the origin and evolution of a family of long interspersed middle repetitive DNA elements. *J Mol Evol* 22: 117–133.

Sperling K (1982). Cell cycle and chromosome cycle: morphological and functional aspects. In: Rao PN, Johnson RT, Sperling K, eds. *Premature Chromosome condensation*. New York: Academic Press, pp. 43–78.

Sperling K, Marcus M (1994). Mapping of genetic activity on mammalian chromosomes. *Chromosomes Today* 8: 169–178. G Allen and Unwin.
Sperling K, Kerem BS, Goitein R, Kottsuch V, Cedar H, Marcus M (1985). DNase I sensitivity in facultative and constitutive heterochromatin. *Chromosoma* 93: 38–42.
Sperling K, Kalscheuer V, Neitzel H (1987). Transcriptional activity of constitutive heterochromatin in the mammal *Microtus agrestis* (*Rodentia, Cricetidae*). *Exp Cell Res* 173: 463–472.
Sperling K, Kalscheuer V, Kottusch-Geiseler V, Neitzel H (1989). Genetic activity of the constitutive heterochromatin in mammals. *Chromosomes Today* 10: 83–89.

5. Plant Cytogenetics

5.1. FISHing Repeated DNA Sequences in *Beta* Genomes

T. Schmidt[1], C. Desel[1], D. Dechyeva[1], B. Fleischer[1], F. Gindullis[2], A. Schmidt[1], J. S. Heslop-Harrison[3] and R. L. Doudrick[4]

[1]*Plant Molecular Cytogenetics Group, Institute of Crop Science and Plant Breeding, Christian Albrechts University, Olshausenstraße 40, D-24118 Kiel, Germany;*
[2]*CellTec GmbH Biotechnologie, Frohmestrasse 110, D-22459 Hamburg, Germany;*
[3]*Department of Biology, University of Leicester, Leicester LE1 7RH, UK;*
[4]*Southern Institute of Forest Genetics, USDA Forest Service, 23332 Highway 67, Saucier MS39574, USA*

Introduction

Nuclear genomes of higher plants vary enormously in size. A substantial proportion of this variation is due to polyploidy, and it is assumed that 50% or more of angiosperms are polyploids. However, although all plants, at the diploid level, require the same number of genes and regulatory DNA sequences for physiological processes like germination, growth, flowering and reproduction, the total amount of the chromosomal DNA still ranges from some 130 Mbp in the model species *Arabidopsis thaliana* L. up to more than 25,000 Mbp in the timber species *Pinus elliottii* Engelm. var. *elliottii*. The differences in complexity of diploid nuclear genomes over several orders of magnitude (Bennett and Leitch, 1995) are caused by repetitive DNA which consists of repeated sequences varying in size from dinucleotides to motifs longer than 10,000 bp and which is present in copy numbers from many hundreds to hundred thousands.

In general, repetitive DNA can be divided into several sequence classes which differ in their organization and localization along the chromosomes, although also intermediate forms of organization may exists. Tandemly repeated DNA occurs mainly on a few preferred positions of the chromosomes (pericentromeric, intercalary and subtelomeric). The repeating units are arranged adjacently to each other; examples for tandemly repeated sequences are satellites and microsatellites. It should be noted here that ribosomal genes and telomeric repeats are also organized as tandemly repeating

M. Schmid and I. Nanda (eds.), Chromosomes Today, Volume 14, 249–265.

units. Satellite DNA is a sequence class which allows interesting insights into the evolution of sequence families, chromosomes and genomes, and the phylogeny of species (Kamm *et al.*, 1995; Hemleben *et al.*, 1992).

In contrast to tandemly repeated sequences, dispersed repetitive DNA sequences, often variable in size, are scattered throughout the genome, interspersed with other sequences and distributed over all or most chromosomes of a complement. Families of mobile DNA elements such as Ty1-*copia*- and Ty3-*gypsy* retrotransposons and LINEs as well as sequences derived from these elements are prominent examples of dispersed sequences of plant genomes (Heslop-Harrison *et al.*, 1997; Schmidt *et al.*, 1999).

Molecular studies including cloning, sequencing and Southern analyses have been used to gain knowledge about the structural characteristics of repetitive DNA in higher plants. However, genetic mapping of multi-locus, highly repeated DNA by molecular methods is often difficult or impossible, and hence *in situ* hybridization approaches are the methods of choice for exact chromosomal localization of repeated DNA. In particular, FISH (fluorescent *in situ* hybridization) is a valuable and accurate method for the physical localization of unique and repetitive sequences along chromosomes.

Although we have used the FISH technology for the physical mapping of DNA sequences in many plants, this paper is focussed on the molecular cytogenetics of *Beta vulgaris* L., sugar beet, a crop grown in temperate climates for sugar production, animal feeding or as a vegetable, and an excellent species to study the genomic and chromosomal organization of repetitive DNA. Diploid forms of sugar beet have $2n = 18$ chromosomes, and some 60% of its 758 Mbp genome (Arumuganathan and Earle, 1991) consist of repeated sequences (Flavell *et al.*, 1974). We have also investigated *Beta* species in three wild beet sections of the genus to enable comparisons of genome organization in closely related taxa. Here, the results obtained by molecular and cytological investigations are compared with data from other plants to highlight common structural features of genome organization. We demonstrate the application of FISH as a tool for the physical mapping of repeats and genes, and for structural analyses of chromosome domains such as centromeres and telomeres.

The molecular cytogenetics of satellite DNA in *Beta vulgaris*

Tandemly repeated DNA sequences form a major component of plant genomes and show species- and chromosome-specific amplification. Some tandem repeats are coding and have a complex organization, e.g. the genes for the ribosomal RNAs, while others are very simple in structure like microsatellites and telomeric DNA. In particular, satellite DNA makes up a large proportion of the heterochromatin, the chromatin fraction, which remains condensed and densely packed throughout the cell cycle. Satellite DNA is a very dynamic component of plant genomes. Amplification may occur by different processes such as unequal cross-over, extrachromosomal amplification via rolling circle molecules or the increase of copy numbers by gene conversion. Repeats of satellite DNA families are constantly subjected to homogenization and fixation within a population and hence characterized by a relatively high genomic turnover. As a result, repeat families can expand or contract rapidly during different evolutionary scales such as species speciation or plant breeding. These genomic changes can be monitored by *in situ* or Southern hybridization experiments, or more sensitively by PCR.

From sugar beet, we have cloned many satellite DNA repeats and investigated their molecular structure (Schmidt and Metzlaff, 1991; Schmidt et al., 1990; Schmidt and Heslop-Harrison, 1993; Kubis et al., 1997). One satellite DNA family, characterized by a conserved recognition sequence for the restriction endonuclease *Bam*HI within the 327 bp repeating units, is located at the centromeres of all 18 chromosomes as revealed by fluorescent *in situ* hybridization (Figure 1A). Despite the differences in copy number between chromosomes, repeat arrays of this DNA family expand over very large regions of the sugar beet centromeres. Complemented by Southern hybridization of PFGE (pulsed-field gel electrophoresis) blots, it has been shown that restriction fragments containing this centromeric sequence family range from several hundred up to 3,000 kb indicating that the *Bam*HI satellite DNA indeed extends over giant regions (Schmidt and Metzlaff, 1991).

The investigation of repetitive DNA provides insights into the evolution of plant genomes and sequence families and supports studies of species phylogeny and relationships. The analysis of the distribution of repeat families within a group of related plant species reveals often that some repeats are amplified in a subset of species only. The genus *Beta* consists of four sections: *Beta, Corollinae, Nanae* and *Procumbentes*; cultivated beet belongs to the section *Beta*. Southern analysis of 22 species of the genus has shown that the *Bam*HI satellite DNA family is exclusively amplified in species of the section *Beta*, indicating that the amplification process took place relatively late in the phylogeny of the genus and after the separation of this section from the remaining species. In contrast, the amplification of another *B. vulgaris* satellite DNA predates the separation of some sections; the repeat has a much wider distribution among *Beta* species. Members of this satellite DNA family are non-homologous to the *Bam*HI satellite, characterized by a conserved endonuclease *Eco*RI site and are only 150 bp long. Both, the *Eco*RI and the *Bam*HI satellite repeats fall into size classes, which are very common for satellite monomers in dicotyledonous plant species (150–180 or 300–360 bp, respectively). These lengths presumably reflect the specific size of the DNA stretch around a nucleosome or nucleosome dimer in order to pack this type of heterochromatin densely.

The *Eco*RI satellite shows a different pattern after FISH analysis. The copies of this DNA family form large heterochromatic clusters at intercalary positions on each chromosome arm of the sugar beet genome (Figure 1B, arrows point to the chromosome pair with the collocalized 5S rRNA genes). It is interesting to note that the size of *Eco*RI repeat blocks on individual arms of a chromosome varies. This chromosome-specific amplification makes it possible to group the nine meta- to submetacentric and morphologically very similar chromosomes of the haploid sugar beet complement into three classes: (i) chromosomes with very large, extended *Eco*RI satellite blocks on both arms, (ii) chromosomes with strong repeat amplification on one arm and a moderate copy number on the opposite arm, (iii) chromosomes with an equal and relatively moderate copy number of *Eco*RI satellite repeats on both arms. These results, in combination with the physical mapping of rRNA genes and other repetitive satellite

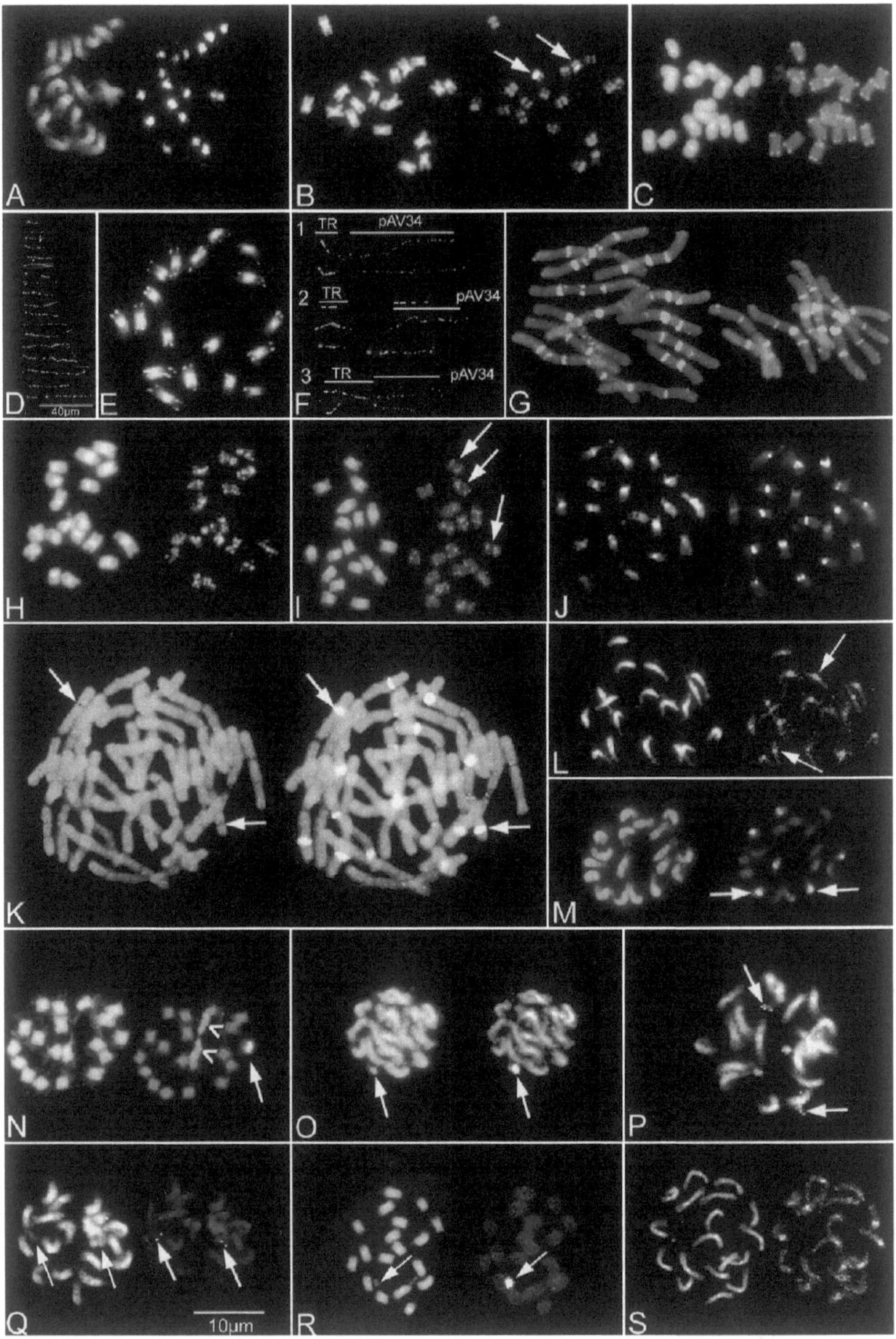
A
B
C
D
40µm
E
1 TR pAV34
2 TR pAV34
3 TR pAV34
F
G
H
I
J
K
L
M
N
O
P
Q
10µm
R
S

Figure 1.

repeats, which already have been cloned and sequenced, and multi-colour FISH, provide a step towards the unequivocal identification of individual sugar beet chromosomes and hence are important for the localization of genes and assignment of linkage groups to chromosomes.

The satellite repeat pAV34 has a length of 360 bp and has been localized in close vicinity to the telomeres which consists, as in most higher plants, of the conserved sequence motif TTTAGGG (Figure 1C). The telomeric repeats are arranged in long arrays, overcome the end-replication problem and protect the chromosome ends from endonucleolytic degradation or fusion to other chromosome ends (Richards and Ausubel, 1988). Telomeric repeats form repeat blocks at the physical ends of the chromosomes, however, the overall array length can vary between individual chromosome

---

Figure 1. *Photomicrographs of plant chromosomes at metaphase after fluorescent* in situ *hybridization. Chromosomes on the left side in each panel (except D–G, K and P) were stained with the fluorochrome DAPI to visualize the chromosome morphology. Bright signals on the right side show sites of probe hybridization detected by UV microscopy. The scale bar in panel Q corresponds to 10 µm. (A) The* Bam*HI satellite pBV1 is detectable around the centromeres of all* Beta vulgaris *chromosomes. (B) Intercalary regions of sugar beet chromosomes contain the* Eco*RI satellite pEV1 in variable copy numbers. Arrows point to the 5S rRNA genes on chromosome IV. (C) Telomeric DNA of sugar beet detected on mitotic metaphase chromosomes and (D) on extended chromatin fibres. (E)* B. vulgaris *chromosomes probed with the subtelomeric satellite pAV34. (F) Analysis of the chromosome ends of* B. vulgaris *chromosomes using the subtelomeric satellite pAV34 and the telomeric repeat TTTAGGG as probes. Different types of sequence organization and interspersion are detectable. (G) Banding pattern of slash pine (*Pinus elliottii *var.* elliottii*) after FISH with the telomeric sequence TTTAGGG. (H) The sugar beet non-LTR retrotransposon (LINE) BNR1 has a clustered chromosomal organization, while (I) Ty1*-copia*-like retrotransposons are almost uniformly dispersed over all chromosomes. Arrows in (I) point to examples of depletion of Ty1*-copia *retrotransposons at centromeres. (J) Centromeric organization of a Long Terminal Repeat (LTR) probe of a Ty3-gypsy retrotransposon on chromosomes of the wild beet* Beta procumbens. *(K) The Ty1*-copia *retrotransposons TPE1 is uniformly dispersed on* Pinus elliottii *var.* elliottii *chromosomes but excluded from 18S-5.8S-25S rRNA genes, detectable as bright signal (examples arrowed). (L) The microsatellite motif $(CA)_8$ is amplified on most centromeres of sugar beet chromosomes (arrows) and weakly dispersed along the chromosome arms. (M) Hybridization of sugar beet chromosomes with $(GATA)_4$ shows genome-wide dispersion of the microsatellite motif with amplification on some centromeres (arrows). (N) Detection of an alien* B. procumbens*-chromosome (arrow) in an interspecific hybrid of* B. vulgaris *and* B. procumbens. *Open arrow heads indicate the sites of the 18S-5.8S-25S rRNA genes in sugar beet. (O) Detection of the wild beet minichromosome (arrow) in the sugar beet fragment addition line PRO1. (P) Physical mapping of two YACs (punctated structure indicated by arrows) on a nematode-resistant sugar beet line carrying a terminal translocation of* B. procumbens. *(Q) Physical mapping of the single copy nematode resistance gene* $Hs1^{pro\text{-}1}$ *with a 684 bp probe. (R) Localization of centromere-specific BAC on the wild beet minichromosome of the fragmnet addition line PRO1. (S) The same BAC hybridizes mainly to the centromeres of all* B. procumbens *chromosomes.*

arms (Figure 1D). In sugar beet, the telomere length ranges from 35 to 170 kb which is agreement with values observed in tomato (Zhong, 1998).

The pAV34 satellite is an ancient component of *Beta* genomes and has been found in the subtelomeric position in species of all four sections of the genus *Beta* (Figure 1E). By FISH to extended DNA fibres isolated from *B. vulgaris* nuclei we estimated the length of the pAV34 satellite arrays and the telomeric repeat blocks. The structural composition of *B. vulgaris* chromosome ends has been determined by fibre FISH with differentially labelled probes for the telomeric DNA and the satellite pAV34. Three different patterns of sequence organization have been observed (Figure 1F):

1. organization of the telomere and the satellite pAV34 in distinct domains separated by a gap,
2. weak interspersion of both sequence repeats with the telomere at the physical end of the chromosomes and a detectable gap between both repeat types,
3. organization in separate repeat blocks without gap or interspersion.

The telomeric DNA is also a probe which can reveal peculiarities of plant genomes. For example, intercalary clusters of telomeric DNA sites were detected in some plant species (Fuchs *et al.*, 1995; Richards *et al.*, 1991). These findings are consistent with fluorescent *in situ* hybridization studies of the extremely large genome of *P. elliottii* var. *elliottii* which revealed that telomeric DNA repeats occur at many intercalary chromosomal regions (Figure 1G). From the strength of the fluorescence signals it is apparent that DNA sequences complementary to telomeric repeats are a major sequence class and contribute considerably to the large genome size of slash pine, which is more than 20,000 Mbp (Ohri and Khoshoo, 1986). The number of intercalary sites ranges from two to five, and the chromosomal position is variable (Schmidt *et al.*, 2000). Therefore, we suggest that FISH with telomere repeats might also provide a useful banding method to distinguish the metacentric chromosomes, to perform karyotype analyses in *Pinus* species as described by fluorochrome banding and rDNA mapping (Doudrick *et al.*, 1995; Doudrick, 1996), and to undertake comparative genome studies in the genus *Pinus*.

Chromosomal organization of retroelements in higher plant genomes

Retrotransposons that proliferate by reverse transcription of RNA intermediates are a feature of all eukaryotic genomes and the major class of mobile genetic elements in plants (Flavell *et al.*, 1992; Voytas *et al.*, 1992; Hirochika and Hirochika, 1993).

Two different classes of retrotransposons, the LTR (long terminal repeats) retrotransposons, also known as viral retrotransposons, and the non-LTR (non-viral) retrotransposons, have been identified in plants. LTR retrotransposons are flanked by long terminal repeats. Their internal domain carries the *pol* gene which encodes protease, RNaseH, reverse transcriptase and integrase. According to the order of genes within the internal domain, LTR retrotransposons can be subdivided in Ty1-*copia*-like and the Ty3-*gypsy*-like retroelements. A fragment of the reverse transcriptase gene of Ty1-*copia* retrotransposons can be amplified by PCR with primers specific for conserved domains. This PCR assay enabled the detection, sequence analyses and the study of evolutionary relationships of Ty1-*copia*-retrotransposons in all major lineages of plants. Furthermore, the method is very useful for the generation and labelling of Ty1-*copia* specific probes from plant genomic DNA. The probes can be directly used for fluorescent *in situ* hybridization, making cloning not necessary.

Until recently, little was known about the chromosomal organization of retroelements in plant genomes. However, during the past two years many investigations were focussed on retroelements as large components of plant genomes and more data are available today (Pearce *et al.*, 1996; Katsiotis *et al.*, 1996; SanMiguel *et al.*, 1996; Heslop-Harrison *et al.*, 1997).

In sugar beet, both non-LTR retrotransposons, from mammalian genome analysis known as LINEs (long interspersed nuclear elements), and LTR retrotransposons of the Ty-*copia* family have been studied in detail (Schmidt *et al.*, 1995). Representative elements of both retrotransposon types have been cloned after PCR isolation from the *B. vulgaris* genome. The BNR1 family is the first non-LTR retrotransposon isolated from a dicotyledonous plant. Members of this sequence family have a highly diverged nucleotide sequence which ranges from 59% to 91% homology. However, the putative amino acid sequence of the cloned reverse transcriptase gene shows a high similarity (42%) with conservation of critical amino acids to LINEs from *Zea mays* (Schwarz-Sommer *et al.*, 1987) and *Lilium speciosum* (Leeton and Smyth, 1993). A considerable similarity (27.5%) of the predicted peptide sequence of BNR1 was found to members of the L1 family from human and rodents (Singer, 1982). Recent results indicate that non-LTR retrotransposons are widespread and ubiquitous components of plants genomes (Wright *et al.*, 1996; Kubis *et al.*, 1998).

Fluorescent *in situ* hybridization of the sugar beet BNR1 family enabled their investigation on the chromosomal level. Most copies of the BNR1 were found in discrete clusters along all chromosomes (Figure 1H). The clusters were often localized in DAPI-positive regions, indicating their occurrence in heterochromatic regions. This picture reflects a significant, non-random distribution and suggests a higher order structure of plant genomes as is known from SINEs (short interspersed repeated sequences) and LINEs in human. No hybridization was detectable in the terminal region of chromosome 1 consisting of the arrays of the 18S-5.8S-25S rRNA genes. This finding contrasts with the LINE family R1Bm and R2Bm which specifically retrotranspose into the rDNA repeats of many insects (Jakubczak *et al.*, 1991).

The genome of *B. vulgaris* harbours a large number of Ty1-*copia* retrotransposons. Sequence analysis and Southern hybridization suggested that the sugar beet Ty1-*copia* retrotransposons can be grouped into related subfamilies. In contrast to the BNR1 family, fluorescent *in situ* hybridization of Ty1-*copia* elements showed a different distribution along the sugar beet chromsomes. Although the exclusion from rDNA sites is similar to BNR1, Ty1-*copia* retrotransposons, represented by the Tbv family, are uniformly dispersed along the chromosomes. Noteworthy is the depletion from the centromeric region of all chromosomes (Figure 1I, arrows). These regions are occupied by large clusters of the *Bam*HI satellite repeat described above. Southern analyses of a wide range of *Beta* species have shown, that the *Bam*HI satellite occurs only in cultivated and wild beets of the section *Beta*. Therefore, we assume that the rapid amplification and homogenization of *Bam*HI tandem repeats is a recent event in the phylogeny of the genus *Beta* and resulted possibly in the displacement of Ty1-*copia* retrotransposons from the centromeres. However, the exclusion of LTR-retrotransposons from *Beta* centromeres has only been observed for Ty1-*copia* retrotransposons. Recently, FISH studies have shown that Ty3-*gypsy* retrotransposons, distinguished from Ty1-*copia* elements by a reversed order of genes within the coding domain, are present in the centromeric region of all *B. vulgaris* chromosomes (Gindullis *et al.*, 2001). Furthermore, we have isolated a complete Ty3-*gypsy*-retrotransposon from the wild beet *B. procumbens* which has a length of approximately 7.2 kb including flanking LTR sequences of 1.1 kb. Using the LTR sequence, which defines this particular retroelement, as a probe demonstrated that it is exclusively localized in the centromeric region of all *B. procumbens* chromosomes (Figure 1J). The contrasting pattern of the chromosomal organization of Ty1-*copia* and Ty3-*gypsy* retrotransposons which differ only in the position of the internal integrase gene, is not clear yet but might be related to differences in transposition frequencies leading to accumulation of *gypsy*-like elements in the centromeric region. Moreover, the sequence conservation of centromeric Ty3-*gypsy* retrotransposons indicates a potential role in the structural and functional composition of *Beta* centromeres. Amplification of Ty3-*gypsy* retrotransposons has also been observed in barley, maize and other cereals (Presting *et al.*, 1998; Ananiev *et al.*, 1998; Aragon-Alcaide *et al.*, 1996).

As a representative conifer species of the gymnosperm divison of plants we have investigated slash pine (*Pinus elliottii* Engelm. var. *elliottii*), an important timber species from the Southern regions of North America. Conifers dominate many ecosystems of temperate climates, but little is known about their major genome constituents and molecular cytogenetics. Chromosome numbers ($2n = 24$) are very conserved in gymnosperms and their genomes are enormous in size. From a genomic library of *Pinus elliottii* var. *elliottii* we have isolated a highly amplified DNA element which has been designated TPE1. Molecular analyses revealed that TPE1 is 1663 bp long and represents a rearranged derivative of a Ty1-*copia* retrotransposon (Kamm *et al.*, 1996). Southern hybridization indicated, that TPE1 is very conserved in many two- and three-needle pine species and its sequence divergence follows the taxonomic grouping. Fluorescent *in situ* hybridization of metaphase chromosomes from slash pine showed that TPE1 is scattered over all chromosome arms. Consistent with the distribution of Ty1-*copia* retrotransposons in *B. vulgaris*, TPE1 is depleted from intercalary and centromeric sites harbouring the major and minor arrays of 18S-5.8S-25S rRNA genes as confirmed by double *in situ* hybridization with a rDNA specific probe (Figure 1K, metaphase spread examined with two different filter sets, arrows point to examples of TPE1 depletion, left, and strong signals of 18S-5.8S-25S rRNA gene arrays, right).

In a comparative study we have analyzed the chromosomal distribution and genomic organization of Ty1-*copia*-like retrotransposons in many species of pteridophytes, gymnosperms and angiosperms (Brandes *et al.*, 1997). A characteristic pattern of the Ty1-*copia* retrotransposon distribution in higher plants emerged: A dispersed distribution throughout most of the chromosomes with reduced hybridization at some regions or with some weak clustering at other regions was found in most species (Heslop-Harrison *et al.*, 1997). This is consistent with database searches which have revealed that Ty1-*copia* retrotransposons or their remnants are scattered throughout the genome and often found in the flanking regions of genes (White *et al.*, 1994). Reduced hybridization occurred mainly at centromeric regions, nucleolus-organizing regions and regions known to be mainly composed of tandemly repeated sequences. In contrast, in species with small genomes like *A. thaliana* and *Cicer arietinum* L., Ty1-*copia* retrotransposons are concentrated in clusters or interspersed within the centromeric heterochromatin (Brandes *et al.*, 1997).

Microsatellites – tandem repeats of simple sequence structure

Tandemly repeated DNA consisting of very short repeating motifs (1–5 bp) and known as microsatellites or simple sequence repeats, has been detected in most eukaryotic genomes. Slipped-strand mispairing during replication is discussed as an important mechanism of amplification and results in a high variability of alleles of a microsatellite locus. Microsatellite repeats are therefore highly informative and polymorphic markers for genetic mapping and genome analysis. Several approaches to use microsatellite polymorphisms as markers have been established. DNA fingerprinting with short synthetic oligonucleotides as probes gives data about the abundance of certain simple sequences motifs within a genome, and can detect polymorphisms between closely related species or varieties, lines or individuals, but the genomic context of the detected repeats remains unknown. Genetic mapping of microsatellites involves the amplification of the repeat arrays using PCR with primers flanking the arrays or also the amplification of DNA stretches between the arrays. PCR primers are often chosen based on databank searches or sequence data from cloned genomic fragments which give selective data about the genomic distribution of microsatellite arrays in the genomes. Furthermore, large arrays of microsatellites are unclonable even in recombination-deficient prokaryotic hosts, mostly because of unusual DNA conformation or instability of the insert due to the formation of secondary structures. Therefore, little is known about the real chromosomal organization and physical localization of microsatellite motifs within plant genomes.

In the genome of sugar beet, different microsatellites are major components of the repeated DNA fraction, and DNA fingerprinting with synthetic oligonucleotides has shown that some are suitable as polymorphic markers for the differentiation of subspecies and cultivars (Schmidt *et al.*, 1993). In a systematic search, we have investigated the chromosomal distribution pattern of di-, tri- and tetranucleotide microsatellites by fluorescent *in situ* hybridization (Schmidt and Heslop-Harrison, 1996b). Repeats representing a range of nucleotide motifs included $(GATA)_4$, $(GGAT)_4$, $(GACA)_4$, $(CAC)_5$, $(CA)_8$, $(GA)_8$, $(TA)_{10}$ which are often used for genetic mapping in plants. To verify that the hybridization pattern of these extremely short probes reflects their distribution along the chromosomes, the microsatellites were chosen so that arrays have 75% homology to one another. The *in situ* hybridization pattern strongly depends on the sequence motifs; contrasting patterns showed that the hybridization conditions are able to discriminate each motif. The dispersed hybridization along chromosome arms out to the telomeres detected as weak fluorescent signals confirms that microsatellite markers give a relatively uniform coverage of the genome.

However, we observed major amplification sites of microsatellites near the centromeres of up to seven chromosome pair or exclusion from particular centromeric, intercalary chromosome regions and 18S-5.8S-25S rRNA genes. Furthermore, the in-gel hybridization pattern of the microsatellite motif to size-separated restriction digests allowed the comparison of the physical and molecular organization of the sequence motifs. The presence, absence and strength of repetitive microsatellite fragments detected by in-gel hybridization tends to correlate with the fluorescent *in situ* pattern on chromosomes: amplification at centromeres is accompanied by major repetitive restriction fragments in gels.

The motif (CA) is a very informative marker in mammalian genome analyses, and human and mouse maps based on PCR polymorphisms of (CA) arrays have been established. In sugar beet, attemps to isolate polymorphic (CA) arrays were not successful. Sequence analysis revealed that positive clones containing (CA) dinucleotide repeats predominantly originate from a simple sequence stretch within the centromeric *Bam*HI satellite monomers (Schmidt *et al.*, 1991; Mörchen *et al.*, 1996). This finding confirms the fluorescent *in situ* hybridization signals of the CA dinucleotide arrays and the *Bam*HI satellite DNA which show that major sites of both repeats are collocalized (Figure 1L, compare also with Figure 1A). At interphase, most (CA) arrays are found within DAPI-positive heterochromatic regions. In contrast, $(GA)_8$ repeats, having the same GC-content as $(CA)_8$, are largely excluded from many centromeres. $(GATA)_4$ is amplified on six centromeres and adjacent regions but is also depleted in some regions (Figure 1M, arrows point to two major centromeric sites). The dispersed hybridization detected on all chromosomes may represent microsatellites flanked by single-copy DNA which can be used for primer design while major sites are either long microsatellite clusters or arrays integrated into other repeats. The chromosomal distribution of the different microsatellites varies widely and has implications for amplification and dispersion mechanisms, and hence genome evolution and utility for genetic mapping. Knowledge of the chromosomal distribution is critical for choosing microsatellites which provide a good genome coverage and hence are suitable as genetic markers in plant genome research.

Application of FISH with tandemly repeated and unique DNA in plant breeding research

The rapid evolution of repetitive DNA within evolutionary short time scales can result in novel and/or diverged repeat variants in a subset of related plants. These species-specific DNA sequences are an excellent source for the isolation of genome-specific DNA probes which can serve as molecular markers in plant breeding research. In plant breeding, the selection of hybrid plants is often based on morphological characters, resistance tests or field trials. These procedures are time-consuming and laborious. Moreover, the assessment of morphological traits and plant performance can sometimes be ambiguous. During the past few years, the development of DNA-based markers has greatly facilitated the breeding process.

Sugar beet is highly susceptible to the beet cyst nematode (*Heterodera schachtii* Schm.). This pest causes severe damage to sugar beet cultivation with considerable loss of yield; hence, breeding of nematode-resistant beet varieties is most desirable. The use of nematicides is essentially impossible, because of the resulting environmental pollution. To introduce nematode resistance into sugar beet, interspecific crosses between wild beets of the section *Procumbentes* (*Beta procumbens* Chr.Sm., *Beta webbiana* Moq. and *Beta patellaris* Moq.) and cultivated beet have been made. Backcrossing and selection resulted in hybrid plants with 18 sugar beet chromosomes and a single wild beet chromosome or chromosome fragment. These hybrids are fully resistant to beet cyst nematodes. In order to introduce marker-assisted selection into beet breeding, genome-specific DNA probes have been isolated from wild beets (Schmidt *et al.*, 1990; Schmidt and Heslop-Harrison, 1996a). It has been shown that nematode-resistant sugar beets can be selected in a rapid screening procedure using a radioactively labelled genome-specific DNA probe which is hybridized to genomic DNA (dot-blot hybridization) or leaf material (squash-dot hybridization). The probe is a *B. procumbens*-specific satellite repeat of 158–159 bp length and was found exclusively in species of the section *Procumbentes*. Analyses of the molecular features of this sequence family have shown that the monomers have a conserved *Sau*3A restriction site, and FISH revealed the occurrence of this satellite DNA in large centromeric clusters on most chromosomes of *B. procumbens* (Schmidt and Heslop-Harrison, 1996a).

The *B. procumbens*-specific satellite repeats, in particular the monomer pTS5, were used to investigate the molecular cytogenetics of nematode-resistant sugar beet hybrids. Investigation of mitotic metaphases of different monosomic addition lines ($2n = 18 + 1$) showed that all resistant plants carry wild beet chromosomes of similar morphology. The wild beet chromosomes conferring nematode resistance are submetacentric to metacentric and not distinguishable from the sugar beet chromosomes (Schmidt *et al.*, 1997). Simultaneous hybridization with pTS5 (Figure 1N, arrow) and a ribosomal probe (Figure 1N, open arrow heads) revealed that none of the *B. procumbens* chromosomes within the three monosomic addition lines harbours the 18S-5.8S-25S rRNA gene arrays. Fluorescent *in situ* hybridization proved also to be a useful technique for the characterization of fragment addition lines which were derived from backcrossing of monosomic addition lines and harbour a small chromosome fragment from *B. procumbens*. We have studied sequence elements that are necessary for chromosome fragment maintenance and segregation in the fragment addition line PRO1. Tandem arrays of the repeat (TTTAGGG), forming the physical ends of plant chromosomes (Richards and Ausubel, 1988), and pTS5 as a sequence specific for the large heterochromatic regions surrounding the centromeres of *B. procumbens* chromosomes were chosen as probes. Hybridization with pTS5 to metaphases of PRO1 showed a strong signal covering most of the *B. procumbens* chromosome fragment. In less condensed prometaphases the signal was detected in a DAPI positive region of the PRO1 fragment representing most probably a part of the centromeric heterochromatin of the original *B. procumbens* chromosome (Figure 1O, arrow). A telomeric hybridization signal was detected at one distal region of the fragment only and was similar in strength to the signals at the sugar beet telomeres (not shown). The results demonstrate that it is not only possible to detect alien chromatin in beet hybrids but also that it is feasible to investigate the structural organization of small chromosome fragments.

Recently, a nematode-resistant sugar beet line carrying a translocation from *B. procumbens* has been used to establish a large insert genomic library. A YAC (yeast artificial chromosome) with an 150-kb insert has been isolated from the wild beet translocation and the nematode resistance gene has been isolated by positional cloning (Cai *et al.*, 1997). Two YACs of the contig spanning the resistance locus were used for the physical mapping of the wild beet translocation, and it could be shown by fluorescent *in situ* hybridization that the investigated resistant plant is homozygous for the translocation and the chromosome fragment from *B. procumbens* is located distal on one chromosome pair (Figure 1P, arrows).

The *in situ* detection of plant single-copy sequences is technically difficult, and the wild beet translocation was used to explore the potential resolution of the FISH approach and to introduce the chromosomal mapping of single-copy genes into genome analysis of *Beta* species. An internal fragment of the nematode resistance gene $Hs1^{pro\text{-}1}$, 684 bp in length, was detected on both chromatids of different *Beta* chromosomes and represents one of the shortest unique DNA sequences localized on mitotic plant chromosomes so far (Figure 1Q, arrows). Comparative chromosomal mapping of the 684 bp long $Hs1^{pro\text{-}1}$ probe in the translocation line, a monosomic addition line and in *B. procumbens* revealed the origin of the wild beet translocation leading to nematode resistant sugar beets.

Centromeric DNA in *Beta* species

In higher eukaryotes, the DNA composition of centromeres displays a high degree of variation, even between chromosomes of a single species. However, the long-range organization of centromeric DNA apparently follows similar structural rules. It has been shown that the centromeric DNA of most higher eukaryotes extends over several hundred kilobase pairs and is predominantly composed of various repetitive sequences such as satellite repeats, retrotransposable elements and simple sequence repeats (Round *et al.*, 1997; Sun *et al.*, 1997; Choo, 1997; Copenhaver *et al.*, 1999). In particular in plants, the organization of the centromeric DNA is still poorly understood. We have recently analyzed the long-range organization of centromeres in the wild beet *B. procumbens* using a set of centromeric repetitive sequences and developed a structural model of a plant centromere (Gindullis *et al.*, 2001). According to this model the two non-homologous satellite repeats pTS4.1 and pTS5 form large arrays directly adjacent to each other and represent the majority of the centromeric DNA.

A similar organization of repeated DNA sequences was found at the centromere of a minichromosome in the chromosome mutant PRO1. PRO1 has originally been selected from interspecific hybrids between *B. vulgaris* and the wild beet *B. procumbens* aiming to establish nematode resistant sugar beets (Jung and Wricke, 1987). The mutant PRO1 carries a single *B. procumbens* chromosome fragment of approximately 6–9 Mbp (Figure 1R, left) and resembles a minichromosome which is functional in sugar beet. For the analysis of a plant centromere this *Beta* system offers several advantages.

Ambiguities in physical mapping due to allelic variations are excluded as the PRO1 centromere can be analyzed on a monosomic minichromosome in a heterologous genetic background. In addition, genome-specific satellite DNAs that can unequivocally distinguish between *B. procumbens* and *B. vulgaris* chromatin are available (Schmidt *et al.*, 1990; Schmidt and Heslop-Harrison, 1996; Dechyeva *et al.*, in preparation). FISH analysis has shown that the satellite arrays reside at the physical end of the minichromosome indicating an acrocentric position of the centromere, and it has been shown that the centromeric satellite arrays of the PRO1 minichromosome occupy approximately 340 kb (Gindullis *et al.*, 2001). The arrays are interspersed with other repetitive sequence elements as revealed by FISH to extended chromatin fibres.

An important step toward the structural analysis of a functional plant centromere was the construction of a PRO1 BAC library with deep genome coverage. The BAC library consists of 50304 clones with an average insert size of 125 kb (Gindullis *et al.*, 2001). The *B. procumbens*-specific sequences pTS4.1 and pTS5 were used for the isolation of BAC clones in order to investigate the molecular structure of the centromeric DNA. The minichromosome origin of these BACs can be shown by FISH (example in Figure 1R). In *B. procumbens*, the BACs hybridize strongly with the centromeric region demonstrating that they contain centromeric DNA (Figure 1S). Additional weak hybridization is caused by dispersed DNA elements which are also present in the large BAC inserts.

Based on the existence and amplification of subsets of centromere-associated repeats the BACs of the PRO centromere can be grouped into four different classes. So far we have isolated six repeated DNA motifs contributing to the complex structure of the PRO1 centromere. These repeats include satellite DNA, Ty3-*gypsy*-like retrotransposons and a dispersed sequences which also exists in other chromosomal regions. Of particular interest are Ty3-*gypsy* retrotransposons which are highly amplified in a single class of PRO1 BACs (Figure 1J) and specifically integrate into the centromeric satellite repeat pTS5. The LTR sequences show less than 2% divergence indicating sequence conservation and a potential role in the formation of the PRO1 centromere.

Regarding the overall structure, *Beta* centromeres show similarities to centromeres from other higher eukaryotes such as human, *Drosophila melanogaster* and *A. thaliana*.

Conclusions

Molecular cytogenetic studies, in particular fluorescent *in situ* hybridization, allows the examination of sequence and genome organization in plants with chromosomes varying in size over a wide range. Such studies have both fundamental implications for genome research and practical use in plant breeding. *In situ* hybridization enables (i) chromosome identification within a complement, (ii) the detection of alien chromatin (chromosomal additions or translocations) in crops, (iii) the physical mapping of unique DNA sequences including single-copy genes, and (iv) the analysis of particular chromosome domains such as telomeres or centromeres.

The comparison of different plant species leads to the definition of common elements of genome organization, and sheds light on general patterns and characteristics of repetitive DNA. It is clear, that repetitive DNA sequences are an important component of the plant genome and affect its structure and function. Hence, the investigation of the genomic and physical distribution of repetitive DNA is critical to gain knowledge about the evolution of sequence families, genomes and plant species. While satellite DNA amplification and diversification reflect phylogenetic relationships between plant species, activation and transposition of different types of retroelements play an important role in the generation of variability of plant genomes and species speciation. Using sugar beet including the chromosomal mutant PRO1 as model systems we were able to build a comprehensive picture of the architecture of a plant chromosome and obtained detailled information about the molecular structure of a plant centromere.

Acknowledgements

T. Schmidt acknowledges the BioFuture grant 0311860 of the Bundesministerium für Bildung und Forschung, financial support by the Deutsche Forschungsgemeinschaft (Schm1048/2-2, Schm1048/2-3) and grants of the USDA Forest Service (SRS 33-G-97-082 and 01-IG-11330126-336). B. Fleischer is supported by a FAZIT fellowship.

References

Ananiev EV, Phillips RL, Rines HW (1998). Chromosome-specific molecular organization of maize (*Zea mays* L.) centromeric regions. *Proc Natl Acad Sci USA* 95: 13,078–13,078.

Aragon-Alcaide L, Miller T, Schwarzacher T, Reader S, Moore G (1996). A cereal centromere sequence. *Chromosoma* 105: 261–268.

Arumuganathan K, Earle ED (1991). Nuclear DNA content of some important plant species. *Plant Mol Biol Rep* 9: 208–218.

Bennett MD, Leitch IJ (1995). Nuclear amounts in angiosperms. *Ann Bot* 76: 113–176.

Brandes A, Heslop-Harrison JS, Kamm A, Kubis S, Doudrick RL, Schmidt T (1997). Chromosomal distribution of *Ty1-copia*-like retrotransposons in pteridophytes, gymnosperms and angiosperms and analysis of new *Ty1-copia*-like families in *Arabidopsis thaliana. Plant Mol Biol* 33: 11–21.

Cai D, Kleine M, Kifle S, Harloff H, Sandal NN, Marcker KA, Klein-Lankhorst RM, Salentijn EMJ, Lange W, Stiekema WJ, Wyss U, Grundler FMW, Jung C (1997). Positional cloning of a gene for nematode resistance in sugar beet. *Science* 275: 832–834.

Choo KHA (1997). *The Centromere*. New York, USA: Oxford University Press Inc.

Copenhaver GP, Nickel K, Kuromori T, Benito M-I, Kaul S, Lin X, Bevan M, Murphy G, Harris B, Parnell LD, McCombie WR, Martienssen RA, Marra M, Preuss D (1999). Genetic definition and sequence analysis of *Arabidopsis* centromeres. *Science* 286: 2,468–2,474.

Desel C, Jung C, Cai D, Kleine M, Schmidt T (2001). High resolution mapping of YACs and single-copy gene $Hs1^{pro\text{-}1}$ on *Beta vulgaris* chromosomes by multi-colour fluorescence *in situ* hybridization. *Plant Mol Biol* 45: 113–122.

Doudrick RL (1996). Genetic recombinational and physical linkage analyses on slash pine. In: Heslop-Harrison JS ed. *Unifying Plant Genomes*. SEB Symposium No. 50. Cambridge, UK: The Company of Biologists Limited.

Doudrick RL, Heslop-Harrison JS, Nelson CD, Schmidt T, Nance WL, Schwarzacher T (1995). The karyotype of *Pinus elliottii* Engelm. var. *elliottii* using *in situ* hybridization and fluorochrome banding patterns. *J Hered* 86: 289–296.

Flavell AJ, Dunbar E, Anderson SR, Pearce R, Hartley R, Kumar A (1992). Ty1-*copia* group retrotransposons are ubiquitous and heterogenous in higher plants. *Nucl Acids Res* 20: 3,639–3,644.

Flavell RB, Bennett MD, Smith JB (1974). Genome size and the proportion of repeated nucleotide sequence DNA in plants. *Biochem Genet* 12: 257–269.

Fuchs J, Brandes A, Schubert I (1995). Telomere sequence localization and karyotype evolution in higher plants. *Plant Syst Evol* 196: 227–241.

Gindullis F, Dechyeva D, Schmidt T (2001). Construction and characterization of a BAC library for the molecular dissection of a single wild beet centromere and sugar beet (*Beta vulgaris*) genome analysis. *Genome* (in press).

Gindullis F, Desel C, Galasso I, Schmidt T (2001). The large-scale organization of centromeric DNA in *Beta* species. *Genome Res* 11: 253–265.

Hemleben V, Zentgraf U, King K, Borisjuk N, Schweizer G (1992). Middle repetitive and highly repetitive sequences detect polymorphisms in plants. In: Kahl G, Appelhans H, Kömpf J, Driesel AJ eds. *BioTechForum-Advances in Molecular Genetics, Vol. 10: DNA-Polymorphisms in Eukaryotic Genomes*. Heidelberg: Hüthig.

Heslop-Harrison JS, Brandes A, Taketa S, Schmidt T, Vershinin AV, Alkhimova EG, Kamm A, Katsiotis A, Doudrick RL, Schwarzacher T, Kubis S, Kumar A, Pearce SR, Flavell AJ, Harrison GE (1997). The chromosomal distribution of retrotransposon-like elements in higher plants and its implication for genome evolution. *Genetica* 100: 197–204.

Hirochika H, Hirochika R (1993). Ty1-copia group retrotransposons as ubiquitoius components of plant genomes. *Jpn J Genet* 68: 35–46.

Jakubczak JL, Burke WD, Eickbush TH (1991). Retrotransposable elements *R1* and *R2* interrupt the rRNA genes of most insects. *Proc Natl Acad Sci USA* 88: 3,295–3,299.

Jung C, Wricke G (1987). Selection of diploid nematode-resistant sugar beets from monosomic addition lines. *Plant Breed* 98: 205–214.

Kamm A, Doudrick RL, Heslop-Harrison JS, Schmidt T (1996). The genomic and physical organization *Ty*1-*copia*-like sequences as a component of large genomes in *Pinus elliottii* var. *elliottii* and other gymnosperms. *Proc Natl Acad Sci USA* 93: 2,708–2,713.

Kamm A, Galasso I, Schmidt T, Heslop-Harrison JS (1995). Analysis of a repetitive DNA family from *Arabidopsis arenosa* and relationship between *Arabidopsis* species. *Plant Mol Biol* 27: 853–862.

Katsiotis A, Schmidt T, Heslop-Harrison JS (1996). Chromosomal and genomic organization of *Ty-copia*-like retrotransposon sequences in the genus *Avena*. *Genome* 39: 410–417.

Kubis S, Heslop-Harrison JS, Schmidt T (1997). A family of differentially amplified repetitive DNA sequences in the genus *Beta* reveals genetic variation in *Beta vulgaris* subspecies and cultivars. *J Mol Evol* 44: 310–320.

Kubis S, Heslop-Harrison JS, Desel C, Schmidt T (1998). The genomic organization of non-LTR retrotransposons (LINEs) from three *Beta* species and five other angiosperms. *Plant Mol Biol* 36: 821–831.

Leeton PJR, Smyth DR (1993). An abundant LINE-like element amplified in the genome of *Lilium speciosum*. *Mol Gen Genet* 237: 97–104.

Meyerowitz EM, Pruitt RE (1985). *Arabidopsis thaliana* and plant molecular genetics. *Science* 229: 1,214–1,218.

Mörchen M, Cuguen J, Michaelis G, Hänni C, Saumitou-Laprade P (1996). Abundance and length polymorphism of microsatellite repeats in *Beta vulgaris* L. *Theor Appl Genet* 92: 326–333.

Ohri D, Khoshoo TN (1986). Genome size in gymnosperms. *Plant Syst Evol* 153: 119–132.

Pearce SR, Harrison G, Wilkinson M, Li D, Heslop-Harrison JS, Flavell A, Kumar A (1996). The Ty1-*copia* group retrotransposons in *Vicia* species: copy number, sequence heterogeneity and chromosomal localization. *Mol Gen Genet* 250: 305–315.

Presting GG, Malysheva L, Fuchs J, Schubert I (1998). A Ty3/*gypsy* retrotransposon-like sequence localizes to the centromeric regions of cereal chromosomes. *Plant J* 16: 721–728.

Richards EJ, Ausubel FM (1988). Isolation of a higher eukaryotic telomere from *Arabidopsis thaliana. Cell* 53: 127–136.

Richards ER, Goodman HM, Ausubel FM (1991). The centromere region of *Arabidopsis thaliana* chromosome I contains telomere-similar sequences. *Nucl Acids Res* 19: 3,351–3,357.

Round EK, Flowers SK, Richards EJ (1997). *Arabidopsis thaliana* centromere regions: Genetic map positions and repetitive DNA structure. *Genome Res* 7: 1,045–1,053.

SanMiguel P, Tikhonov A, Jin Y-K, Motchoulskaia N, Zakharov D, Melake-Berhan A, Springer PS, Edwards KJ, Lee M, Avramova Z, Bennetzen JL (1996). Nested retrotransposons in the intergenic regions of the maize genome. *Science* 274: 765–768.

Schmidt A, Doudrick RL, Heslop-Harrison JS, Schmidt T (2000). The contribution of short repeats of low sequence complexity to large conifer genomes. *Theor Appl Genet* 101: 7–14.

Schmidt T, Heslop-Harrison JS (1993). Variability and evolution of highly repeated DNA sequences in the genus *Beta. Genome* 36: 1,074–1,079.

Schmidt T, Heslop-Harrison JS (1996b). The physical and genomic organization of microsatellites in sugar beet (*Beta vulgaris*). *Proc Natl Acad Sci USA* 93: 8,761–8,765.

Schmidt T, Metzlaff M (1991). Cloning and characterization of a *Beta vulgaris* satellite DNA family. *Gene* 101: 247–250.

Schmidt T, Boblenz K, Metzlaff M, Kaemmer D, Weising K, Kahl G (1993). DNA-fingerprinting in sugar beet (*Beta vulgaris*) – identification of double-haploid breeding lines. *Theor Appl Genet* 85: 653–657.

Schmidt T, Jung C, Metzlaff M (1991). Distribution and evolution of two satellite DNAs in the genus *Beta. Theor Appl Genet* 82: 793–796.

Schmidt T, Jung C, Heslop-Harrison JS, Kleine M (1997). Detection of alien chromatin conferring resistance to the beet cyst nematode (*Heterodera schachtii* Schm.) in cultivated beet (*Beta vulgaris* L.) using *in situ* hybridization. *Chrom Res* 5: 186–193.

Schmidt T, Junghans H, Metzlaff M (1990). Construction of *B. procumbens*-specific DNA probes and their application for the screening of *B. vulgaris* = *B. procumbens* ($2n = 19$) addition lines. *Theor Appl Genet* 79: 177–181.

Schmidt T, Kubis S, Heslop-Harrison JS (1995). Analysis and chromosomal localization of retrotransposons in sugar beet (*Beta vulgaris* L.): LINEs and *Ty1-copia*-like elements as major components of the genome. *Chrom Res* 3:335–345.

Schmidt T, Schwarzacher T, Heslop-Harrison JS (1994). Physical mapping of rRNA genes by fluorescent *in situ* hybridization and structural analysis of 5S rRNA genes and intergenic spacer sequences in sugar beet (*Beta vulgaris*). *Theor Appl Genet* 88: 629–636.

Schmidt, T, Heslop-Harrison JS (1996a). High resolution mapping of repetitive DNA by *in situ* hybridization – molecular and chromosomal features of prominent dispersed and discretely localized DNA families of the wild beet species *Beta procumbens. Plant Mol Biol* 30: 1,099–1,119.

Schwarzacher T, Leitch AR, Bennett MD, Heslop-Harrison JS (1989). *In situ* localization of parental genomes in a wide hybrid. *Ann Bot* 64: 315–324.

Schwarz-Sommer Z, Leclercq L, Goebel E, Saedler H (1987). Cin4, an insert altering the structure of the *A1* gene in *Zea mays*, exhibits properties of nonviral retrotransposons. *EMBO J* 13: 3,873–3,880.

Singer M (1982). Highly repeated sequences in mammalian genomes. *Int Rev Cytol* 76: 67–112.

Sun X, Wahlstrom J, Karpen G (1997). Molecular structure of a functional *Drosophila* centromere. *Cell* 91: 1,007–1,019.

Voytas DF, Cummings MP, Konieczny FM, Ausubel FM, Rodermel SR (1992). *copia*-like retrotransposons are ubiquitous among plants. *Proc Natl Acad Sci USA* 89: 7,124–7,128.
White SE, Habera LF, Wessler SR (1994). Retrotransposons in the flanking regions of normal plant genes: A role for copia-like elements in the evolution of gene structure and expression. *Proc Natl Acad Sci USA* 91: 11,792–11,796.
Wright DA, Ke N, Smalle J, Hauge BM, Goodman HM, Voytas DF (1996). Multiple non-LTR retrotransposons in the genome of *Arabidopsis thaliana. Genetics* 142: 569–578.
Zhong X-B (1998). Tomato genome mapping by fluorescence *in situ* hybridisation. PhD Thesis, University of Wageningen.

5.2. Interphase Cytogenetics in Plants

I. Schubert, Z. Jasencakova, A. Meister, P. Fransz and M. Lysak
Institute of Plant Genetics and Crop Plant Research (IPK), D-06466 Gatersleben, Germany

Since the eighties of nineteenth century, when Flemming and others discovered the mitotic and meiotic cycle and Waldeyer (1888) coined the term "chromosome", chromosome research was performed for one century mainly on mitotic and meiotic chromosomes. During this period cytogeneticists accumulated broad knowledge on species-specific karyotypes and on the structural features of chromosomes in their highly condensed "transport" from. The development of non-radioactive *in situ* hybridization techniques during the 1980s together with facilities to isolate specific chromosomes allowed to study defined chromosomes and chromosome domains also during interphase when essential processes such as transcription, replication and DNA repair take place. In particular "chromosome painting" developed by Pinkel *et al.* (1988) and Lichter *et al.*, (1988), enabled to address and to investigate individual chromosome territories within cell nuclei.

While chromosome painting has been worked out for more than 40 mammalian species and later on for birds, reptiles and insects, chromosomes of euploid plants were, in spite of intense efforts, not accessible by that approach. The reason for this failure is the high number of dispersed repeats which are highly homogeneous between the chromosomes of a given plant species with a large genome. Therefore, it is difficult to remove such sequences from chromosome-derived probes or to prevent efficiently their hybridization (Fuchs *et al.*, 1996; Schwarzacher *et al.*, 1997; Schubert *et al.*, 2001).

Only chromosomes and interphase territories of alien chromosomes/chromosome regions within interspecific hybrids and their progenies could be distinguished by genomic *in situ* hybridization (GISH) as developed by Trude Schwarzacher and co-authors in 1989. Painting by GISH is based on dispersed repeats that evolve much faster than coding sequences and thus allow in most cases to distinguish even closely related genomes during nuclear divisions as well as in interphase nuclei. However, it remains an open question whether the painted alien chromosomes in fact behave like

M. Schmid and I. Nanda (eds.), Chromosomes Today, Volume 14, 267–274.

the host chromosomes and in the same way as in their own genetic background. Recently, tools have been developed to address individual chromosomes and chromosome regions in euploid plants along the cell cycle.

Chromosome painting in plants

One approach suitable for plant interphase cytogenetics has been provided by the development of chromosome painting in Arabidopsis. This technique allows for the first time to investigate individual chromosome territories of a euploid plant during all cell cycle and developmental stages (Figure 1). Although the application of chromosomes painting by use of labelled pools of contiguous large insert clones is still restricted to small plant genomes with a low content of (dispersed) repetitive sequences, it enables to address important biological questions. For instance: (i) what is the spatial and temporal sequence of chromosome synapsis during meiosis I; (ii) does homologue association occur during somatic interphase or at specific developmental stages, and if yes, what is its significance; (iii) which chromosomes are involved in spontaneous or induced chromosome mutations; (iv) which chromosomes of closely related species are homologous to each other and what types of chromosome rearrangements have occurred during the evolution of the respective karyotypes. For details see Lysak *et al.*, this volume.

Interphase cytogenetics on flow-sorted plant nuclei

Another approach of plant interphase cytogenetics combining FISH with various probes and immuno-detection of various antigens on isolated nuclei sorted according to their DNA content into fractions of defined cell cycle stages has also been developed at the IPK. We have used this approach to follow histone acetylation within rDNA-, eu- and heterochromatin-containing chromosome domains during the cell cycle and in correlation with transcription and replication processes in interphase nuclei and have compared this dynamic chromatin modification in dicot and monocot plants as well as with corresponding data from non-plant organisms.

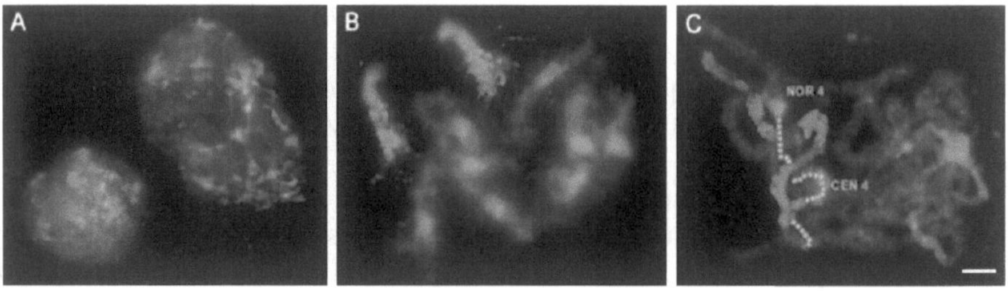

Figure 1. *Painting of the long arm of Arabtdopsis chromosome 4 by FISH with alternatively labelled pools of 113 BACs. (A) Interphase nuclei showing associated (left) and separated (right) territories for that arm; (B) prometaphase; (C) pachytene of meiosis I. Bar = 5 μm.*

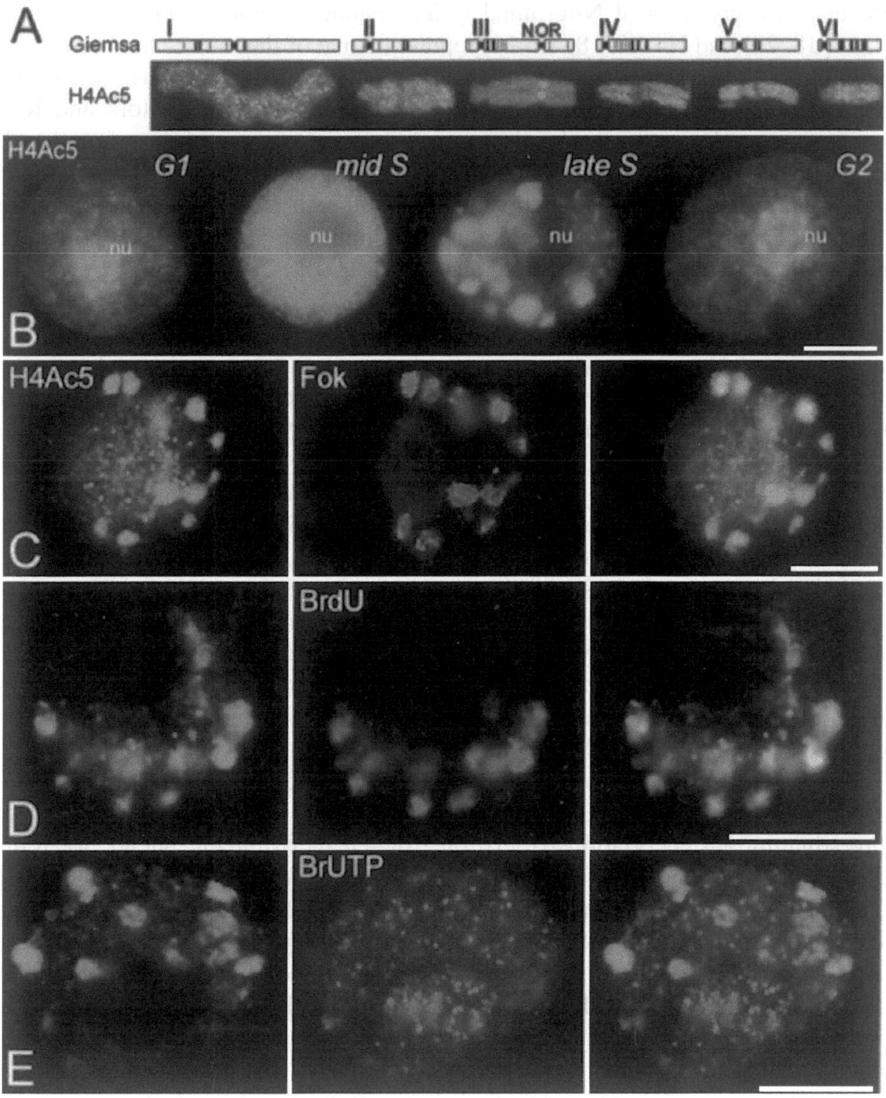

Figure 2. *Immunolabelling of histone H4 acetylated at lysine 5 at distinct chromatin domains of the field bean during different cell cycle stages. (A) Metaphase chromosomes, (B) nuclei in G1, mid S, late S and G2, (C) nucleus in late S after H4Ac5 detection and FISH with the heterochromatin-specific Fok-element, (D) nucleus in late S after pulse treatment with BrdU for 30 min (red) and detection of H4Ac5, (E) nucleus in late S after incorporation of BrUTP for 10 min (red) and detection of H4Ac5. For details see text. Bars = 10 μm.*

Reversible acetylation of N-terminal lysine residues of nucleosomal histones H3 (K9, 14, 18, 23) and H4 (K5, 8, 12, 16) represents one of the major chromatin modifications in all eukaryotes studied so far. Histone acetylation mediates a less condensed chromatin structure and easier access of transcription factors and RNA polymerase complexes. At the level of genes, histone acetylation was found to be linked with the potential activity in transcription and recombination processes. It is therefore considered as an epigenetic code which together with other modifications, for instance methylation, phosphorylation and ubiquitination, in a concerted manner regulates downstream functions (Strahl and Allis, 2000).

Field bean

The intensity of histone acetylation varies along mitotic chromosomes. In general, transcriptionally inactive heterochromatin is less acetylated than euchromatic domains. This is true for animals and for plants. In plants the NOR is often the most intensely acetylated region of the karyotype. In the field bean this is the case for all acetylatable lysines of H4 and H3, except for H3Ac23. Contrary to mammalian chromosomes, field bean metaphase chromosomes showed different acetylation intensity of H3 and H4 in that the heterochromatin fraction containing a tandemly repeated so-called Fok-element is less acetylated than euchromatin, while DAPI-positive, Fok-element-free heterochromatin (adjacent to the centromeres) proved to be even more strongly acetylated at lysines 9/18 and 14 than euchromatin (Belyaev *et al.*, 1998). When deacetylases were blocked by trichostatin A for >2 h before mitosis, the acetylation of H4 within heterochromatin switched from underacetylation toward hyperacetylation (comparable to the NOR), while the acetylation pattern of H3 remained constant. This suggested a cell cycle dependent variation at least for H4 acetylation (Belyaev *et al.*, 1997, 1998).

Investigating H4 acetylation pattern on isolated meristematic nuclei, we found indeed clearly distinguishable labelling patterns, with different acetylation intensity at nucleoli, eu- and heterochromatic compartments. On nuclei, sorted according to their DNA content into G1, early S, mid S, late S and G2 fractions, a clear ceil cycle-dependent variation of these acetylation patterns was detectable. Strongly labelled nucleoli and weaker labelled chromatin with some unlabelled areas occurred most frequently during G1 and G2 (reflecting the labelling pattern of mitotic chromosomes). During S phase, the chromatin (except for some unlabelled areas) was most strongly labelled while nucleoli were unlabelled. Very intensely labelled spots instead of unlabelled regions appeared only during late S, when heterochromatin is replicating and in early G2, when also nucleoli became again strongly labelled (Figure 2).

To address the differently labelled chromatin domains, we performed FISH with Fok-elements after immunostaining of acetylated H4. This has shown that under-acetylated chromatin in G1 and G2 nuclei as well as hyperacetylated chromatin in late S corresponded to heterochromatin domains (Figure 2d). Combined immunostaining of acetylated H4 and of BrdU after a 30 min pulse treatment revealed clear colocalization of both immunosignals during different stages of S-phase (Figure 2f).

To test for a possible correlation of H4 acetylation and transcriptional activity, run-on transcription on isolated nuclei was performed. The strongest transcriptional activity was found in nucleoli independent of their acetylation status. As expected, no transcription was found within heterochromatic domains even when strongly acetylated (Figure 2h).

Immunostaining of acetylated isoforms of H3 also revealed different labelling patterns on interphase nuclei. However, these did not vary along the cell cyle. As on metaphase chromosomes, the strongest acetylation did not colocalize with the Fok-element-containing heterochromatin domains, but with DAPI-positive proximal heterochromatin.

In conclusion these data demonstrate that in field bean eu- and heterochromatin are strongest acetylated at H4 during replication, while nucleolus organizers are strongest acetylated during mitosis and are apparently free of acetylated H4 during S phase. The incorporation of the acetylated isoforms H4Ac5 and 12 seemed to be conserved for insects, mammals and plants, while K16 but not K8 of H4 is also highly acetylated during/after replication and K16 showed a delayed deacetylation as compared to K5 and 12 in V. faba. H3 acetylation remains fairly constant throughout the mitotic cycle with a sequence-dependent overacetylation at Fok-element-free heterochromatic domains (Jasencakova *et al.*, 2000).

Barley

Comparative studies were performed for the monocot barley. Two-colour FISH with centromere-specific retroelements and telomere-specific subtelomere repeats on sorted nuclei revealed maintenance of "Rabl orientation", i.e., clustering of centromeres at one pole and of telomeres at the other, throughout interphase. The centromere-adjacent heterochromatin, visualized by FISH with GAA-oligos, was also found at the centromeric pole. After BrdU pulse the following order of replication could be established: rDNA during very early S; euchromatin from early to mid S; centromeres from mid S on; and heterochromatin at the centromeric pole during late S.

The patterns of H4 acetylation of chromosomes and nuclei and their variation along the cell cycle were similar to the situation observed in V. faba for K5 and 12: strong acetylation of the nucleoli and moderate acetylation of the remaining chromatin appeared most frequently in G1 and G2 nuclei, while during early and mid S phase nucleoli were unlabelled and most of the chromatin revealed the strongest labelling. The centromeric/heterochromatic pole showed strongest acetylation during late S, when these domains are replicating, and in early G2, when nucleoli became labelled again. This could be verified by immunolabelling of acetylated H4 and by FISH with centromere- and heterochromatin-specific probes.

During mitosis, most chromosomes revealed strongest acetylation at their distal euchromatic parts, however, in a few cases deacetylation of K5 and K12 was delayed up to late telophase, resulting in a uniform acetylation, sometimes even more pronounced within the pericentromeric regions. NORs were stronger acetylated than euchromatin during mitosis independent of their nucleolus forming activity.

Strong acetylation of K16 was constantly restricted to the euchromatic pole. A strong acetylation of K8 was limited to (late) S phase. Acetylation of H3 at K14 and 9/18 was weakest at the centromeric regions throughout the cell cycle (Figure 3).

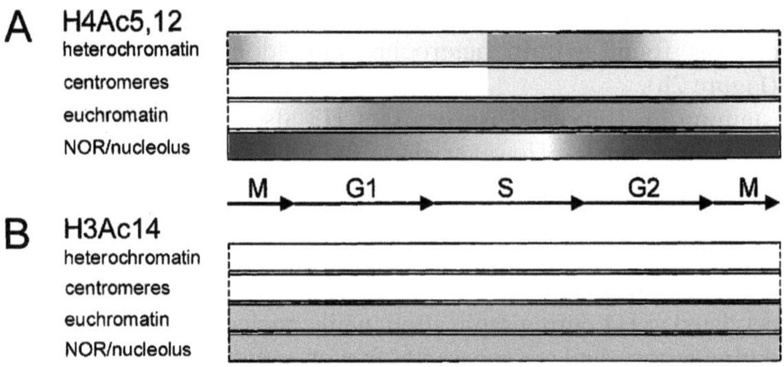

Figure 3. *Scheme of histone acetylation at different chromatin domains of barley along the mitotic cell cycle.*

Arabidopsis

Similar investigations are now in progress for the model plant A. thaliana. In Arabidopsis interphase nuclei, chromosomes do not show a bipolar Rabl orientation; instead, the ten DAPI-positive heterochromatic chromocentres around each centromere revealed a dispersed location (in unsquashed nuclei mainly at the periphery) and telomeres were clustered around nucleoli. Interestingly, most of the 5S rDNA and even the terminally positioned 45S rDNA were associated with the heterochromatic chromocentres and allowed to identify most of them individually.

Immunodetection of 5′-methyl cytosine has shown that the most heavily methylated domains in Arabidopsis nuclei are the chromocentres. These and the nucleolus proved to be underacetylated at K5 of H4 as compared to the euchromatin in most nuclei (Fransz *et al.*, submitted).

Surprisingly, and in contrast to the situation in field bean and barley, H4 acetylation patterns proved to be constant in Arabidopsis nuclei along the cell cycle except for K16 of H4 which showed stronger acetylation at pericentromeric chromocentres (excluding rDNA) during their replication at the end of S-phase. Instead, K9 of H3 showed a preferential labelling of all heterochromatic chromocentres during late S when chromocentres replicate and later on in endopolyploid nuclei.

We are now testing how this pattern correlates with methylation of K9 of H3 which proved to be a prerequisite for heterochromatin assembly and epigenetic silencing mediated by binding of HP1 homologues to methylated lysine 9 in mice (Lachner *et al.*, 2001) and fission yeast (Nakayama *et al.*, 2001; Bannister *et al.*, 2001).

Summary and conclusions

Novel approaches of interphase cytogenetics enabled to study (i) chromosome territories of Arabidopsis by painting and (ii) chromosomal acetylation patterns of H4 (K5, 8, 12, 16) and H3 (K14, 9/18, 23) along the cell cycle for monocot and dicot species by immunostaining and FISH on flow-sorted nuclei. The main conclusion from the latter experiments is that the global acetylation of H4 within eu- and heterochromatin is correlated with replication rather than with transcription, masking the fine-tuning of acetylation at active promoters. The strong acetylation at the time of replication, which becomes reversed for heterochromatin towards mitosis, might be linked with ubiquitous post-replication repair processes. Deposition-related acetylation of K5 and K12 of H4 is apparently conserved from insects and mammals (Sobel *et al.*, 1995) to field bean and barley. In plants, NORs tend to be strongest acetylated during mitosis while nucleoli show no detectable H4 acetylation during S phase. Contrary to animals, the chromosomal acetylation patterns of H4 and H3 in plants may differ as to their cell cycle-dependency and/or due to sequence-specificity. Arabidopsis nuclei deviate from those of field bean and barley since acetylation of H3 (K9) instead of H4 (K5, 12) showed cell cycle-dependent intensity (most strongly at heterochromatin during late S). Thus, histone acetylation patterns reveal evolutionarily conserved as well as group- or species-specific features.

Acknowledgement

This work has been supported by grants of the Land Sachsen-Anhalt (3233A/0020B; 3035A/0088B) and the Deutsche Forschungsgemeinschaft (FR1497/1-1).

References

Bannister AJ, Zegerman P, Partridge JF, Miska EA, Thomas JO, Allshire RC, Kouzarides T (2001). Selective recognition of methylated lysine 9 on histone H3 by the HP1 chromo domain. *Nature* 410: 120–124.

Belyaev ND, Houben A, Baranczewski P, Schubert I (1997). Histone H4 acetylation in plant heterochromatin is altered during cell cycle. *Chromosoma* 106: 193–197.

Belyaev ND, Houben A, Baranczewski P, Schubert I (1998). The acetylation patterns of histones H3 and H4 along *Vicia faba* chromosomes are different. *Chromosome Res* 6: 59–63.

Fransz P, de Jong JH, Lysak M, Ruffini Castiglione M, Schubert I (2001). Chromosomes in interphase nuclei of *Arabidopsis thaliana* are organized as euchromatin loops around heterochromatin domains. *Science*: submitted.

Fuchs J, Houben A, Brandes A, Schubert I (1996). Chromosome 'painting' in plants – a feasible technique? *Chromosoma* 104: 315–320.

Jasencakova Z, Meister A, Schubert I (2001). Chromatin organization and its relation to replication and histone acetylation during the cell cycle in barley. *Chromosoma* 110: 83–92.

Jasencakova Z, Meister A, Walter J, Turner BM, Schubert I (2000). Histone H4 acetylation of euchromatin and heterochromatin is cell cycle dependent and correlated with replication rather than with transcription. *Plant Cell* 12: 2,087–2,100.

Lachner M, O'Carroll D, Rea S, Mechtier K, Jenuwein T (2001). Methylation of histone H3 lysine 9 creates a binding site for HP1 proteins. *Nature* 410: 116–120.

Lichter P, Cremer T, Borden J, Manuelidis L, Ward DC (1988). Delineation of individual human chromosomes in metaphase and interphase cells by *in situ* suppression hybridization using recombinant DNA libraries. *Human Genet* 80: 224–234.

Lysak MA, Fransz PF, Ali HBM, Schubert I. Chromosome painting in *Arabidopsis thaliana. Plant J*, submitted.

Nakayama J-i, Rice JC, Strahl BD, Allis CD, Grewal SIS (2001). Role of histone H3 lysine 9 methylation in epigenetic control of heterochromatin assembly. *Science* 292: 110–113.

Pinkel D, Landegent J, Collins C, Fuscoe J, Segraves R, Lucas J, Gray J (1988). Fluorescence *in situ* hybridization with human chromosome-specific libraries: Detection of trisomy 21 and translocations of chromosome 4. *Proc Natl Acad Sci USA* 85: 9,138–9,142.

Schubert I, Fransz PF, Fuchs J, de Jong JH (2001). Chromosome painting in plants. *Methods Cell Sci.*

Schwarzacher T, Leitch AR, Bennett MD, Heslop-Harrison JS (1989). *In situ* localization of parental genomes in a wide hybrid. *Ann Bot* 64: 315–324.

Schwarzacher T, Wang ML, Leitch AR, Miller N, Moore G, Heslop-Harrison JS (1997). Flow cytometric analysis of the chromosomes and stability of a wheat cell-culture line. *Theor Appl Genet* 94: 91–97.

Sobel RE, Cook RG, Perry CA, Annunziato AT, Allis CD (1995). Conservation of deposition-related acetylation sites in newly synthesized histones H3 and H4. *Proc Natl Acad Sci USA* 92: 1,237–1,241.

Strahl BD, Allis CD (2000). The language of covalent histone modifications. *Nature* 403: 41–45.

MIX
Papier aus verantwortungsvollen Quellen
Paper from responsible sources
FSC® C105338
FSC
www.fsc.org

If you have any concerns about our products,
you can contact us on
ProductSafety@springernature.com

In case Publisher is established outside the EU,
the EU authorized representative is:
Springer Nature Customer Service Center GmbH
Europaplatz 3, 69115 Heidelberg, Germany

Printed by Libri Plureos GmbH
in Hamburg, Germany